GENOMICS FOR BIOSAFETY IN PLANT BIOTECHNOLOGY

NATO Science Series

A series presenting the results of scientific meetings supported under the NATO Science Programme.

The series is published by IOS Press and Kluwer Academic Publishers in conjunction with the NATO Scientific Affairs Division.

Sub-Series

I.	Life and Behavioural Sciences	IOS Press
II.	Mathematics, Physics and Chemistry	Kluwer Academic Publishers
III.	Computer and Systems Sciences	IOS Press
IV.	Earth and Environmental Sciences	Kluwer Academic Publishers
V.	Science and Technology Policy	IOS Press

The NATO Science Series continues the series of books published formerly as the NATO ASI Series.

The NATO Science Programme offers support for collaboration in civil science between scientists of countries of the Euro-Atlantic Partnership Council. The types of scientific meeting generally supported are "Advanced Study Institutes" and "Advanced Research Workshops", although other types of meeting are supported from time to time. The NATO Science Series collects together the results of these meetings. The meetings are co-organized by scientists from NATO countries and scientists from NATO's Partner countries – countries of the CIS and Central and Eastern Europe.

Advanced Study Institutes are high-level tutorial courses offering in-depth study of latest advances in a field.
Advanced Research Workshops are expert meetings aimed at critical assessment of a field, and identification of directions for future action.

As a consequence of the restructuring of the NATO Science Programme in 1999, the NATO Science Series has been re-organized and there are currently five sub-series as noted above. Please consult the following web sites for information on previous volumes published in the series, as well as details of earlier sub-series:

http://www.nato.int/science
http://www.wkap.nl
http://www.iospress.nl
http://www.wtv-books.de/nato_pco.htm

Series I: Life and Behavioural Sciences - Vol. 359 ISSN: 1566-7693

Genomics for Biosafety in Plant Biotechnology

Edited by

Jan-Peter Nap

Plant Research International, Wageningen University and Research Centre & Centre for BioSystems Genomics, Wageningen & Hanzehogeschool Groningen, Groningen, The Netherlands

Atanas Atanassov

AgroBioInstitute, Sofia, Bulgaria

and

Willem J. Stiekema

Centre for BioSystems Genomics, Wageningen & Plant Research International & Laboratory of Bioinformatics, Wageningen University and Research Centre, The Netherlands

IOS Press

Amsterdam • Berlin • Oxford • Tokyo • Washington, DC

Published in cooperation with NATO Scientific Affairs Division

Proceedings of the NATO Advanced Research Workshop on
Genomics for Biosafety in Plant Biotechnology
15–19 October 2003
Bansko, Bulgaria

ISBN 1 58603 432 4
Library of Congress Control Number: 2004109514

Publisher
IOS Press
Nieuwe Hemweg 6B
1013 BG Amsterdam
Netherlands
fax: +31 20 620 3419
e-mail: order@iospress.nl

Distributor in the UK and Ireland
IOS Press/Lavis Marketing
73 Lime Walk
Headington
Oxford OX3 7AD
England
fax: +44 1865 750079

Distributor in the USA and Canada
IOS Press, Inc.
4502 Rachael Manor Drive
Fairfax, VA 22032
USA
fax: +1 703 323 3668
e-mail: iosbooks@iospress.com

PRINTED IN THE NETHERLANDS

Preface

In mainstream plant biotechnology research, attention is rapidly shifting towards genomics and its "omics" offspring. The various genomics approaches (transcriptomics, proteomics, metabo-lomics, phenomics etc.) aim at the large-scale characterisation of DNA, RNA, proteins, metabolites and phenotypes of plant cells during growth and development. This will result in a better explanation and utilisation of the variability in the functioning of cells, organisms and ecosystems. At the same time, this approach will generate knowledge and technology applicable for the overall assessment of safety of the products made by breeding and biotechnology, of food and feed. The advances of genomics in plant research will influence notably the safety assessment of genetically modified (GM) crops around the world. Appropriate organisation and safety regulation of GM crops is of particular interest for the economies in Central and Eastern Europe.

Right now, however, is the relationship between genomics and biosafety a largely unexplored area. There is a clear gap between the issues of biosafety and its regulation on the one hand and the technological developments in genomics on the other. The purpose of this volume of the NATO Science Series is to start bridging that gap. This book contains the papers of the main lectures that were presented at the NATO Advanced Research Workshop "Genomics for Biosafety in Plant Biotechnology: New Challenges", held in Bansko, Bulgaria, 15-19 October 2003. The event was organised by the AgroBioInstitute, Sofia, Bulgaria in close collaboration with the Centre for BioSystems Genomics and Plant Research International, Wageningen, The Netherlands.

This book is meant as the start of a discussion on the use and need of plant genomics in biosafety. It will not present definitive answers, but it offers a wide variety of viewpoints, developments and issues to consider. Contributions range from the development of validated protocols for NMR in metabolomics to the use of metabolomics for the prediction of unintended effects in GM crops, from the philosophical implications of the precautionary principle - or approach - to the *in silico* prediction of allergenicity of proteins, or from the ecological genomics of non-target effects of insect resistance in GM crops to novel strategies for obtaining drought tolerance. Leitmotiv is that plant genomics could and should learn from all discussions on GM crops. A major issue of the discussion has been and still is establishing the limits of omics technology in biosafety assessments. Does the technological feasibility that "everything" can be measured automatically imply that "everything" must be measured? Who will pay? How to organise?

This book will therefore be of major interest to all researchers in academia, industry or agriculture with an interest in life sciences or social sciences, genomics and biosafety, as well as to graduate and (advanced) undergraduate students in topics as broad as plant biotechnology, genetics, food science, bioinformatics, philosophy, regulatory science and law. Moreover, the book should appeal to policy makers in governments and non-governmental organisations as well as to interest groups that are, or want to be, involved in shaping agriculture and food science for the future.

We thank the Public Diplomacy Division of NATO for largely funding this scientific meeting and the organising bodies for their financial support. We would like to thank the members of the international scientific and local organising committees for their contributions. A special thanks goes to Ms. Denitsa Marinova who was efficient in organising and excellent in troubleshooting. We thank all discussion leaders and chairpersons for their involvement. We feel particularly indebted to all participants that submitted chapters for this book and contributed to the success of the meeting. We trust that this book will reflect that success.

April 2004, Wageningen/Sofia

Jan-Peter Nap
Atanas Atanassov
Willem J. Stiekema

Contents

1. Profiling Technology in the Assessment of GM Crops

Potential of Integrated Functional Genomics in Biosafety Assessments

Aaron M. ELMER, Corey D. BROECKLING, Fang CHEN, Richard A. DIXON, Bryna E. DONNELLY, Anthony L. DURAN, David V. HUHMAN, Zhentian LEI, Bonnie S. WATSON and Lloyd W. SUMNER*
The Samuel Roberts Noble Foundation, Ardmore, Oklahoma, United States of America
**email: lwsumner@noble.org*

Abstract. Integrated functional genomic technologies offer comprehensive profiling at the mRNA, protein, and metabolite levels. These transcriptomic, proteomic, and metabolomic approaches are critical to evaluating biochemical aspects of genetic engineering. Based on the comprehensive nature of functional genomics, these technologies offer powerful tools in comparative analyses and safety assessments of genetically engineered plants.

Introduction

It is estimated that the world's population will nearly double over the next 50 years to approximately 10 billion people [1]. This population expansion has the potential to generate a global crisis unless agricultural and political infrastructures work cooperatively to ensure adequate food quantity and quality. Meeting these goals will require organized improvements in land management, alterations in cultural and regional farming practices, increased financial resources, education of the work force, improved post-harvest technologies, maintenance of international distribution routes, efficient diplomatic channels, global agricultural policies, an educated public, and dramatically increased crop productivity (Figure 1).

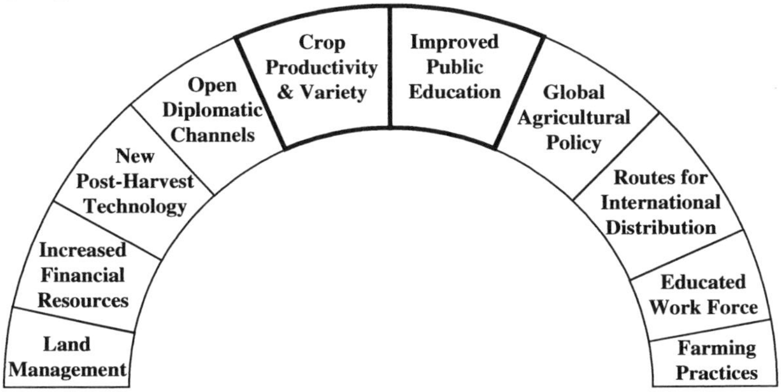

Figure 1. The ability of agriculture to feed the world's population in the 21st century demands improvements in the scientific, political, and agricultural arenas, as well as cooperation among all of the components listed.

At exactly the same time that agricultural output is required to increase in quality and quantity, environmental and ecological demands necessitate decreasing the use of pesticides, herbicides, fertilizers, and water. Identifying beneficial traits available in current germplasms will be essential to fulfilling future demands on agriculture due to the absence of necessary performance traits in current cultivated crops. It will be vital to incorporate as many of these traits as possible within the next 50 years using both selective breeding programs and recombinant DNA (rDNA) technologies for genetic modification (GM) [2].

1. GM Technology: Developments and Discussions

Traditional selective breeding techniques have been responsible for molding the form and composition of modern crops, fruits, and vegetables. Over the past 10,000 years, domestication and selective breeding have brought about large changes in plant variety, morphology, and composition [3]. Examples include modern corn (*Zea mays*) from *Zea mexicana* (teosinte) [4,5], modern bread wheat (*Triticum aestivum*) from a hybrid of *Aegilops sp.* and *Triticum turgidum* (Emmer wheat), and modern soybean (*Glycine max*) from *Glycine soja*. At the same time, many advantageous traits have been lost over the course of artificial selection [6], and reinserting some of these traits could dramatically improve productivity and value. Introgression of these quality traits into new crop varieties using traditional breeding may take as long as 5 to 15 years depending on the plant species and the trait(s) being incorporated. Given these time constraints, it is unlikely that traditional breeding programs could meet the agricultural demands of the world population within 50 to 100 years. A quick glance at the timeline presented by a rapidly increasing world population, combined with the throughput for developing new plant varieties and the fact that insufficient natural genetic variation may be present in some species, presents researchers with an impetus for advancement and acceptance of rDNA technology.

The use of rDNA technologies has fuelled a political debate over the past decade, primarily in Europe. Since the first plant transformations in the early 1980's, the power and potential benefit of these technologies are still being realized. Utilization of rDNA technologies facilitates a variety of genetic modifications including the introduction of foreign genes, introduction of modified native genes, altered expression patterns, or modulated expression levels. These advances have resulted in the ability to genetically modify crop species resulting in enhanced desirable traits or decreased undesirable traits.

In addition to improvement of classical agronomic traits, the full potential of GM technology also includes ecological applications such as in bioremediation [7,8], biosensors [9], and "pharmaceutical" applications such as development of edible vaccines [10,11] or expression of recombinant proteins having potential pharmaceutical value (*i.e.* insulin) in plants. Two important traits that have already been introduced using GM technology include herbicide and insect resistance. First, plants transformed with enolpyruvylshikimate-3-phosphate synthase are resistant to the herbicide glyphosate isopropylamine, the active ingredient in Roundup®. Roundup Ready® crops increase productivity by permitting herbicide use during crop growth and maturation. The herbicide eliminates competitive plants while not affecting the Roundup Ready® varieties. While some argue that this will actually increase the use of herbicides, proper application of Roundup® results in a reduction in ecological impact due to the low toxicity, negligible bioaccumulation and biodegradability of glyphosate isopropylamine [12,13] compared with alternative herbicides. Second, the decreases in crop productivity and the agricultural losses associated with insect damage have been addressed by insertion of insecticidal crystal (Cry) proteins into corn and cotton. These enterotoxic Cry proteins are naturally found in the soil bacterium, *Bacillus thuringiensis* (Bt), and have been extensively used as pesticides for 35 years in the form of foliar sprays. The increased crop productivity in Bt corn is tied directly to the pressure exerted by the pest [14, 15]. Increased yields in Bt crops occur only during growing seasons when the pest pressure is high.

As with all new technologies, it is important to balance benefit with safety. Initial concerns about the effects of Bt corn on non-pest insects, such as the Monarch butterfly (*Danaus plexippus*) [16], have been shown negligible [17]. The potential for development of Cry resistant pests remains a concern [18,19], but it is unlikely to become more problematic than pesticide resistance [20]. Insect resistance management (IRM) strategies are being used to alter current farming practices to maximize benefits and minimize potential for Cry resistant pests. According to IRM strategies, co-cultivating Bt crops and non-Bt crops has the potential to increase yields during growing seasons with high pest pressure and maintain yields during growing seasons with low pest pressure, while reducing the incidence of pest resistance and the ecological impact to non-pest species [21]. The benefits of both Roundup Ready® and Bt crops go beyond increased yield, reduced pesticide application, decreased post-harvest processing, and reduced use of ecologically unfriendly herbicides. These benefits also have the secondary effect of saving millions of gallons of gasoline due to fewer herbicide applications, fewer pesticide applications, and decreased post-harvest processing [22]; reduced worker exposure to pesticides during broadcast applications [23]; and reduced health related complications associated with herbicide application due to the low volatility of glyphosate isopropylamine.

Public acceptance of novel technology is typically associated with a perceived benefit. Foremost, GM crops with enhanced productivity should equate to lower cost foods that will have a major market influence. In addition, added benefits in nutritional or quality traits are highly beneficial. "Golden rice" was one of the first GM enhanced crops offering a beneficial nutritional trait for vitamin deficient populations dependent on rice as staple food [24-31]. On the global scale, this GM food has the potential capacity to eliminate blindness due to vitamin A deficiencies in up to half a million individuals per year [24, 32]. Other quality traits such as "no cry" onions that have modified enzymes associated with the synthesis of lachrymator compounds will eliminate the unpleasant experience often associated with onions [33]. Continued metabolic engineering is expected to produce food sources with added nutritional benefits in the very near future.

Although GM crop technology has been adopted at a rate faster than any other major new technology this century, it has nevertheless been the subject of controversy. Both the public and select members of the scientific community have concerns about GM crop safety related to environmental impact and consumption of food products. To help address these concerns, the Plant Journal published a series of articles on GM crop plants that are freely accessible via the Internet. Articles include reviews on the ecological impact [34], regulations [35] and food safety [36] of GM plants.

Geographical inconsistencies currently exist with respect to the acceptance and use of GM crops. The United States, Argentina, Canada, and China together account for 99% of GM crops planted [37]. Other large countries, such as India and Brazil, have recently instituted governmental policy to implement GM crops. Globally, over 175 million hectares of GM crops were planted in 16 different countries between 1996 and 2001 [35]. In contrast, many other countries, including those in the European Union, have either placed a moratorium on, or actively discouraged, the use of GM crops. Some of the controversy over the use and safety of GM crops is counterintuitive since the fundamental and applied research that produces the medicines consumed by these countries uses similar technology. Although some individuals may be willing to take medicines produced by genetically modified organisms (GMOs) on a short-term basis, consumers appear to be more cautious about their long-term dietary intake.

In some cases, the distrust of GMOs by the general public can be correlated to a lack of scientific understanding. For example, surveys have shown that some members of the public are unaware that non-GM plants contain DNA or chemicals. The lack of knowledge about the chemical nature of life is linked to a fear of the unknown, and exacerbated by a distrust of governmental agencies. Some of these fears may be addressed by quantifying the changes in GM crops induced by rDNA technologies relative to known and trusted products (i.e. non-GMOs). New technologies now enable more comprehensive methods for comparative global analyses of mRNA, protein, and metabolites, and therefore defining the GM crop in terms of exactly what it does or does not contain when compared with all available non-GM varieties of the same crop.

2. Assessing Safety

There are multiple US agencies responsible for the assessment of GMO safety. These include the US Department of Agriculture (USDA), Environmental Protection Agency (EPA), and the Food and Drug Administration (FDA). The USDA evaluates field testing and environmental safety issue to assure that the new plant varieties do not present a "plant pest" risk or the potential for significant impact on the environment through non-target activity, outcrossing, or protein expression. EPA regulates pesticide applications and the pesticidal properties of plants. The FDA is responsible for food and feed safety, nutritional equivalence, compositional equivalence, and allergenicity potential.

2.1 Substantial Equivalence

Although there is no *a priori* rationale, concerns about the safety of GM crops include increased anti-nutrient content due to changes in metabolism, allergic reactions due to introduced foreign proteins, and potential DNA transfer to the consumer. Quelling public and scientific concern about GM crops has led to the development of decision trees for safety assessment. Currently, the only standardized criteria for safe GM crops is the concept of substantial equivalence. A GM crop and its parent line are considered to be substantially equivalent if the quantified characteristics for the two demonstrate a difference of less than 20% [38]. The GM crop may also be categorized as substantially equivalent except for the inserted gene as long as the only difference between the two crop lines is the inserted gene or gene product. It is also possible to be categorized as not substantially equivalent, and thus require further testing based upon the nature of the gene insert or the identified non-equivalence [39].

Even with substantial equivalence established, concerns about DNA transfer still remain. Recombinant DNA concentration is approximately 1-10 ppb in most GM crops. The low concentration of rDNA alone makes the probability of direct gene transfer improbable. In addition, ingested DNA is rapidly digested and considered to be safe [40]. In contrast, small fragments of soybean leaf DNA (from a non-GM variety) have been found in mouse liver and spleen [41]. Direct feeding studies using green fluorescent protein (GFP) DNA have produced similar results. The small fragments of soybean leaf and GFP DNA present in the liver and spleen are not integrated into the mouse DNA, nor were they expressed or passed to the next generation [41], and thus posed no adverse risks.

The potential for allergic reaction remains the most significant concern associated with GM crop consumption. This, however, is very unlikely to be a problem unless the product of the transgene itself is directly allergenic. Standard protocols call for bioinformatic analysis to compare the translation product of the inserted rDNA to known allergens. Any gene product from rDNA that shares >35% homology or more than 6 contiguous amino acids with a known allergen must be isolated and subjected to further testing to determine the level of risk involved in releasing the GM crop for public consumption. Testing may require purifying the protein and subjecting it to further testing using animal models and test sera from individuals with known allergies to the homologous allergen [42]. Such tests are not required for plants produced by classical breeding, which can, by its nature, introduce unknown allergenic products. For example, an estimated 1 to 2% of adults and approximately 5% of children are allergic to non-GM products currently on the market [43]. These products are generally considered safe and include tree nuts, peanuts, cereal grains (wheat, rye and barley), in addition to non-plant products such as cow's milk, eggs, fish, and shellfish.

2.2 Unintended Changes

Our ability to characterize potential non-target changes induced by rDNA technologies can benefit from the use of global analytical techniques to adequately characterize the differences between a GM crop and parental lines. Further, the future of genetic engineering in plants will most likely proceed with the introduction of nutritional enhancements. These traits are expected to result in both quantitative and qualitative phenotypic changes, with potential for inducing both intended and unintended effects [36]. Global profiling will thereby be necessary for understanding the biochemistry as well as the safety assessment of these lines. The development of large-scale quantitative and qualitative profiling techniques for measuring gene expression products at the mRNA, protein, and metabolite levels coincides with the demands for GM crop safety assessment.

Transcriptomics, proteomics, and metabolomics will play key roles in evaluating the success of the intended genetic engineering, elucidating unintended effects, and determining substantial equivalence. Determination of substantial equivalence employs standard statistical analyses to determine if the quantified characteristics are statistically different between the crops being compared. This type of analysis is helpful for determining if the single gene or multiple gene insertions present in new varieties of GM crops have resulted in unintended changes.

2.3 Safety Assessments

The three stages at which safety assessment should be used are gene discovery, line selection, and commercialization [44]. The best time to assess the safety of gene products harboring homology to known allergens or toxins is during the gene discovery phase. Positive results indicating a toxicity or allergenicity potential at this stage prevent loss of time and money necessary for line selection and commercialization. Using transcriptomics, proteomics, and metabolomics to determine substantial equivalents during the line selection stage will provide information about the maximum level of expression that maintains substantial equivalency and minimum level of expression that provides the desired trait. Identifying specific lines demonstrating both substantial equivalence as well as the desired traits will not only save time and money bringing beneficial products to market, but will also ensure that the final product will comply with GM crop regulations.

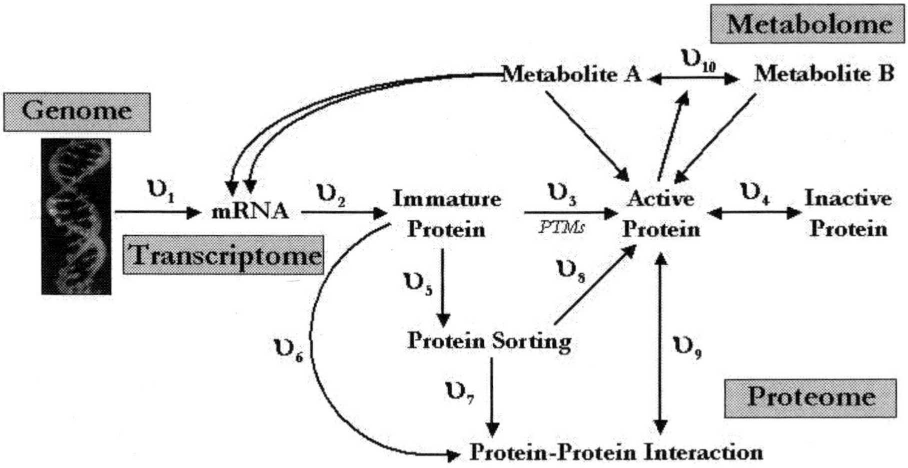

Figure 2. Holistic view of the multi-level reaction networks which comprise the dynamic biochemistry of life. These networks can be visualized in four general categories including the static genome, the transcriptome, the proteome, and the metabolome.

3. Integrated Functional Genomics

Cellular biochemistry consists of a series of reaction networks beginning with a static genome (Figure 2). The genome is transcribed to form mRNA that must be transported to the cytosol and loaded onto ribosomes before being translated into protein. The pool of mRNA present in the cytosol represents both free mRNA and polysomal mRNA. Polysomal mRNA represents transcripts that are being actively translated into protein. These proteins may or may not be active enzymes following translation. Often proteins must undergo further post-translational modifications prior to enzymatic activation. In addition, many proteins contain signaling sequences targeting them to specific sub-cellular locations where the signal sequences are cleaved and the enzyme activated. It may be necessary for proteins to engage in protein-protein interactions before resulting in enzymatically active oligomeric complexes. The active enzymes and enzyme complexes then catalyze the chemistry of life. Flux through these metabolic pathways is governed by multiple parameters including reaction kinetics, mRNA stability, protein turnover rates, as well as feed-forward and feedback regulatory mechanisms. If one is to comprehend the impact of genetic perturbations and successfully engineer complex metabolic pathways, then global views of transcription, translation, and metabolism are critical.

3.1 Functional Genomics and Systems Biology

The scale upon which biology can now be queried has rapidly evolved toward the large-scale quantitative and qualitative profiling of gene expression products at the mRNA, protein, and metabolite levels. These large-scale approaches are generally referred to as functional genomics or systems biology (Figure 3). High-density cDNA and oligonucleotide "GeneChips" are currently available for a large number of species including Arabidopsis, rice, *Medicago truncatula*, and tomato which enable simultaneous analysis of the entire genome for sequenced organisms or many thousands of potential transcripts from

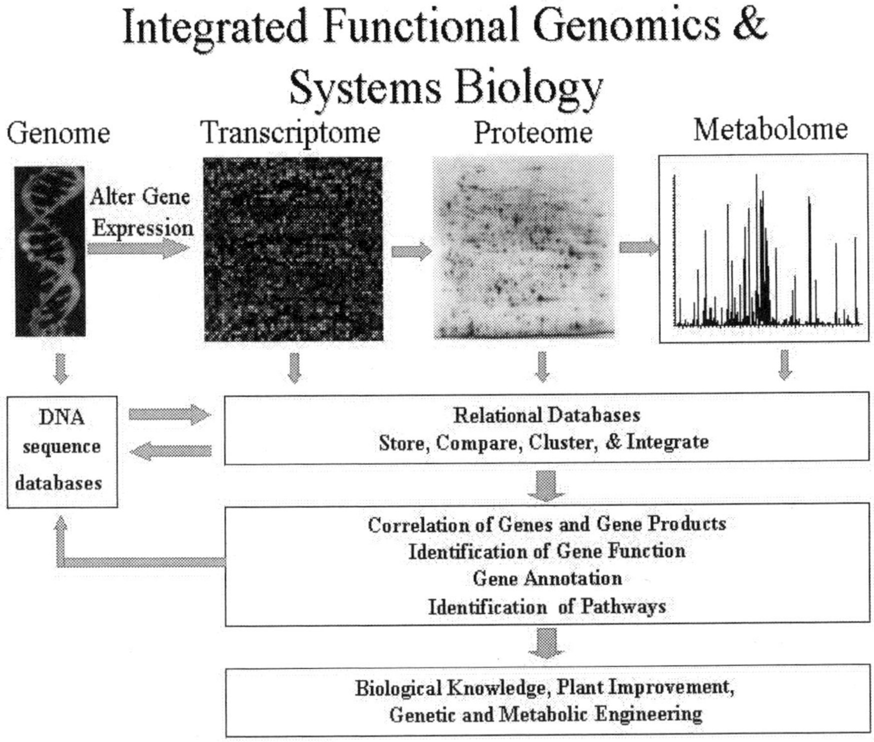

Figure 3. Integrated functional genomic data collection and down-stream processing. Transcript profiling is achieved using high-density microarrays; protein profiling is achieved using 2-DE, and metabolite profiling is accomplished using multiple hyphenated techniques such as GC-MS and LC-MS. Processed data are deposited into relational databases capable of storage and interrogation through correlation analysis, cluster analysis, or analysis of variance. The results of these analyses are used for gene annotation and the identification of novel pathways. Ultimately, the integration of the data will provide the underlying knowledge of biological processes that will enable plant improvement through metabolic engineering.

unsequenced organisms. In addition, two-dimensional electrophoresis (2-DE) and other multidimensional separation technologies now exist that are capable of analyzing thousands of proteins simultaneously. More recently, technologies such as NMR, GC-MS, and LC-MS have been recruited to simultaneously profile large numbers of metabolites. Each of these integrated functional genomics technologies are discussed in more detail below. The important point is that the comprehensive or large volume data generated from these experiments represent global snapshots of the dynamic biochemistry of life. When these snapshots are compared before and after genetic perturbations, differences can be used to decipher gene function or to evaluate the holistic response of a biological system to both biotic and abiotic stimuli. By combining analyses of the transcriptome, proteome, and metabolome changes in biochemical pathways can be characterized at all levels providing a comprehensive understanding of the intricacies of biochemistry never before possible. Due to the comprehensive nature of these global data sets, they are also highly useful in evaluating the intended and unintended consequences of transgene insertion, and thus offer a high level of confidence in the safety assessment of GM crops.

3.2 Transcriptomics

Transcriptome analysis has the ability to quantify changes in global expression of messenger RNA (mRNA) following genetic transformation. This technology employs DNA microarrays (GeneChips) designed from either genomic data or from tentative consensus (TC) sequences created using aligned expressed sequence tags (ESTs). Neither approach to GeneChips guarantees representation of all possible mRNA transcripts. GeneChips designed from genomic data rely on bioinformatic approaches to predict open reading frames (ORF) and splicing sites, whereas GeneChips designed from TCs have only *in silico*-assembled transcripts as their basis. Despite these shortcomings, transcriptomics remains a widely used technique for studying global gene expression. A GeneChip may only contain a hundred oligonucleotides in a directed analysis of specific genes, or may contain tens of thousands of oligonucleotides for genome wide experiments involving Arabidopsis, rice, tomato, wheat, barley, or *M. truncatula* (Figure 4). Applications for GeneChips technologies include basic measurements of transcript abundance and stability as well as the identification of novel genes, metabolic pathways, and biological responses.

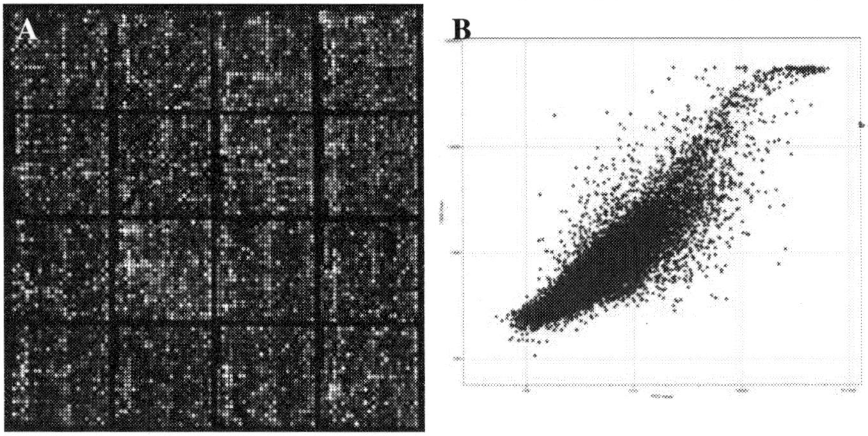

Figure 4. The *Medicago truncatula* 16,000 70-mer oligo set (Version 1.0; Qiagen-Operon). (A). One half of the microarray showing hybridization of *M. truncatula* cell cultures: reference RNA and RNA from methyl jasmonate (MJ)-elicited cell culture. (B). A log-log plot of the Cy3 vs. Cy5 signal ratio showing genes up and down regulated in response to MJ application (N. Aziz and G.D. May, unpublished results).

3.3 Proteomics

The goal of proteomics is global analysis of all proteins present within an organism, tissue, or cell type being studied. This includes extraction, separation, quantification, and identification of the individual proteins. The most common method employed for this type of analysis is 2-D gel electrophoresis (2-DE), which separates proteins based on their isoelectric point (pI) and their molecular weight (MW). The proteins are then fixed, and visualized using stains such as coomassie brilliant blue G-250 (Figure 5).

Protein spots of interest are dissected from the gel and enzymatically digested using a protease such as trypsin (EC 3.4.21.4). Peptides eluted from the gel piece may be analyzed by matrix assisted laser desorption ionization time-of-flight mass spectrometry (MALDI-TOFMS) to generate a peptide mass fingerprint (PMF), or can be subjected to liquid chromatography tandem mass spectrometry (LC-MS/MS) to generate sequence tag data (STD) in combination with the PMF (Figure 6).

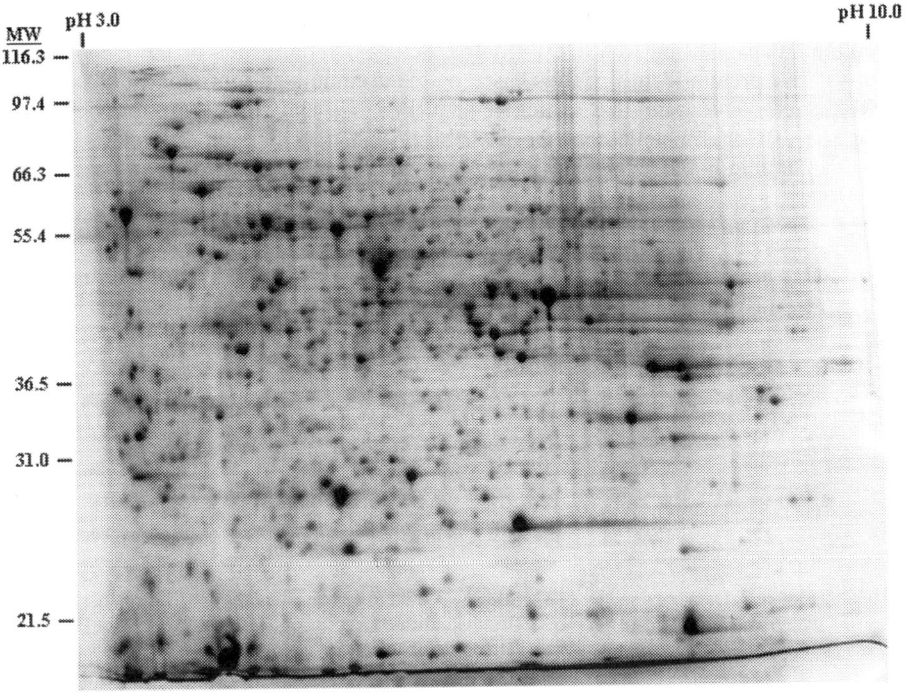

Figure 5. Protein extracted from *Medicago truncatula* cell culture separated on a 24 cm, pH 3-10 nonlinear, Immobiline Dry Strip™ gel (Amersham Biosciences). Staining the gel with coomassie brilliant blue G-250 reveals approximately 1600 proteins.

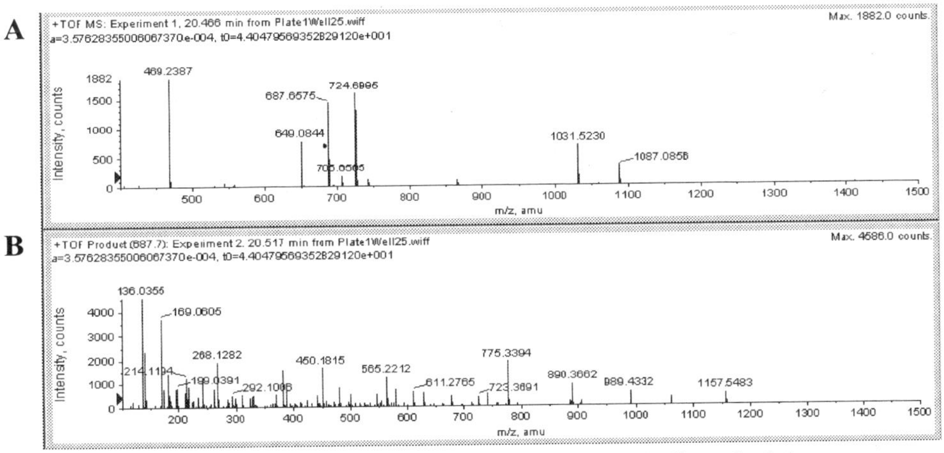

Figure 6. LC-MS/MS data of peptides extracted from a trypsin-digested gel plug. (A) Peptide mass fingerprint (PMF) at 20.466 minutes. (B) Collision-induced fragmentation of the 687.7 m/z peptide. This data was used to generate the sequence tag (STD).

GC-MS has been used to resolve temporal and spatial differences in metabolite concentration during development of the model legume *Medicago truncatula* [51]. Data obtained from these studies indicate that GC-MS techniques can resolve approximately 300 components in each of the polar and non-polar chromatograms with minimal redundancy. Combining the number of components in the polar and nonpolar chromatograms with a detailed 10-week developmental study of the leaves, stems, and roots, resulted in a data set containing approximately 500,000 data points. Differences among developmental stages were then determined using multivariate analyses, correlation coefficients, and metabolic networks. This same technology can easily be used to compare new varieties or GM crops with currently consumed crops.

Genetic modification of alfalfa (*Medicago sativa* L.) lignin pathway intermediates has been performed in control and corresponding transgenic lines. These lines were then studied using metabolic profiling and principal component analysis (PCA) to evaluate lignin biosynthesis [52]. The transgenic lines, down-regulated in either caffeic acid 3-*O*-methyltransferase (COMT) or caffeoyl CoA 3-*O*-methyltransferase (CCoAOMT) using anti-sense constructs, were derived from *Medicago sativa* cv. Regen SY. PCA analysis of phenolic profiles obtained from leaf tissue suggested that the transgenic lines were indistinguishable from the Regen SY control line (Fig 9, I.A) suggesting substantial equivalence in leaf tissue. Interestingly, the PCA data indicate that the transgenic leaf profiles were more similar to the parent line than to those from another cultivar (alfalfa C349) (Fig. 9.I.A and 9.I.B). Thus, it is clear that different cultivars of a crop species generated by classical breeding are not, by the strict definition, substantially equivalent. The PCA analysis of stem tissue reveals clear differentiation of the alfalfa C349, Regen SY control, COMT, and CCOMT lines (Fig 9.I.D-9.I.F). This differentiation in PCA space can be attributed to variations in levels of caffeoyl glucoside.

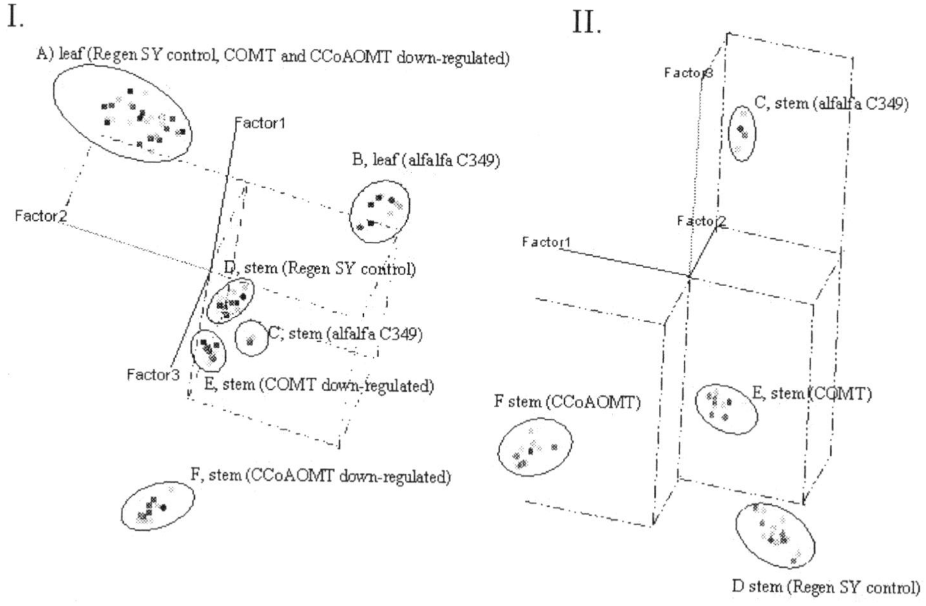

Figure 9. Principal component analysis (PCA) of soluble phenolics data obtained for leaves and stems from four alfalfa genotypes by HPLC with UV detection measured at three wavelengths (254, 280, 310 nm). Panel I includes both leaf and stem data; whereas, panel II contains an expanded view of only stem data. Clustering was observed for A, leaf (Regen SY control, COMT and CCoAOMT down-regulated); B, leaf (alfalfa C349); C, stem (alfalfa C349); D, stem (Regen SY control); E, stem (COMT down-regulated); F, stem (CCoAOMT down-regulated) datasets.

4. Conclusions and Future Outlook

Currently, genetic transformation methods are available for a large number of plant species and have generated crops with enhanced productivity and nutritional content. Although GM crops are not yet globally accepted, farmers are adopting them at an amazing rate. The examples of Bt and Round-up Ready crops represent only a fraction of the potential utility of GM crops as field trials have been conducted on 52 species of field crops, vegetables, herbaceous, and woody plants [53]. Given the current international production levels, numerous field trails, and substantial research associated with GM organisms, it is not only evident that GMOs are here to stay but also that these organisms are moving into main-stream agriculture. Although GM crops have great potential and value, it is important that we clearly understand and define the impacts of genetic engineering on the biochemical composition and safety of these products. The comprehensive data generated through integrated functional genomics technologies are tailor-made for assisting in the assessment of GM crop safety. The depth to which these technologies can currently evaluate substantial equivalence is impressive, and their efficacy will continue to improve as the technologies associated with transcriptomics, proteomics, and metabolomics evolve.

Although functional genomic technologies are critical to the actual process of genetic engineering, these technologies will also play an important role in the 'GM debate' through the qualitative and quantitative definition of intended and unintended effects. Definition of both the intended and unintended biochemical effects in a comprehensive manner will dispel the fear of the unknown and provide greater reassurance to the consumer. Functional genomic tools will provide continual evidence that natural variation in compositional traits of classically generated plant varieties is significantly greater than that associated with GM lines, and thereby document the greater precision of genetic engineering of traits relative to traditional breeding approaches. This should provide additional understanding and positive reinforcement of the benefits and safety of GM plants.

Acknowledgements. Work in the authors' laboratories was supported by the Samuel Roberts Noble Foundation, the National Science Foundation, and the US Department of Energy.

References

[1] Cohen, J.E. (2003). Human population: the next half century. Science 302, 1172-1175.

[2] Lee, M. (1998). Genome projects and gene pools: new germplasm for plant breeding? Proc. Natl Acad. Sci. USA 95, 2001-2004.

[3] Lev-Yadun, S., Gopher, A. and Abbo, S. (2000). Archaeology. The cradle of agriculture. Science 288, 1602-1603.

[4] Benz, B.F. (2001). Archaeological evidence of teosinte domestication from Guila Naquitz, Oaxaca. Proc. Natl Acad. Sci. USA 98, 2104-2106.

[5] Piperno, D.R. and Flannery, K.V. (2001). The earliest archaeological maize (*Zea mays* L.). from highland Mexico: new accelerator mass spectrometry dates and their implications. Proc. Natl Acad. Sci. USA 98, 2101-2103.

[6] Tanksley, S.D. and McCouch, S.R. (1997). Seed banks and molecular maps: unlocking genetic potential from the wild. Science 277, 1063-1066.

[7] Watanabe, M.E. (2001). Can bioremediation bounce back? Nat. Biotechnol. 19, 1111-1115.

[8] Ruiz, O.N., Hussein, H.S., Terry, N. and Daniell, H. (2003). Phytoremediation of organomercurial compounds via chloroplast genetic engineering. Plant Physiol. 132, 1344-1352.

[9] Manak, M.S., Paul, A.L., Sehnke, P.C. and Ferl, R.J. (2002). Remote sensing of gene expression in Planta: transgenic plants as monitors of exogenous stress perception in extraterrestrial environments. Life Supp. Biosph. Sci. 8, 83-91.

[10] Mason, H.S., Haq, T.A., Clements, J.D. and Arntzen, C.J. (1998). Edible vaccine protects mice against *Escherichia coli* heat-labile enterotoxin (LT): potatoes expressing a synthetic LT-B gene. Vaccine 16, 1336-1343.

[11] Giddings, G., Allison, G., Brooks, D. and Carter, A. (2000). Transgenic plants as factories for biopharmaceuticals. Nat. Biotechnol. 18, 1151-1155.
[12] WHO (1994). Glyphosate. Environmental Health Criteria 159. Geneva: World Health Organization, International Programme on Chemical Safety.
[13] WHO (1998). Guidelines for Drinking. Water Quality. Vol. 2, pp. 53-59.
[14] Catangui, M.A. and Berg, R.K. (2002). Comparison of *Bacillus thuringiensis* corn hybrids and insecticide-treated isolines exposed to bivoltine European corn borer (Lepidoptera: Crambidae) in South Dakota. J. Econ. Entomol. 95, 155-166.
[15] Catangui, M.A. (2003). Transgenic *Bacillus thuringiensis* corn hybrid performance against univoltine European corn borer (Lepidoptera: Crambidae) in South Dakota. J. Econ. Entomol. 96, 957-968.
[16] Losey, J.E., Rayor, L.S. and Carter, M.E. (1999). Transgenic pollen harms monarch larvae. Nature 399, 214.
[17] Sears, M.K., Hellmich, R.L., Stanley-Horn, D.E. et al. (2001). Impact of Bt corn pollen on monarch butterfly populations: a risk assessment. Proc. Natl Acad. Sci. USA 98, 11937-11942.
[18] Gonzalez-Cabrera, J., Herrero, S. and Ferre, J. (2001). High genetic variability for resistance to *Bacillus thuringiensis* toxins in a single population of diamondback moth. Appl. Environ. Microbiol. 67, 5043-5048.
[19] Griffitts, J.S., Huffman, D.L. and Whitacre, J.L. (2003). Resistance to a bacterial toxin is mediated by removal of a conserved glycosylation pathway required for toxin-host interactions. J. Biol. Chem. 278, 45594-45602.
[20] Georghiou, G.P. and Lagunes-Tejeda, A. (1991). The occurrence of resistance to pesticides in arthropods. Rome: FAO.
[21] Glaser, J.A. and Matten, S.R. (2003). Sustainability of insect resistance management strategies for transgenic Bt corn. Biotechnol. Adv. 22, 45-69.
[22] Fawcett, R. and Towery, D. (2002). Conservation Tillage and Plant Biotechnology: How new technologies can improve the environment by reducing the need to plow. West Lafayette, Indiana: Conservation Technology Information Center.
[23] Pray, C.E., Huang, J., Hu, R. and Rozelle, S. (2002). Five years of Bt cotton in China - the benefits continue. Plant J. 31, 423-430.
[24] Potrykus, I. (2003). Nutritionally enhanced rice to combat malnutrition disorders of the poor. Nutr. Rev. 61, S101-104.
[25] Chong, M. (2003). Acceptance of golden rice in the Philippine 'rice bowl'. Nat. Biotechnol. 21, 971-972.
[26] Thomson, J.A. (2002). Research needs to improve agricultural productivity and food quality, with emphasis on biotechnology. J. Nutr. 132, 3441S-3442S.
[27] King, J.C. (2002). Biotechnology: a solution for improving nutrient bioavailability. Int. J. Vitam. Nutr. Res. 72, 7-12.
[28] Beyer, P., Al-Babili, S. and Ye, X. (2002). Golden Rice: introducing the beta-carotene biosynthesis pathway into rice endosperm by genetic engineering to defeat vitamin A deficiency. J. Nutr. 132, 506S-510S.
[29] Potrykus, I. (2001). Golden rice and beyond. Plant Physiol. 125, 1157-1161.
[30] Normile, D. (2000). Agricultural biotechnology. Monsanto donates its share of golden rice. Science 289, 843-845.
[31] Ye, X., Kloumlti, A., Zhang, J. et al. (2000). Engineering the provitamin a (-carotene) biosynthetic pathway into (carotenoid-free) rice endosperm. Science 287, 303-305.
[32] Sommer, A. (1989). New imperatives for an old vitamin (A). J. Nutr. 119, 96-100.
[33] Imai, S., Tsuge, N. and Tomotake, M. (2002). Plant biochemistry: an onion enzyme that makes the eyes water. Nature 419, 685.
[34] Conner, A.J., Glare, T.R. and Nap, J.P. (2003). The release of genetically modified crops into the environment. Part II. Overview of ecological risk assessment. Plant J. 33, 19-46.
[35] Nap, J.P., Metz, P.L., Escaler, M. and Conner, A.J. (2003). The release of genetically modified crops into the environment. Part I. Overview of current status and regulations. Plant J. 33, 1-18.
[36] Kuiper, H.A., Kleter, G.A., Noteborn, H.P. and Kok, E.J. (2001). Assessment of the food safety issues related to genetically modified foods. Plant J. 27, 503-528.
[37] NASS (2002). Prospective Plantings. March 2002.
[38] Nordic Council (2000). Safety Assessment of Novel Food Plants. Chemical analytical approaches to the establishment of substantial equivalence. Copenhagen, Denmark: Nordic Council of Ministers.
[39] OECD (1993). Safety Evaluation of Foods Derived by Modern Biotechnology. Paris, France: OECD.
[40] Beever, D.E. and Kemp, C.F. (2000). Safety issues associated with the DNA in animal feed derived from genetically modified crops. A review of scientific and regulatory procedures. Nutr. Abstr. Rev. (Series B: Livestock Feeds and Feeding) 70, 175-182.
[41] Hohlweg, U. and Doerfler, W. (2001). On the fate of plant or other foreign genes upon the uptake in food or after intramuscular injection in mice. Mol. Genet. Gen. 265, 225-233.

[42] FAO/WHO (2001). Allergenicity of genetically modified foods. Report of a joint FAO/WHO expert consultation on foods derived from biotechnology. Rome, Italy: FAO
[43] Kimber, I. and Dearman, R.J. (2002). Factors affecting the development of food allergy. Proc. Nutr. Soc. 61, 435-439.
[44] Taylor, S.L. (2003). Safety assessment of foods produced through agricultural biotechnology. Nutr. Rev. 61, S135-140.
[45] Watson, B.S., Asirvatham, V.S., Wang, L. and Sumner, L.W. (2003). Mapping the proteome of barrel medic (*Medicago truncatula*). Plant Physiol. 131, 1104-1123.
[46] Washburn, M.P., Wolters, D. and Yates III, J.R. (2001). Large-scale analysis of the yeast proteome by multidimensional protein identification technology. Nat. Biotechnol. 19, 242-247.
[47] Yates, III J.R. (1998). Mass spectrometry and the age of the proteome. J. Mass. Spectrom. 33, 1-19.
[48] Gygi, S.P., Rist, B., Gerber, S.A. et al. (1999). Quantitative analysis of complex protein mixtures using isotope-coded affinity tags. Nat. Biotechnol. 17, 994-999.
[49] Smith, R.D., Anderson, G.A. and Lipton, M.S. (2002). An accurate mass tag strategy for quantitative and high-throughput proteome measurements. Proteomics 2, 513-523.
[50] Ferguson, P.L. and Smith, R.D. (2003). Proteome analysis by mass spectrometry. Annu. Rev. Biophys. Biomol. Struct. 32, 399-424.
[51] Duran, A.L., Broeckling, C.D., Sumner, L.W. et al. (2003). Temporal and spatial metabolomics of *Medicago truncatula* development. Potsdam, Germany: Proc. 2nd Int. Conf. Plant Metabol.
[52] Chen, F., Duran, A.L. and Blount, J.W. (2003). Profiling phenolic metabolites in transgenic alfalfa modified in lignin biosynthesis. Phytochem. 64, 1013-1021.
[53] Dunwell, J.M. (2000). Transgenic approaches to crop improvement. J. Exp. Bot. 51, 487-496.

Genomics for Biosafety in Plant Biotechnology
J.P.H. Nap, A. Atanassov and W.J. Stiekema (Eds.)
IOS Press, 2004

NMR Approaches to Detect Unintended Effects of Genetic Modification in Plants

Marianne DEFERNEZ* and Ian J. COLQUHOUN
*Institute of Food Research, Norwich Research Park, Colney, Norwich NR4 7UA, United Kingdom; *email: marianne.defernez@bbsrc.ac.uk*

Abstract. NMR is introduced as a method for metabolite profiling. The benefits of the technique, typical sample preparation for plant extract and methodology for signal assignments are presented, before the usefulness and shortcomings of various approaches for data analysis (univariate and multivariate) are discussed. Finally a case study is used to discuss ways to better interpret and assess the significance of the observations we make when analysing genetically modified products. The objective is to provide a comparative assessment of their composition, therefore assisting the evaluation of their safety.

Introduction

One important requirement for the safety assessment of genetically modified foods is that for the analysis of their composition. The reasoning behind this is that in terms of safety and nutritional quality, GM foods can be considered acceptable if their composition is shown not to be significantly different from that of existing comparable foods that are already regarded as safe. This is the concept of 'substantial equivalence', which has been largely adopted by regulatory bodies as a key step towards approval of GM crops [1]. Food found not to be substantially equivalent may require further testing and safety assessment. Great attention is given to any new components that are introduced deliberately by the modification. They obviously fall outside the substantial equivalence framework and are treated separately.

The comparison between new and existing crops is carried out across a range of analytes, and in practice this and the necessary analyses are decided in advance using the best current knowledge in nutrition, toxicology etc. However there is always the difficulty with any set of predefined or targeted analyses that unintended effects of the genetic modification may escape detection. This is true for all types of genetic modification, but particularly so with direct metabolic perturbations aimed at single enzymes or reactions: these may lead to the modification of further pathways and increase or decrease in the amount of metabolites other than those intended, through the further regulation of metabolic pathways [2]. This problem is the main motivation for the development of 'non-targeted' molecular profiling methods for safety assessment of GMOs [3], which together with targeted approaches, can contribute to the safety assessment of crops. In our work we use ^{1}H-NMR, GC/MS, and LC/MS, as shown schematically in Figure 1. We report here on the methodology used for the application of ^{1}H-NMR spectroscopy. We present the technique and its usage in the context of comparative compositional analysis for safety assessment, and conclude with thoughts on the questions to be asked to make such assessment fair and complete.

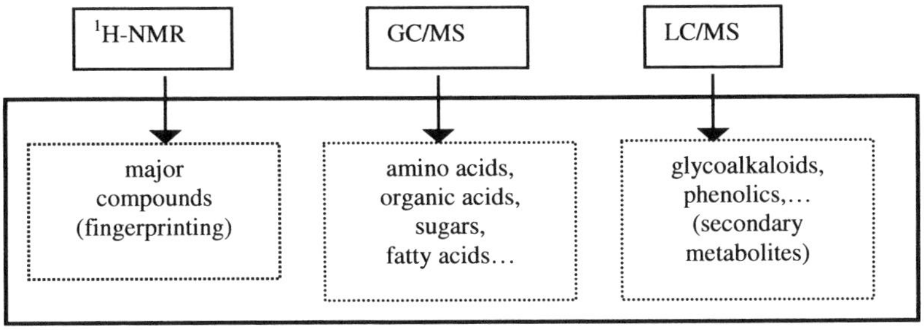

Figure 1. Metabolite profiling techniques and examples of compounds detected

1. Benefits of NMR for Metabolite Profiling

^{1}H-NMR spectroscopy is one of the techniques whose potential is currently being explored in the emerging field of metabolomics. The reasons for this are the advantages that this technique offers. These include:

- A broad range of compounds can be analysed: it is a non-targeted method, in that in principle it produces signals for all proton-containing chemical species.
- It is a relatively rapid technique: a spectrum can be obtained in a few minutes. In our research, a run of typically 20 min gives an acceptable signal-to-noise ratio [4,5]. There is also the possibility to use an autosampler, which increases the potential throughput as samples can be queued to run automatically when the operator is not present.
- The sample preparation can be kept quite simple. There is, of course, the possibility to purify samples prior to analysis. For metabolite profiling, crude extracts only are prepared. These give rise to complex, but good quality NMR spectra (see below).
- Features in the spectra carry information on the structure of the compounds giving rise to the signals. The structure of unknown compounds can be inferred from the spectra.

The main drawback of NMR is its low sensitivity, which restricts its use to the study of major compounds in our metabolite profiling work.

2. Sample Preparation and NMR Parameters for Plant Extracts

The main types of samples that have been the subject of our studies include potato tubers, tomato fruits, and Arabidopsis leaves. In this paper we present the general methodology, but include more specific details on the study on potato tubers. A typical protocol would be for each extract to be prepared as follows:
(1) addition of 70% methanol-d_4 / 30% buffer to freeze-dried powder of the material;
(2) stirring of the mixture;
(3) centrifugation;
(4) collection of the supernatant and storage at -18° C until required for analysis.
Further details for specific applications can be obtained from relevant publications [4,5].

The ^{1}H-NMR spectra of extracts of e.g. potato tubers are recorded at 27°C on a 400MHz JEOL GX spectrometer, with methanol-d_4 as the internal lock. Typically each spectrum consists of 100-500 scans of 8192 complex data points with a spectral width of 5000 Hz, an acquisition time of 1.64 s and a recycle delay of 2 s per scan. The pulse angle is 50°. A presaturation sequence is used to suppress the residual water signal with low power selective irradiation at the water frequency during the recycle delay. Spectra are Fourier transformed with 1Hz line broadening, phased and baseline corrected.

yet have a chemical assignment. There are many ways in which such data can be analysed and displayed, for instance calculating distribution parameters for populations of samples (for instance two populations could be a group of genetically modified and a group of control samples), drawing histograms or making comparisons between different populations using box-and whiskers plots, showing the fold increase/decrease (in genetically modified samples compared to controls for instance) in a number of selected compounds.

4.2 Multivariate Analysis

The purpose of multivariate analysis is to examine many measurements simultaneously. With NMR spectra what constitutes a measurement can be a number of things. First of all it could be the integrated value, as used in univariate analysis, only here a technique able to handle many such measurements would be used (see below). Alternatively, it could simply be the NMR intensity at a given frequency; indeed each of these is in effect a measurement, and there are as many measurements as there are digitised points in the spectrum (which for our spectra would typically be 8192, although the baseline at high and low field is usually removed and leaves us with ~5000 points). The input in the latter case would be the whole of the NMR trace.

There are a number of multivariate (also called chemometric) techniques. In the context of evaluating genetic modification most useful are those that allow to carry out data exploration, simplify the data, look for differences between groups of samples, detect outliers, etc. The most commonly used technique for such purposes is principal component analysis (PCA) [6]. This is an unsupervised technique, which means that the origin or status of the samples (e.g. whether they are genetically modified or not) is not used in the analysis. Other unsupervised techniques include cluster analysis, whilst supervised techniques include canonical variates analysis and discriminant analysis, as well as partial least squares (often used for quantification but can also be used for classification). Because for these the group membership (i.e. GMO/ non-GMO) is used to carry out the analysis, the danger of overfitting (resulting in over-optimistic results or non-genuine clustering) is higher [7]. Some techniques can be used in both ways, particularly the natural computation methods such as genetic algorithms and artificial neural networks. The advantages and potential problems associated with such approaches are illustrated below.

4.3 General Elements of Principle Component Analysis (PCA)

PCA is a mathematical operation that allows to rearrange the information contained in e.g. a series of spectra and to display it in a much-simplified manner. Hence it constitutes a convenient way to gain an overall picture of the relative composition within a set of samples. Essentially the input data used in this analysis is a series of measurements on a number of variables. PCA defines new variables and the data is then expressed as the co-ordinates (or *scores*) of the samples on this new set of variables (or PC axes). Typically, most of the information in the data set is condensed in the first few PC axes, because the way these are defined is such that the variance of the scores on successive PCs is maximised. In addition, the axes are orthogonal to one another and the data is thus uncorrelated. The relationship between the initial variables and the new axes is expressed through the *loadings*, which are the weights each of the initial variables have in the linear combination that defines the new axes. As mentioned above a possible input for such analysis could be a series of integrated values, or alternatively the whole NMR trace.

4.4 Specific Issues for Multivariate Analysis of Whole NMR Spectra

An underlying assumption of techniques such as PCA is that each of the original variables corresponds to the same measurement throughout the series of samples. Our work has shown that the analyses are affected by shifts on the frequency axis and by variation in line shape [8]. Our conclusions were that the sample preparation and the careful adjustment of the resolution were important in optimising the repeatability of the spectra and the suitability of methods such as PCA to analyse them. Even with the buffer, very slight shifts due to pH differences are sometimes observed. These are not easy to deal with, but we have used a procedure based on that developed by Vogels et al. [9,10] with relatively good success, to adjust the alignment prior to multivariate analysis. A number of researchers also use as input deresolved spectra, particularly when using the "bucketing" procedure offered by software packages associated with Bruker spectrometers. Our own experience of deresolving spectra has not usually met with success in avoiding altogether the aforementioned issues, when a problem existed in the original data. The critical stages at which care ought to be taken are shown in Figure 5.

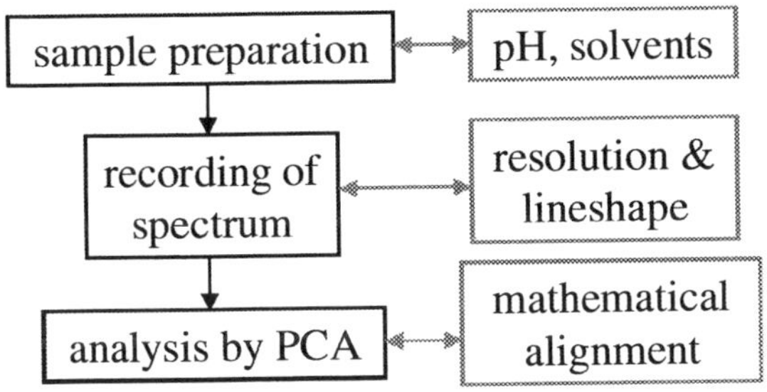

Figure 5. Critical stages in NMR experiments used in conjunction with PCA

A summary of the main advantages and disadvantages of the different approaches is given in Table 1. A combination of approaches is generally most effective to best define the magnitude of the variation that may be observed in a data set.

Table 1. Summary of advantages (+) and disadvantages (–) of the different approaches

	Univariate analysis on each separate 'integrated' value	**Multivariate analysis on all 'integrated' values**	**Multivariate analysis on whole traces**
+	Conceptually easy	All accountable compounds analysed simultaneously	All detected compounds analysed simultaneously
–	No account is taken of interactions between compounds Only applicable to signals that can be integrated, not all recorded compounds (difficult for small S/N ratio, overlapping signals)	Only applicable to signals that can be integrated, not all recorded compounds (difficult for small S/N ratio, overlapping signal)	Care required of instrumental- and sample- related artefacts

5. Case Study: Development of a Methodology for Assessing Significance of Changes

A good starting point when assessing the safety of genetically modified crops is to compare a given modification with a control. What we would like to demonstrate here are ways where this can be taken a step further. In particular, the true significance of any metabolite changes that may be observed can only be assessed when these are put into context by examining all or part of the following variables:

- Plant to plant variability
- Tuber to tuber variability
- Differences between controls and modified lines
- Differences between different types of control
- Persistence of metabolite profile characteristics in successive generations

One of our largest studies concerns the occurrence of unintended effects in genetically modified potato tubers. This work has so far encompassed 4 successive harvests, for which we are developing ^{1}H-NMR, HPLC(-MS), and GC(-MS) methods of analysis. Without getting in too much detail, we highlight here the harvesting plan and the factors that we were able to test.

GM and control lines from 2 varieties were grown. The modification to one variety (cv. Record) involved up- or down-regulation of fructokinase (FK), an enzyme that phosphorylates fructose and is involved in the starch biosynthesis pathway. The other variety (cv. Desiree) had undergone one of three types of modification. The first one (W2GBSS) involved a fungal gene believed to encode a glucan branching enzyme. The second one (MAL1) concerned a potato gene which encodes a glycoprotein processing enzyme [11]. The third one was aimed at manipulating the expression of the S-adenosylmethionine decarboxylase (SAMDC) gene, a key enzyme involved in the biosynthesis of polyamines [12]; SAMDC samples had a 35S promoter which caused this gene to be expressed in the whole plant, whereas SAMPAT constructs had the same transgene but a tuber-specific promoter.

For each of the Desiree modifications, control lines created at the same time by transformation with an 'empty' pBIN19 vector were grown in parallel with the modified plants in each year's crop. For the SAMDC lines, there was in addition a tissue culture grown line, and 5 lines (tet-repressor) which had an inducible promoter that had not been activated and hence could also be considered as controls. From the second year there were also a number of additional controls obtained separately from any of the modifications. The controls for the Record samples were constituted by wild-type samples.

Table 2 shows the number and description of the samples obtained, as well as details on the harvesting plan. On the basis of these data, the following parameters can be identified and analysed:

- Year 1: the variation between plants;
- Year 2: variations between as well as within plants (tuber to tuber);
- Year 3: differences between plots (environmental) and variation within plot;
- Year 4: differences between plots;
- All years: Persistence of metabolite profile characteristics in successive generations.

We are focussing our analyses on these different parameters, using the various analytical approaches outlined above.

Table 2. Sample description and main characteristics of harvesting plan. Abbreviations used: s = sense, as = anti-sense, ev = empty vector control, tc = tissue culture control, tet-rep = tet-repressor. The numerical indicates the number of independent lines available for analysis

Cultivar	Year 1 (1999)	Year 2 (2000)	Year 3 (2001)	Year 4 (2002)
Record FK	3s 6as 1wt	3s 6as 1wt	-	-
Desiree				
W2GBSS	4as 2ev	4as 2ev	4as 2ev	4as 2ev
MAL1	7as 1s 1ev	7as 1s 1ev	7as 1s 1ev	7as 1s 1ev
SAMDC	4as 5 tet-rep 1ev 1tc	5 tet-rep 1ev 1tc	5 tet-rep 1ev 1tc	2as 1ev 1tc
SAMPAT	-	2s 1as 1ev 1tc	2s 1as 1ev 1tc	2s 1as 1ev 1tc
Other cultivars	-	-	-	32 non GM
Trial conditions	Containment (polytunnels)	Containment (Tygan netting)	Field	Field
Replicates	4 plants per line x 1 subset of tubers per plant	5 plants per line x 3 subsets of tubers per plant	4 plots in the field (6 plants per line) x 3 subsets of tubers	4 plots in the field (3 plants per line) x 1 subset of tubers

6. Conclusions and Future Outlook

The study of genetically modified potato tubers over four successive harvests will allow us to demonstrate possible approaches to try and assess the safety of GMOs. This is being carried out by examining the composition of the plant under study and comparing GMOs to non-GMOs/ controls of the same or other variety. The strengths and limitations of NMR as a metabolite profiling technique and the appropriateness of univariate and multivariate statistical methods were demonstrated here. The NMR results are necessarily based on a set of the most abundant metabolites: amino acids, sugars, organic and fatty acids, major phenolics such as chlorogenic acid and a few miscellaneous compounds (choline, trigonelline). Nevertheless the experimental simplicity enabled us to obtain a rapid overview of the numerous samples in each year's data set, to identify obvious outliers, to examine the variation between 'replicates', to show that one of the four plots (in year 3) was different from the others, and to uncover some more subtle differences between the various groups and between individual lines within a group. We could not use NMR to analyse the less abundant secondary metabolites and initially employed reverse phase HPLC and subsequently LC/MS, which detects the important glycoalkaloids. Although NMR and HPLC cover largely different sets of molecules the two methods were in close agreement on which GM lines showed the greatest difference from the controls.

Acknowledgements. We would like to acknowledge funding from the EU 5th Framework Programme (GMOCARE, QLK1-1999-00765), the UK Food Standards Agency Novel Foods Programme (GO2001), and from a Biotechnology and Biological Sciences Research Council's competitive strategic grant. We thank Prof. Howard Davies, Dr. Louise Shepherd and colleagues (Scottish Crop Research Institute, Invergowrie, Dundee, Scotland, UK) for providing potato samples.

References

[1] Kuiper, H.A. and Kleter, G.A. (2003). The scientific basis for risk assessment and regulation of genetically modified foods. Trends Food Sci. Tech. 14, 277-293.

[2] Firn, R.D. and Jones, C.G. (1999). Secondary metabolism and the risks of GMOs. Nature 400, 13-14.

[3] Kuiper, H.A., Kok, E.J. and Engel, K-H. (2003). Exploitation of molecular profiling techniques for GM food safety assessment. Curr. Opin. Biotechnol. 14, 238-243.

[4] Le Gall, G., Puaud, M. and Colquhoun, I.J. (2001). Discrimination between orange juice and pulp wash by ^{1}H NMR nuclear magnetic resonance spectroscopy. J. Agric. Food Chem. 49, 580-588.

[5] Le Gall, G., Colquhoun, I.J., Davis, A.L., Collins, G.J. and Verhoeyen, M. E. (2003). Metabolite profiling of tomato using ^{1}H NMR spectroscopy as a tool to detect potential unintended effects following a genetic modification. J. Agric. Food Chem. 51, 2447-2456.

[6] Krzanowski, W.J. (1988). Principles of Multivariate Analysis: a user's perspective. New York, USA: Oxford University Press.

[7] Defernez, M. and Kemsley, E.K. (1997). The use and misuse of chemometrics for treating classification problems. Trends Anal. Chem. 16, 216-221.

[8] Defernez, M. and Colquhoun, I.J. (2003). Factors affecting the robustness of metabolite fingerprinting using ^{1}H-NMR spectra. Phytochemistry 62, 1009-1017.

[9] Vogels, J.T.W.E., Tas, A.C., van den Berg, F. and van der Greef, J. (1993). A new method for classification of wines based on proton and carbon-13 NMR spectroscopy in combination with pattern recognition techniques. Chem. Intell. Lab. Sys. 21, 249-258.

[10] Vogels, J.T.W.E., Terwel, L., Tas, A.C., van den Berg, F., Dukel, F., and van der Greef, J. (1996). Detection of adulteration in orange juice by new screening method using proton NMR spectroscopy in combination with pattern recognition techniques. J. Agric. Food Chem. 44, 175-180.

[11] Taylor, M.A., Ross, H.A., McRae, D., Stewart, D., Roberts, I., Duncan, G. et al. (2000). A potato α-glucosidase gene encodes a glycoprotein-processing α -glucosidase II-like activity and effects of down-regulation in transgenic plants. Plant J. 24, 305-316.

[12] Kumar, A., Taylor, M.A., Mad Arif, S.A. and Davies, H.V. (1996). Potato plants expressing antisense and sense S-adenosylmethionine decarboxylase (SAMDC) transgenes show altered levels of polyamines and ethylene: antisense plants display abnormal phenotypes. Plant J. 9,147-158.

Genomics for Biosafety in Plant Biotechnology
J.P.H. Nap, A. Atanassov and W.J. Stiekema (Eds.)
IOS Press, 2004

Biochemical Profiling in Plant Systems: Phenotypes beyond Chemical Equivalence

Wolfram WECKWERTH
Max-Planck-Institute of Molecular Plant Physiology
14424 Potsdam, Germany;
email: weckwerth@mpimp-golm.mpg.de

Abstract. At the systems level, gene function will depend on multiple parameters: the developmental stage, the environmental and experimental conditions, as well as the expression levels of many other genes. This results in dynamic changes in transcript, protein and metabolite profiles. Therefore, to achieve a next level of biological understanding, cells, tissues and organs should be described in detail at different levels to provide the complete molecular phenotype. Not only the level of gene expression (transcripts) or activity (proteins), but also all metabolites should be taken into account. Moreover, such a description should include or give consideration to the dynamic interactions of all different components involved. Providing a genomics technology platform of differential profiling strategies to pinpoint alterations or demonstrate the absence of change in genetically modified (GM) compared to non-GM crops will significantly improve the judgement and use of the concept of 'substantial equivalence', especially in the environmental context.

Introduction

In 1993, the concept of 'substantial equivalence' was introduced: "The concept of substantial equivalence embodies the idea that existing organisms used as foods, or as a source of food, can be used as the basis for comparison when assessing the safety of human consumption of a food or food component that has been modified or is new." [2]. Although this definition of substantial equivalence is not yet very clear and may need refinement, once equivalence is demonstrated it may support acceptance of genetically modified (GM) crops.

Another potential of the concept of substantial equivalence may be that it will help to demonstrate that natural variation and conventional breeding strategies result in phenotypes which are as far -or further- from substantial equivalence to known food as GM crops can be. For example, the allergenic potential of conventionally bred crops can exceed that from GM crops. Also the culture conditions may influence the chemical equivalence of food and feed, when comparing biologically cultivated crops and conventional cropping. A major challenge is that once such knowledge has become science fact, it must be taken up in public communication and understanding. However, the analysis can not be restricted to chemical equivalence. We are not yet able to reliably predict any biochemical of toxicological effect of a GM crop solely from its chemical composition. Integrative approaches towards chemical composition, biochemical behaviour and environment-dependent changes are required. The integrative and comprehensive genomics-based analyses of plant material [1] will support and improve existing technologies for defining and establishing substantial equivalence of GM and new crops.

1. Biochemical Profiling of Plant Systems to Connect Genotype and Phenotype

Various examples illustrate that the analyses of plant systems require more than establishing chemical equivalence. For example, despite detailed knowledge of the chemistry and flavour of grape and wine, the relationship between the genetics of grape and the flavour of wine is not well understood [2]. Another example is that GM glyphosate-tolerant soybean (*Glycine max*) was concluded to be substantially equivalent to non-GM soybean. This conclusion was based on the analysis of a fairly restricted set of variables such as amount of protein, carbohydrates, vitamins and minerals, amino acids, fatty acids, fibre, ash, isoflavones and lecithins, comparing untreated GM soybean to untreated non-GM soybean [2]. Yet it is known that upon glyphosate treatment plants significantly change their levels of for example phenolic compounds such as flavones [2]. Therefore there is a need for a more comprehensive analysis of plant material, also taking the importance of actual growth and environment parameters into account.

Similar to other biological systems, any plant system reflects a genotype-to-phenotype relationship. The environment and particular developmental stage of the plant will interact with the genotype to result in a constantly changing phenotype (Figure 1). This phenotype can be described on the morphological/phenomenological level and on a molecular level. The aim of non-targeted biochemical profiling is to detect all the molecular components of a biological system, plant tissue under clearly defined environmental conditions, as systematic, comprehensive and accurate as possible in a complete unbiased way thereby revealing key points of regulation, candidate genes or candidate metabolites.

With the help of various multivariate statistical methods and alternative methods of analysis, one is able to classify different genotypes and phenotypes based on their biochemical characteristics. Most promising is the application of this concept in the differential analysis of control and treated/genetically modified plant.

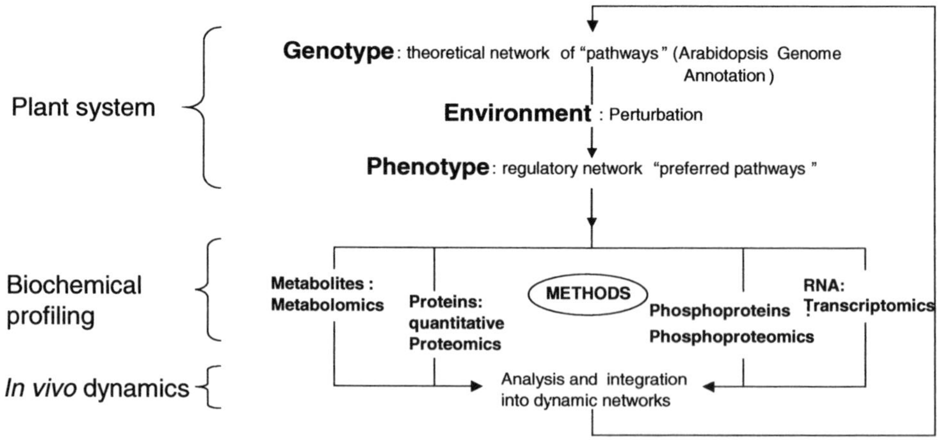

Figure 1. Efforts to describe the *in vivo* dynamics of plant systems with the help of biochemical profiling

2. Metabolomics to Analyse *In Vivo* Dynamics and Pleiotropic Effects

The approach of systems biology is supported by the development of novel technologies such as metabolomics. Metabolomics aims to describe the complete metabolic state of a plant based on accurate metabolite identification and quantitation. This level of molecular information is now regarded as a highly valuable complement to the established approaches of transcriptomics and proteomics.

Plant extracts contain several thousand metabolites, comprising carbohydrates, amino acids, fatty acids, lipids and an unclear amount of secondary metabolites. Only the combination of various analytical techniques is able to deal with mixtures of such a high complexity. A 'golden standard' of non-targeted detection, identification and quantitation of metabolites in complex mixtures is gas chromatography (GC) or liquid chromatography (LC) coupled to mass spectrometry (MS). In Figure 2, a typical gas chromatography mass spectrometry (GC/MS) chromatogram is shown, using time-of-flight (TOF) mass spectroscopy. An advantage of such instrumentation over GC quadruple systems is a markedly improved peak deconvolution. As a consequence, a much higher number of peaks can be identified. This improves the accuracy of the analyses.

In comparison to other genomics technologies, such as RNA microarrays or protein two-dimensional gel electrophoresis, GC/MS- and LC/MS-based metabolite assays in crude extracts combine robustness with very high accuracy and sample throughput. This is essential to analyse and define the dynamics of a biological system.

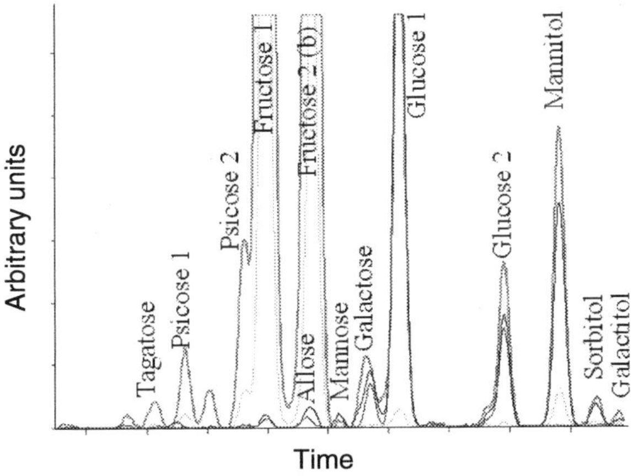

Figure 2. Deconvolution in a complex GC/MS(TOF) chromatogram. Different mass ion traces are extracted from the total ion chromatogram to identify co-eluting sugars.

2.1 The in vivo *Dynamics of Plant Cell Metabolism*

The combination of metabolite profiling and quantitation allows to reconstruct the *in vivo* dynamics of a cell's metabolism. Starting point for such an analysis is the typical glycolytic reaction network occurring in a cell (Figure 3A). Glucose uptake of a cell can be modelled as a fluctuating influx. Such influx is considered to be a typical phenomenon in biological systems. Its fluctuations propagate through the whole reaction network. Samples are taken from such a cell, metabolite levels are determined and plotted in relationship to each other (Figure 3B). This resulted in the scatter plots shown for glucose-6-phosphate (G6P) in relation to fructose-6-phosphate (F6P), fructose-1,6-bisphosphate (FBP), pyruvate (Pyr) and acetaldehyde (ACA) using arbitrary units (a.u.). Each dot represents a single sample.

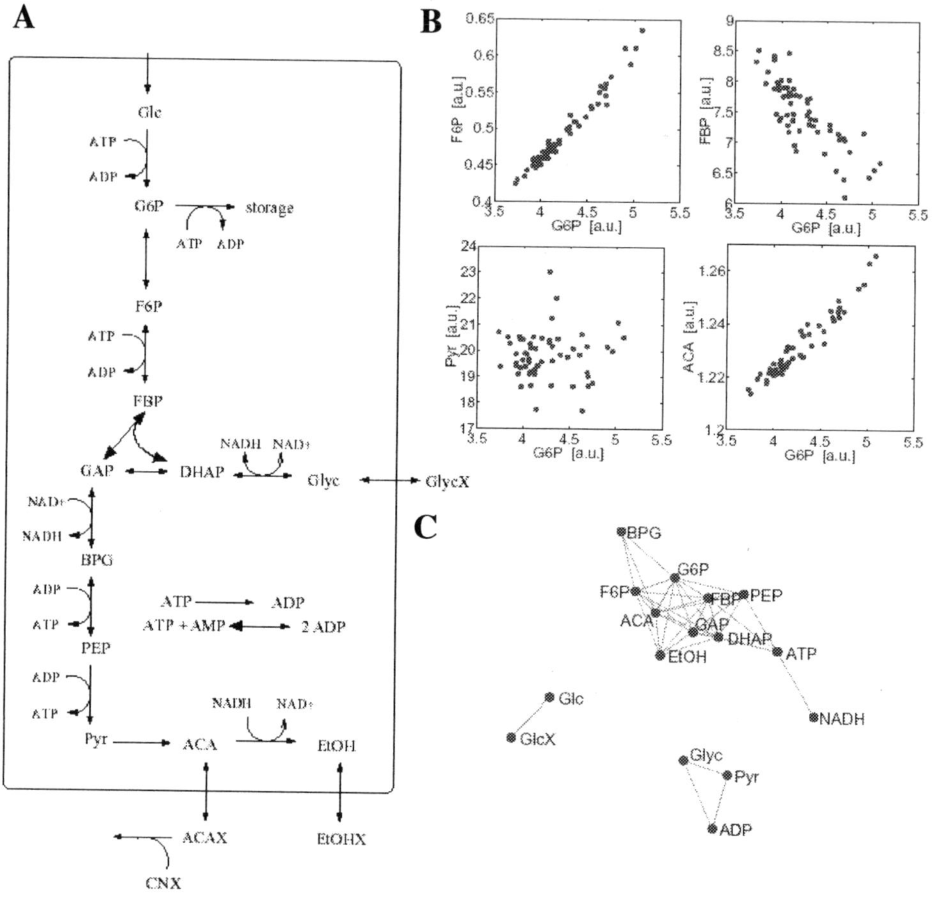

Figure 3. Reconstruction of a metabolic network by metabolomics. (A) Glycolysis as example of experimental metabolomic network and underlying biochemical pathway. (B) Scatter plot of the relationships between metabolite levels. (C) Correlation matrix of metabolite pairs. For details see text and [1].

The interactions or reactions between individual metabolites are reflected as linear correlation of metabolite pairs. Figure 3 shows clear correlation between G6P/F6P and G6P/FBP. There is no correlation between G6P and Pyr. This way, it is possible to correlate all metabolites and identify the high correlation coefficients. The resulting correlation matrix of all metabolite pairs is a fingerprint of the reaction network.

It is further possible to visualise the correlation matrix and compare the resulting "measured" network with the underlying reaction network (Figure 3). In the example of glucose influx in glycolysis, it is interesting is that a strong correlation is observed between ACA/G6P, whereas pyruvate is not correlated to G6P. This indicates, that it is not straightforward to predict what kind of network a reaction will produce at a systems level. The equivalence of a reaction network and the resulting correlation matrix is the foundation of the connection between the measured metabolomic networks and the underlying regulatory biochemical network *in vivo*. Any alteration in the reaction network, for instance

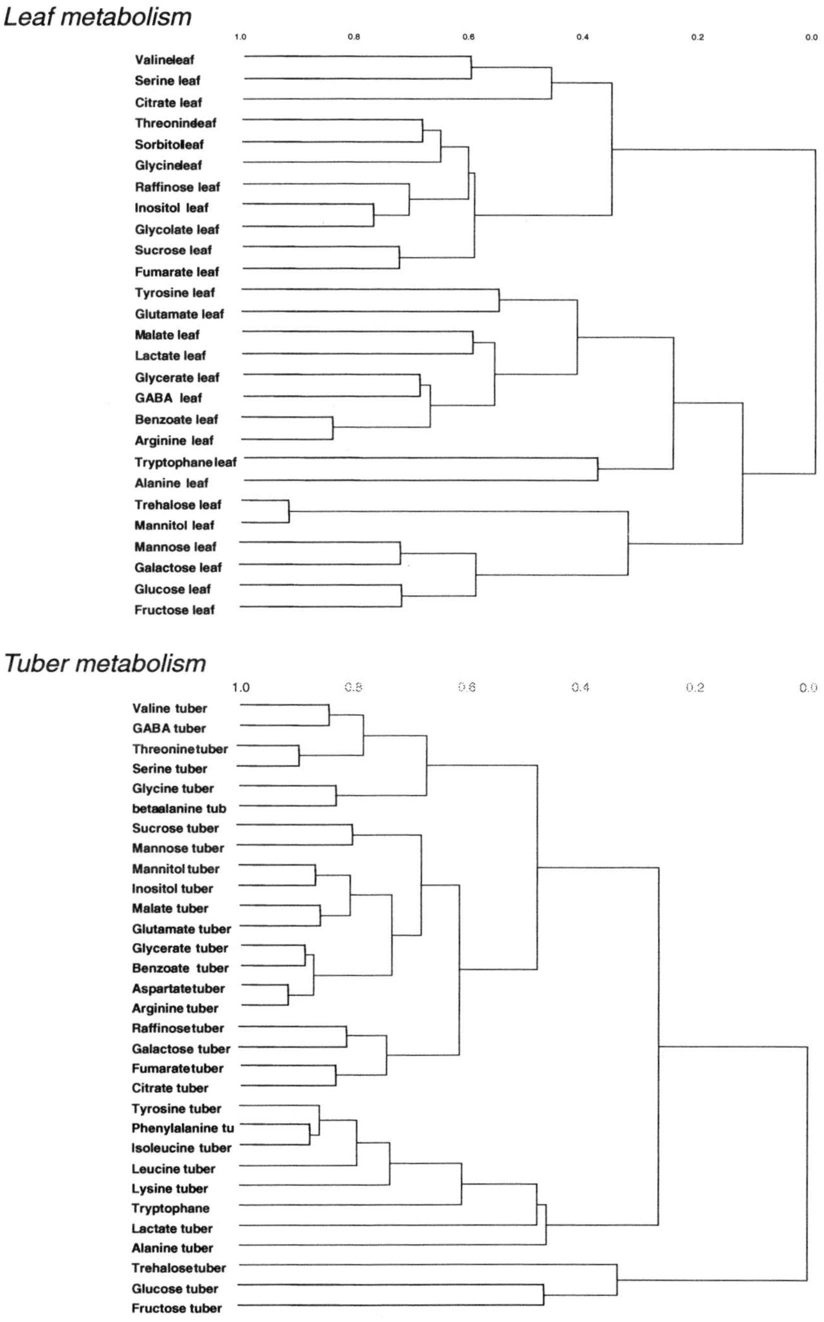

Figure 4. Metabolite network topologies for two different plant tissues, potato tuber as sink tissue and potato leaf as source tissue, show the expected differences. For details see text and [1].

gene suppression, enhancement of a reaction or adding new pathways will result in a changed correlation matrix. Therefore the correlation matrices can be considered true fingerprints of the actual metabolic state of a cell, organ or plant.

2.2 Metabolomics for Fingerprinting Pleiotropic Effects

Correlation matrices can be used to visualise potential differences between GM and non-GM crops, even if variances in average metabolite levels are (too) small. Such differences can be visualised using various statistical methods such as hierarchical clustering analysis (HCA), multidimensional scaling or principal component analysis (PCA) [1,3]. As example, there are clear differences between the metabolite networks in potato leaf and tuber tissue (Figure 4). The metabolite profile comprises the same set of compounds. The metabolite correlations are calculated as Euclidean distances using hierarchical clustering analysis.

As expected, the profiles differ considerably. Non-green potato tuber tissue is biochemically very different from photosynthetically active leaf tissue. This results in different metabolic network topologies (Figure 4). For example, the sugars result in different clusters in leaf tissue compared to tuber tissue. Such changes in metabolic network topologies may also be observed in case of a change in genotype. This way, changes in biochemical characteristics or behaviour of a GM crop relative to its parental non-GM crop may be detected. Any alteration of the genotype may produce pleiotropic effects. These are effects that cannot be *a priori* predicted and may affect one or more biochemical pathways or networks. Metabolomics is able to give a fingerprint of the occurrence of such pleiotropic effects.

3. Linking Metabolomics and Proteomics

A major methodological challenge is combining the dynamics in protein and metabolite expression. Both components are ultimately connected in a biochemical network, metabolite levels as indicators of enzyme activity and pathway fluxes. We are exploring the possibilities to measure protein expression via so-called shotgun proteomics and integrate metabolite and protein data. The principle of shotgun proteomics is outlined in Figure 5. The heterogeneous complex mixture of proteins is proteolytically digested, resulting in a more homogeneous sample of tryptic peptides with comparable sizes and chemical properties. These characteristics are exploited in sophisticated LC/MS.

Using this approach we are currently profiling the *Arabidopsis* proteome in well-defined and well-controlled developmental stages and environmental conditions. Aim is to combine and integrate quantitative protein data and metabolomic networks to constitute biochemical networks. This should allow to define and detect co-regulation at a systems level depending on controlled experimental conditions (Figure 6). An example is the analysis of two *Arabidopsis* varieties Col2 and C24. A subset of quantitative metabolite and protein data analysed by PCA shows clear clustering of the two varieties (Figure 7). On the basis of this type of analysis, individual metabolite and protein components can be analysed for their importance and impact and corresponding system network dynamics [4]. This kind of analysis can be easily transferred to the comparison of a GM crop and its control. Clustering reveals biochemical alteration. Parallel identification of the relevant components, metabolites, proteins, and notably their combination, is a promising strategy to define biomarkers representative for genetic engineering.

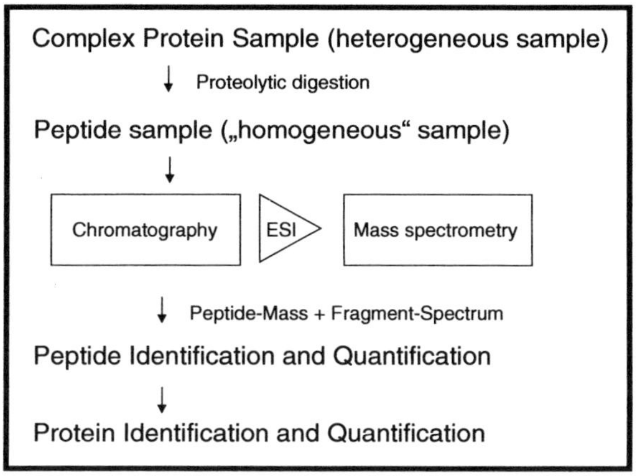

Figure 5. Outline of shotgun proteomics

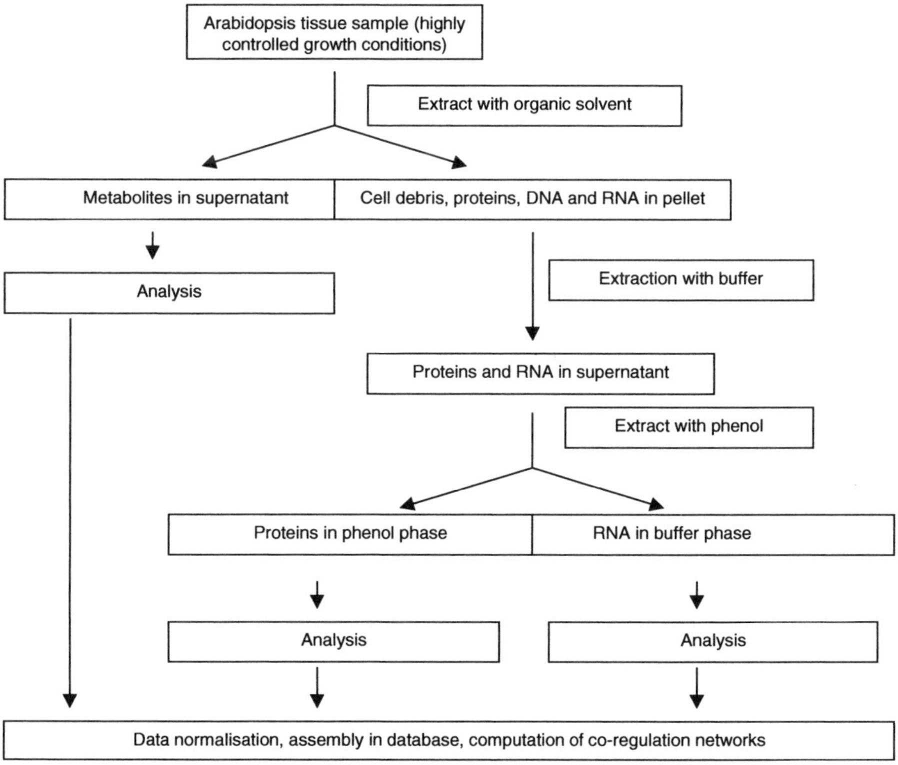

Figure 6. Scheme of integrated extraction and combination of quantitative metabolite and protein data. For details, see [4]

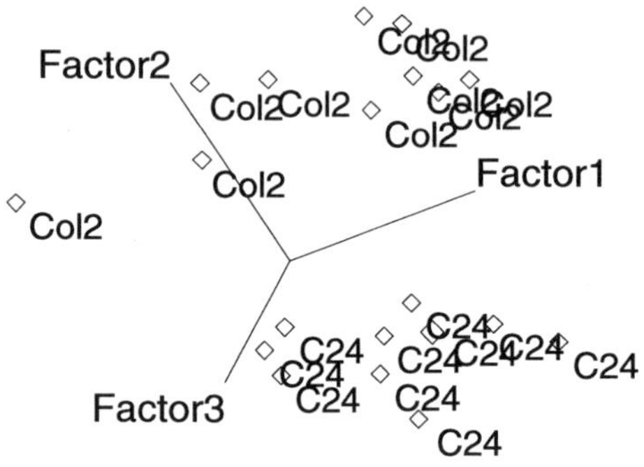

Figure 7. Clustering of *Arabidopsis* varieties C24 and Col2 using quantitative metabolite and protein data.

4. Systematic Profiling and the Assessment of GM Crops

Development of appropriate integrative profiling strategies will enable a far more extensive understanding of metabolism and molecular phenotypes than ever before. Each 'omics' technology alone is able to detect specific biomarkers in relation to genetic engineering in plants. Linking the different 'omics' will give more possibilities and further insight, notably in the occurrence and potential impact of unexpected (pleiotropic) effects in GM crops.

Using these biological markers and subsequent mechanistic biological knowledge will contribute to the understanding of the behaviour of a biochemical network in its genomic and environmental context. Predictions can then be made. These predictions and extrapolations are much more comprehensible than those based on the measurement of a single molecule or a few targeted biomarkers at one time. However, there is a huge gap between the amount of data generated via omics profiling methods and their biological interpretation in molecular mechanism. More work is required to fill this gap and to distinguish between biological noise and biological relevance and significance. Provided that the future omics data are accurate, technically precise and reliable, there may be no need for understanding the molecular mechanisms to define substantial equivalence. It will be possible to classify substantial equivalence of different genotypes and their corresponding environmental phenotypes without full understanding of the causal relationships at the molecular level.

References

[1] Weckwerth, W. (2003). Metabolomics in systems biology. Ann. Rev. Plant Biol. 54, 669-689.
[2] Millstone, E., Brunner, E. and Mayer, S. (1999). Beyond 'substantial equivalence'. Nature 401, 525-526.
[3] Steuer, R., Kurths, J. Fiehn, O. and Weckwerth, W. (2003). Observing and interpreting correlations in metabolomic networks. Bioinformatics 22, 1019-26.
[4] Weckwerth, W., Wenzel, K. and Fiehn, O. (2004). Process for the integrated extraction, identification and quantification of metabolites, proteins and RNA to reveal their co-regulation in biochemical networks. Proteomics 4, 78-83.

Genomics for Biosafety in Plant Biotechnology
J.P.H. Nap, A. Atanassov and W.J. Stiekema (Eds.)
IOS Press, 2004

The Biosafety of Metabolic Change: Altering Plant Terpenoid Metabolism

Teun W.J.M. van HERPEN[1], Wilfried SCHWAB[2], Maarten A. JONGSMA[1], Harro J. BOUWMEESTER[1] and Asaph AHARONI[1*]

[1]Plant Research International, Wageningen University and Research Centre, P.O. Box 16, 6700 AA Wageningen, The Netherlands; [2]Biomolecular Food Technology, Technical University München, Freising, Germany;
**email: asaph.aharoni@wur.nl*

Abstract. Metabolic diversity in plants increases with enzymes generating multiple products from a single substrate. This way, the diversity of metabolites, both endogenous and as well as introduced by genetic modification (GM) is influenced. Transgenic *Arabidopsis* and potato plants altered in mono- and sesquiterpene metabolism show this phenomenon. Overexpression of the *FaNES1* gene encoding a dual linalool (mono-terpene)/ nerolidol (sesquiterpene) synthase in plastids of Arabidopsis resulted in high levels of free linalool and substantial levels of linalool derivatives. Levels were at least 10 fold higher than the aglycones. Preliminary trials showed that the increased levels of linalool and derivatives altered the attractivity to insects. Transgenic potato plants overexpressing the same gene produced various linalool derivatives, some of which differed from those present in Arabidopsis. Potato produced some nerolidol, suggesting that its precursor, farnesyl diphosphate, is present in plastids. In conclusion, the outcome of plant metabolic engineering strategies in terms of the metabolic fate of target compounds is difficult to predict. This should be taken into account when considering the biosafety of GM plants engineered to produce new metabolites.

Introduction

It is now more than two decades since the first report on plant transformation was published. After this breakthrough, it was demonstrated in a large number of cases that the outcome of experiments aiming at engineering plants can be surprising and is difficult to predict. Our ability to predict the outcome of experiments with transgenic plants is largely dependent on our understanding of the biological system under study. The system includes networks of genes, proteins and metabolites and the higher order networks, which consist of the interplay between various genetic pathways and between them and the environment. Our knowledge although still limited is growing at an exponential rate, mainly as a result of the past years advances in technologies, which accelerate gene discovery and characterisation.

Metabolic engineering of plants is one of the most promising fields in the development of transgenic crops. Plants are metabolically engineered for resistance to biotic and abiotic stresses, the production of natural, industrial and pharmaceutical products and for improved quality as food and feed. For example, vaccine production against anthrax and cholera in tobacco [1,2] and the expression of anthocyanins and proanthocyanins in alfalfa [3]. Metabolic engineering of plants is associated with the potential perturbation of the endogenous metabolism and the possibility that endogenous enzymes unintentionally modify the newly produced metabolite. Such changes can in some cases lead to a change in plant morphology [4].

We have chosen the terpenoid pathway as our model to test the outcome of metabolic engineering in terms of metabolic fate of the introduced metabolites. Terpenoids are the most structurally varied class of plant natural products, and our specific goal in this study was to alter the metabolism of the C10 monoterpenes. Although many terpenoids are associated with primary metabolism (e.g. the phytol side chain of chlorophyll, carotenoid pigments, and the plant hormone gibberellin) others are typical plant secondary metabolites. They are commercially important due to their wide application in a vast number of industrial products such as flavouring agents, pharmaceuticals, perfumes, insecticides and anti-microbial agents. In nature, they play significant roles in plant-environment interactions, plant-plant communication, and in plant-insect and plant-animal interactions. For example, linalool is an acyclic monoterpene alcohol that has a peculiar creamy floral, sweet taste [5]. In *Clarkia breweri*, linalool, among other compounds, is responsible for the attraction of pollinating moths [6].

All terpenoids, including the mono- and sesquiterpenes, are synthesised from isopentenyl diphosphate (IPP) and dimethylallyl diphosphate (DMAPP), originating from either the mevalonate pathway active in the cytosol-endoplasmic reticulum or from the plastidial 2-*C*-methyl-D-erythritol 4-phosphate (MEP) pathway [7] (Figure 1). Condensation of IPP and DMAPP will yield geranyl diphosphate (GDP), the precursor of monoterpenes. The condensation of two molecules of IPP and one DMAPP generates farnesyl diphosphate (FDP) the precursor for sesquiterpene biosynthesis. Following the formation of GDP and FDP the different mono- and sesquiterpenes are generated through the action of terpenoid synthases (TPSs). The action of various enzyme classes, such as the cytochrome P450 hydroxylases (Figure 1), dehydrogenases, and glycosyl and methyl

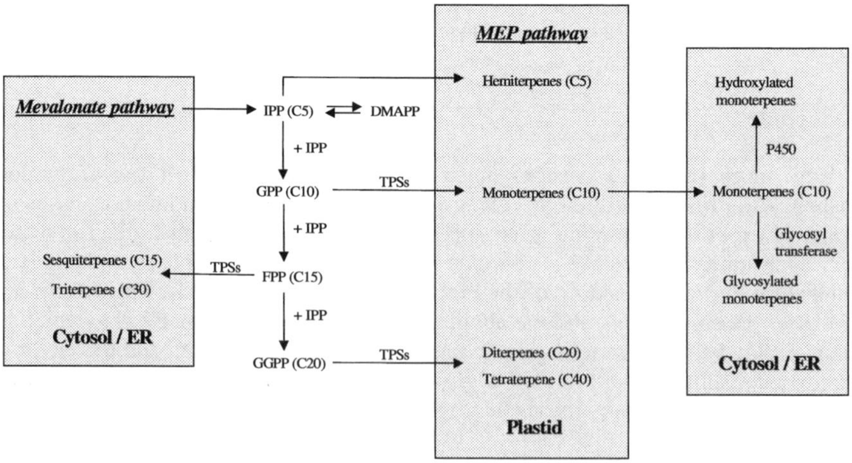

Figure 1. Scheme of the terpenoid biosynthetic pathway. The pathway in outside the blocks is common to both cytosol and plastids. ER, endoplasmatic reticulum; IPP, isopentenyl diphosphate; DMAPP, dimethylallyldiphosphate; GPP, gergeranyl diphosphate; FPP, farnesyl diphosphate; GGPP, geranyl geranyl diphosphate; TPSs, terpenoid synthases; P450, cytochrome P450.

transferases [8] can further modify primary terpene skeletons formed by TPSs. Although there is increasing evidence that there is exchange of intermediates between compartments [9,10], it could be generalised that the cytoplasmic mevalonate pathway is responsible for the production of sesquiterpenes (C_{15}). In plastids, the MEP pathway is responsible for the production of monoterpenes.

In recent years, a number of examples of the production of transgenic plants expressing monoterpene synthases have been reported. Constitutive expression of the *Clarkia breweri* *S*-linalool synthase (*LIS*) gene in *Petunia hybrida* demonstrated that linalool is produced in a number of organs [11]. It is also efficiently converted to the non-volatile *S*-linalyl-ß-D-glucopyranoside, probably by the action of endogenous glucosyl-transferase activity [11]. Expression of *LIS* in tomato under the control of a late fruit ripening promoter (E8 promoter) resulted in the accumulation of *S*-linalool and 8-hydroxy linalool in ripening fruit [12]. The same gene was also constitutively expressed in carnation (*Dianthus caryophyllus*) and both free linalool and its *cis*- and *trans*-oxides could be detected by headspace sampling of leaves and flowers of transgenic plants [13]. In an extract of transgenic flowers only *trans*-linalool oxide could be detected. In another study three different monoterpene synthase genes from lemon (+)-limonene, γ-terpinene and (-)-ß-pinene synthase were introduced into one tobacco plant [14]. These plants produced limonene, γ-terpinene and ß-pinene as main products, also α-thujene, α-pinene, sabinene and *p*-cymene were identified as side products.

The *Perilla frutenscens* limonene synthase (*LS*) gene was expressed in tobacco in three different compartments namely plastids, cytosol and endoplasmatic reticulum (ER) [15]. Plastid and cytosol expressed *LS* produced 143 and 40 mg limonene g^{-1} fresh weight respectively, no enzyme activity was detected for the ER expressed *LS*. Two sesquiterpene synthase genes have also been expressed in transgenic plants. Tobacco plants were transformed by Hohn and Ohlrogge [16] with a fungal trichodiene synthase and by Wallaart et al. [17] with the *Artemisia annua* amorpha-4,11-diene synthase. In both cases, only very low levels of the expected sesquiterpene products could be detected. The levels of trichodiene in cell-suspension cultures of the transgenic tobacco expressing trichodiene synthase could be increased 10-fold after elicitation [18]. The elicited transgenic suspension culture also accumulated an oxygenated trichodiene derivative.

In this paper we present the investigation of the fate of the newly introduced monoterpene linalool to the model plant *Arabidopsis thaliana* and the crop potato. Engineered monoterpenes in Arabidopsis are extensively modified. However, the type of modification and the relative amounts of each derivative are different between plant species and most probably depend on their enzymatic background. We therefore call attention to the fact that such alteration in metabolism are likely to occur and should be taken into account when conducting metabolic engineering in plants.

1. *FaNES1* from Strawberry

Previous work using cDNA microarrays in strawberry identified more than 200 clones showing differential expression during fruit development [19]. Data analysis revealed a putative terpene synthase homologue (*FaNES1*) with a strongly up-regulated expression during fruit ripening. The recombinant FaNES1 protein produced in *E. coli* cells catalyses the biosynthesis of the monoterpene *S*-linalool from GDP and its sesquiterpene counterpart (3S)-(E)-nerolidol from FDP. Enantioselective GC-MS analysis showed that the enantiomeric composition of the products formed by the recombinant FaNES1 enzyme is identical to that produced by ripe strawberry fruit (Figure 2).

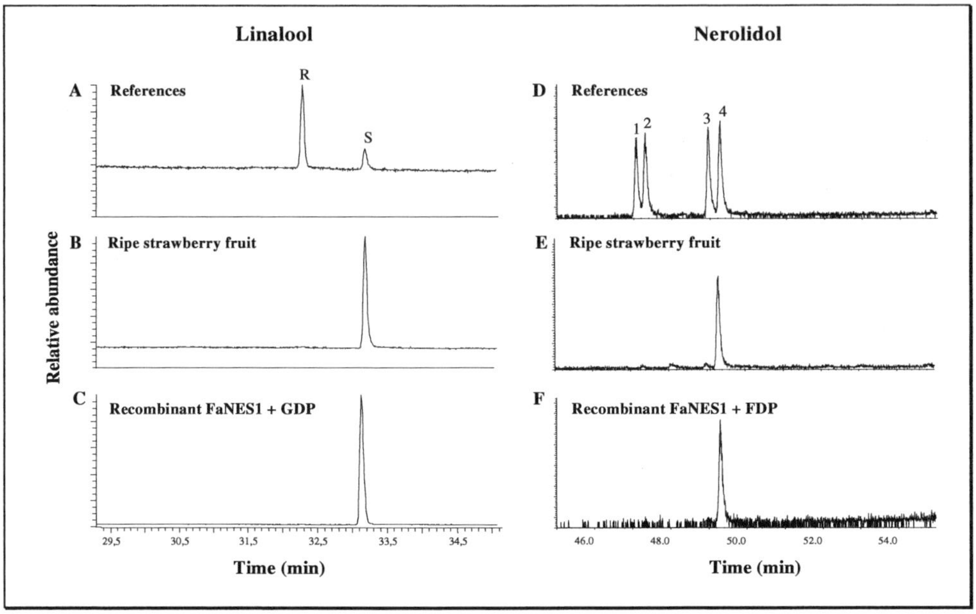

Figure 2. Enantioselective GC-analysis of linalool and nerolidol produced by recombinant FaNES1 protein and ripe fruit of the cultivated strawberry. In (A) references of R and S linalool enantiomers and in (D) references of nerolidol enantiomers: 1, 2- (3*R*)- and (3*S*)-(*Z*)-nerolidol (elution order is not known), 3- (3*R*)-(*E*)-nerolidol, 4- (3*S*)-(*E*)-nerolidol.

2. Linalool and Nerolidol in Arabidopsis Expressing *FaNES1*

Strawberry *FaNES1* under the control of the double 35S promoter of the cauliflower mosaic virus (CaMV) was used for transformation of Arabidopsis. The protein encoded by *FaNES1*, normally localised in the cytosol, was directed to the plastids by the fusion of a plastid targeting signal to its N-terminus. Over 120 putative transgenic plants were generated and two rosette leaves from each individual plant were screened for terpene emission by the SPME method [20]. Leaves of wild-type Arabidopsis did not emit any detectable level of linalool, however transgenic lines showed different levels of linalool emission (data not shown). Furthermore, plants emitting high levels of linalool were also emitting low levels of the sesquiterpene nerolidol.

The headspace volatiles emitted by intact transgenic plants during 24 hours were analysed using the Tenax trapping/elution method for vegetative plants (Figure 3). Six primary transformants were examined in three separate experiments. These showed a similar pattern of linalool and nerolidol emission in the SPME analysis. The high emission of linalool was usually accompanied by detectable amounts of nerolidol. MDGC-MS analysis revealed that the transgenic Arabidopsis plants produced exclusively the enantiomer *S*-linalool. This matched earlier observations on the recombinant FaNES1 enzyme and the linalool produced by ripe strawberry fruit (Figure 2). Quantification of linalool emission showed that the levels of linalool released amounted to 20-50 $\mu g.day^{-1}.g^{-1}$ fresh weight (fwt). This is at least 100 fold higher than the levels of monoterpene limonene produced by wild-type Arabidopsis leaves.

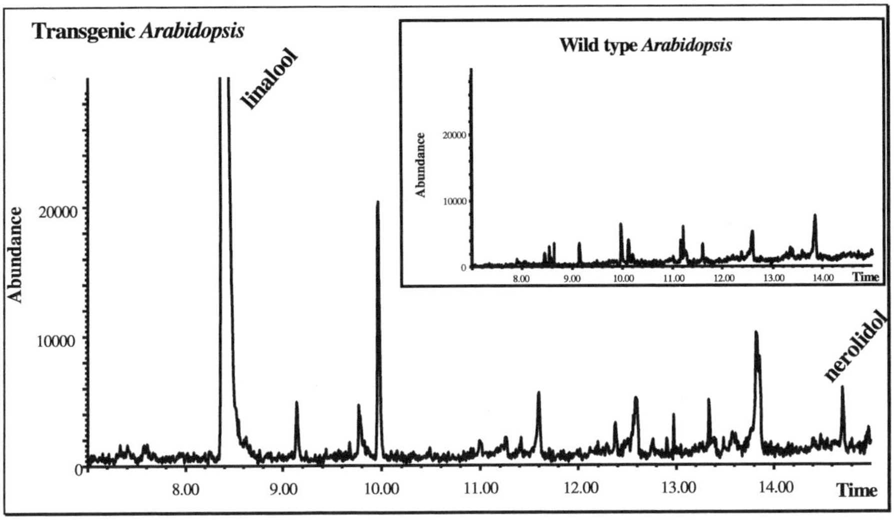

Figure 3. Linalool and nerolidol emission by the *FaNES1*-expressing Arabidopsis plants. Headspace Tenax trapping and subsequent GC-MS analysis show emission of linalool and nerolidol from vegetative parts of the transgenic Arabidopsis and their absence in the headspace profile of the wild type plant.

3. Hydroxylated and Glycosylated Linalool Derivatives of Linalool in *FaNES1* Plants

To investigate the fate of linalool *in planta*, volatiles of Arabidopsis rosette leaves were isolated by solid phase extraction and subjected to GC-MS analysis. In all of the extracts obtained from wild type plants linalool and nerolidol could not be detected but transgenic plants contained up to 1.5 $\mu g.g^{-1}$ fwt of linalool, nerolidol, *E*-8-hydroxylinalool (largest amount) and *E*-8-hydroxy-6,7-dihydrolinalool (Figure 4A). Plants used for the analysis were derived from the progeny of the self-pollinated primary transformant (T1). Thus, production of linalool and nerolidol was inherited to the next generation. The presence of glycoside-bound terpenoids was determined after enzymatic hydrolysis of glycosides. GC-MS analysis of released alcohols showed that wild type plants contained some glycosylated *E*-8-hydroxylinalool, but much higher amounts of E-8-hydroxylinalool were released from the glycosidic extract obtained from transgenic plants (Figure 4B). In addition, glycosides of *Z*-8-hydroxylinalool (not detected in the aglycon analysis) and *E*-8-hydroxy-6,7-dihydro-linalool were observed in transgenic plants but not in the wild type plants.

In some transformants, the sum of glycosidically bound volatiles was interestingly 40- to 60-fold higher than the sum of aglycons. Linalyl ß-D-glucopyranoside, which was previously demonstrated to be produced in transgenic petunia plants expressing the *Clarkia breweri* linalool synthase gene [11], could not be detected by LC-MS analysis in either wild type or transgenic Arabidopsis plants (data not shown). However, the ion trace m/z 391 of the glycosidic extracts obtained from transgenic Arabidopsis plants showed a compound eluting before linalyl ß-D-glucopyranoside under reversed phase condition. This compound is more polar than linalyl ß-D-glucopyranoside and displays the pseudomolecular ion mass m/z 391, which makes it likely that it represents 8-hydroxylinalyl glucoside. Although linalool and nerolidol are present in transgenic Arabidopsis leaves, glycosylated tertiary alcohols were not observed. Thus, we assume that the carbohydrate moiety in 8-hydroxylinalyl glycoside is attached to the primary hydroxyl group.

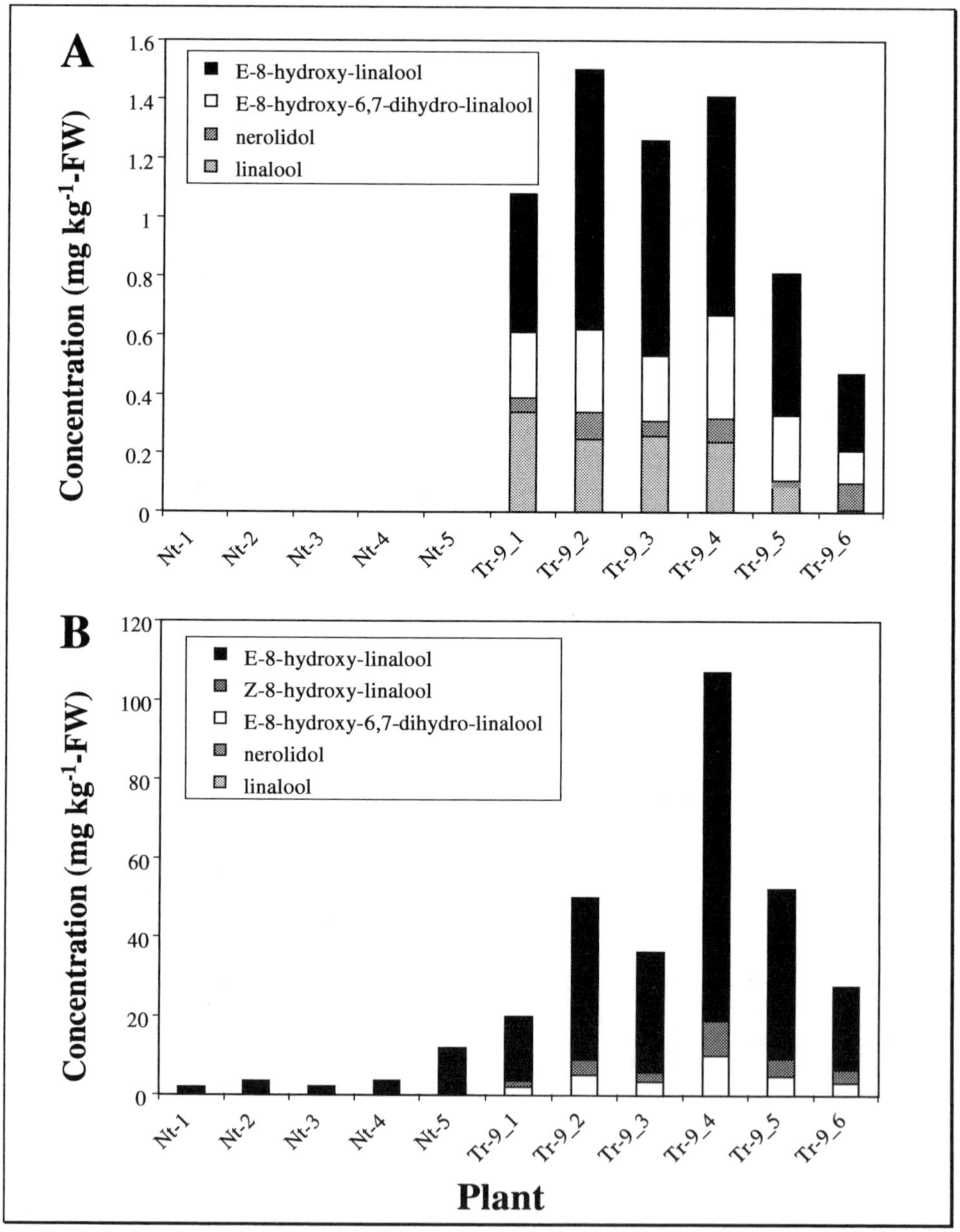

Figure 4. Production of linalool, linalool derivatives and nerolidol by leaves of wild type and *FaNES1* expressing Arabidopsis plants. Free (panel A) and glycosidically bound (panel B) terpenoids of Arabidopsis rosette leaves were isolated by solid phase extraction, subjected to GC-MS analysis and quantified. Plants are progeny (T2) of the selfed Tr-9 primary transformant (T1). Nt, not transformed, Tr, transgenic.

A scheme of the linalool derivatives detected in the transgenic Arabidopsis plants is shown in Figure 5.

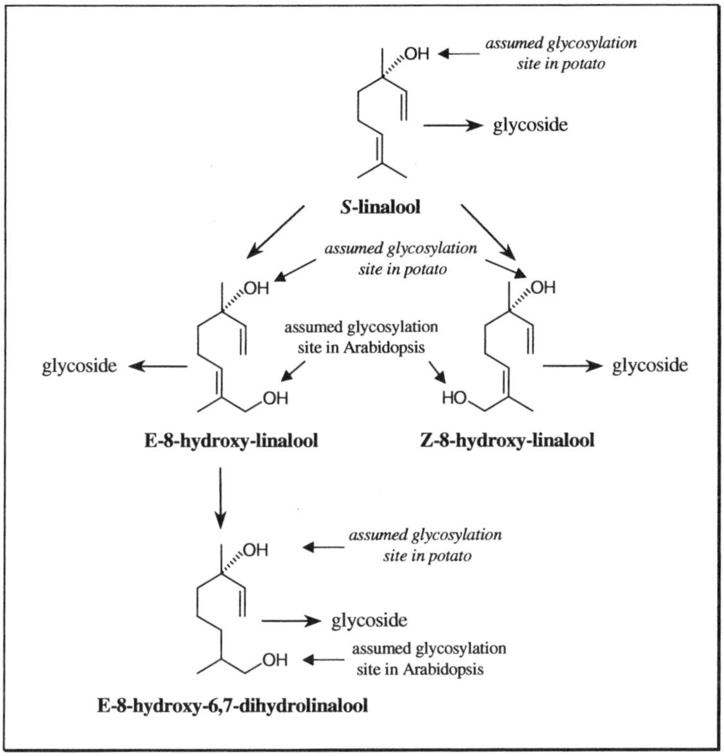

Figure 5. Production of linalool and its derivatives by *FaNES1* expressing Arabidopsis plants. The assumed glycosylation sites in linalool and its derivatives produced by leaves of transgenic Arabidopsis and potato plants are indicated. Potato plants were transformed with the same construct used for expressing *FaNES1* in Arabidopsis (unpublished data).

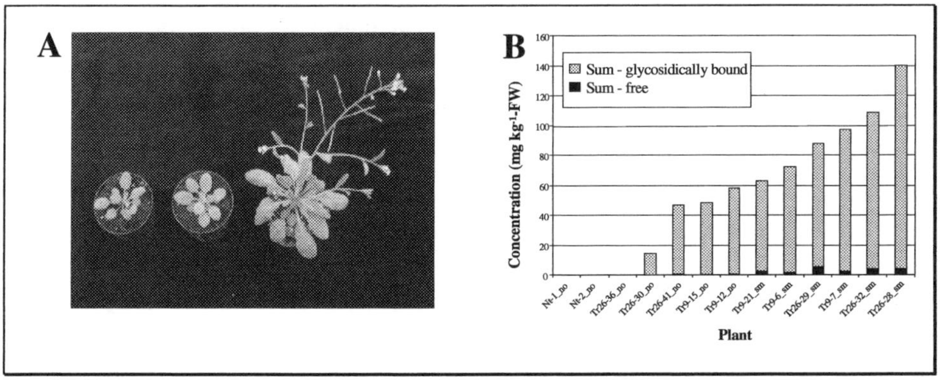

Figure 6. *FaNES1*-expressing Arabidopsis plants show growth retardation. (A), two transgenic lines (left and middle) and a wild type plant (right) of the same age; (B), correlation between the size of *FaNES1* expressing Arabidopsis plants and production of free and glycosylated linalool, linalool derivatives and nerolidol. Nt, not transformed; Tr, transgenic, no, normal size inflorescence stems of above 16 cm; sm, small size inflorescence stems of below 5 cm.

4. *FaNES1*-expressing Plants

Transgenic Arabidopsis plants overexpressing *FaNES1* grew less vigorously than their non-transgenic counterparts (Figure 6A). Both the basal rosette and the flowering stems were retarded in growth. Due to the delayed development of the transgenic plants they had a much younger appearance, including lighter green leaves, but they did eventually produce flowering stems and viable seeds. We also analysed volatile levels in transgenic lines showing growth retardation. The severity of the phenotype correlated with the levels of linalool and its derivatives that were produced (Figure 6B), thus confirming the relationship between *FaNES1* gene expression levels and the retarded growth of the transgenic plants.

To investigate the fate of linalool in another plant species, we generated transgenic potato plants using the same construct used for expressing *FaNES1* in Arabidopsis (unpublished data). Volatiles and glycosidically bound terpenoids produced by young potato leaf tissue were isolated and subjected to gas chromatography mass spectrometry (GC-MS) analysis. Wild type leaves already produce linalool, nerolidol and *E*-8-hydroxylinalool. However, in leaves of *FaNES1* transgenic plants, levels of linalool and *E*-8-hydroxylinalool are considerably higher than in the wild type leaves and *Z*-8-hydroxylinalool as well as *E*-8-hydroxy-6,7-dihydrolinalool were detected. The concentration of nerolidol remained unchanged in leaves of transgenic plants. Glycosidically bound terpenoids were determined after enzymatic hydrolysis of glycosides. GC-MS analysis of released alcohols showed that glycosylated linalool, *E*-, and *Z*-8-hydroxylinalool were detected in wild type plants but levels were considerably higher in leaves of transgenic plants.

HPLC-MS analysis confirmed the presence of linalyl-ß-D-glucopyranoside in wild and transgenic potato leaves. Transgenic plants also produced glycosylated *E*-8-hydroxy-6,7-dihydrolinalool, not detected in wild type plants. The different assumed glycosylation sites in linalool and its derivatives produced by leaves of transgenic Arabidopsis and potato plants are indicated in Figure 5. As described above the FaNES1 recombinant protein catalyses the bio-transformation of GDP to *S*-(+)-linalool and FDP to *S*-(+)-trans-nerolidol. To investigate the contribution of FaNES1 to the production of linalool in transgenic potato plants we performed chiral analyses of linalool isolated from transgenic and wild type plants. Wild type potato leaves already contain linalool with an average enantiomeric ratio of 98:2 (R:S). However, in transgenic potato leaves the enantiomeric ratio of linalool changed in preference of the *S*-enantiomer to an average of 8:92 (R:S). The same trend was observed for linalool released by enzymatic treatment of the glycosidic extracts.

5. Behaviour of Aphids Towards *FaNES1*-expressing Plants

Plant volatiles including monoterpenes such as linalool may influence the behaviour of insects, either by attracting or repelling them. Transgenic Arabidopsis generated in the course of this study, emitting volatile terpenoids which are normally not released by wild type Arabidopsis leaves presented therefore a good experimental material to test the effect of the newly emitted volatiles on insect behaviour. To do so, we investigated the preference of *Myzus persicae* aphids (an important agricultural pest) towards detached leaves of either transgenic *FaNES1* expressing plants or wild-type plants in a dual-choice assay. The amount of aphids on leaves was counted at seven time points up to 50 hours after the start of the experiment and statistical analysis was applied to determine the significance of the differences observed. The results show that from 20 hours onwards aphids significantly preferred leaves from wild-type plants (Figure 7). Linalool and/or its derivatives in the transgenic plant leaves act as repellent of *Myzus persicae* aphids.

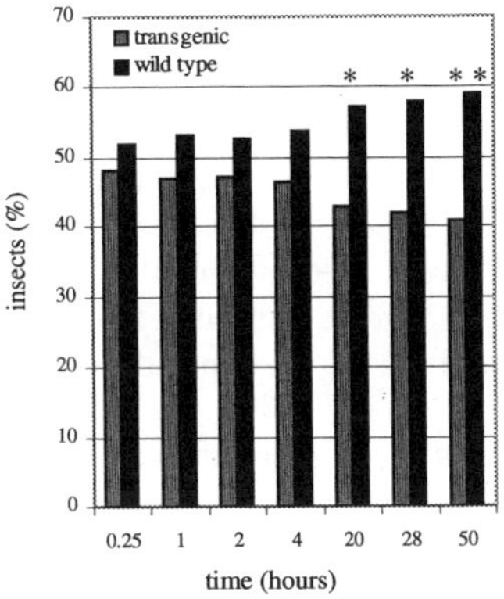

Figure 7. The behaviour of *Myzus persicae* aphids is influenced by volatiles produced in transgenic plants expressing the *FaNES1* gene. The results from a dual-choice assay examining the preference of *Myzus persicae* aphids (represented by percentage) towards detached leaves of either wild type plants and transgenic *FaNES1* expressing plants during time. The calculated p-values of the Wilcoxon Signed-Rank Test [21] showed significant evidence for repellency by the transgenic lines in the last three time points ($p < 0.05$; $p < 0.01$).

6. Effects of Linalool and Derivatives in Transgenic Plants

In this paper the overexpression of the strawberry *FaNES1* gene in transgenic Arabidopsis plants is described. This approach resulted in the production of significant levels of both free, hydroxylated and glycosylated linalool derivatives. The levels emitted by young plants (at 20-50 $\mu g.day^{-1}$ per plant) surpassed the levels produced by flowering wild-type Arabidopsis as reported by Chen et al. [22] (0.5 $\mu g.day^{-1}$ per plant) 40-100 fold.

Previous studies of various transgenic plants expressing the Clarkia linalool synthase gene (*LIS*) demonstrated that the linalool formed is further metabolised to either glycosylated or hydroxylated forms. It was shown that transgenic *LIS* petunia plants emit only little amounts of free linalool from flowers while no free linalool could be detected in leaves [11]. All the linalool produced was converted to the non-volatile, linaloyl-ß-D-glucoside [11]. Carnation plants expressing the same gene, produced and emitted linalool and linalool oxides from flowers (13) whereas transgenic tomato produced linalool and 8-hydroxylinalool from ripe fruit (12). Transgenic potato plants overexpressing *FaNES1*, also produced an array of different linalool derivatives.

In transgenic Arabidopsis plants we identified a different pattern of linalool derivatisation than reported for the other species mentioned above. Both free and glycosylated linalool and linalool derivatives were identified in leaf extracts, with total levels of glycosides at least 10-fold higher (up to more than 60-fold in some transgenic lines) than those of the aglycons. *E*-8-hydroxy-linalool was the predominant component in both the aglycon as well as the glycosidic fraction. This specific pattern of linalool hydroxylation is most probably a result of an endogenous Arabidopsis hydroxylase, an assumption that is further supported by the presence of only *E*-8-hydroxy-linalool but not linalool or other linalool derivatives in leaves of wild-type Arabidopsis. Apparently, metabolism of the monoterpene introduced in a transgenic plant is determined by the background of enzymatic activity present in a given plant species.

Transgenic Arabidopsis plants expressing *FaNES1* were delayed in their growth compared with wild type plants grown under equal conditions. This phenotype was transmitted to further generations and correlated with levels of linalool production. One, major explanation for such a phenotype might be the reduction in the availability of substrates for other metabolites produced in the plastids that play an important role in plant growth and development. These include the growth regulators gibberellins and abscisic acid and other vital components such as carotenoids, chlorophyll and quinones. Transgenic tomato plants constitutively expressing the carotenogenic enzyme phytoene synthase showed a dwarf phenotype [23]. This phenotype was shown to result from a decrease in the levels of gibberellin A1, synthesised from GGDP the common precursor to gibberellins and carotenoids. A similar effect was reported for transgenic tobacco plants expressing high levels of the same enzyme [24]. A second explanation for the phenotypic alteration we observed in transgenic *FaNES1* plants might be the toxic effect of the high levels of linalool produced, since several monoterpenes, including linalool, have been demonstrated to be harmful to plant cells [25,26]. It should be noted that transgenic petunia, tomato and carnation plants producing high levels of linalool and its derivatives were not described to show a similar phenotype.

Mono and sesquiterpenes have been shown to be of ecological impact in plant defence [27,28]. Linalool for example has been shown to be one of the volatile compounds released after herbivore attack in some plant species, and might be involved in attracting predators of the herbivores [28,29]. In this study we showed by dual choice assays that volatiles produced by the *FaNES1* expressing plants could influence the behaviour of *Myzus persicae* aphids. The same aphid species were reported earlier to be repelled by linalool (as we report here) when tested using an olfactometer [30].

7. Predicting the Metabolic Fate of an Engineered Metabolite

When engineering metabolic pathways in plants, the metabolic fate of the newly introduced metabolite is difficult to predict. The composition of plant genomes in terms of gene function and the activity of enzymes, especially those associated with secondary metabolism, are major factors effecting the outcome of metabolic engineering. Metabolism is the largest functional category in the Arabidopsis genome [31]. This category contains large gene families such as the cytochrome P450 family, which contains 286 members. Cytochrome P450 enzymes catalyse mainly NADPH and O2 dependent hydroxylation reactions and participate in a myriad of biosynthetic pathways for example of phenylpropanoids, alkaloids, terpenoids, lipids, cyanogenic glycosides, glucosinolates, gibberellins, jasmonic acid and brassino-steroids. The enzyme activity of only a few members of this family is known. It is therefore not inconceivable that one newly introduced compound might be hydroxylated by one or more of the cytochrome P450 family members.

In most examples of metabolic engineering of terpenoid biosynthesis up to date, the introduced metabolite (mostly the monoterpene linalool) was glycosylated and/or hydroxylated, with highly efficient glycosylation in petunia, Arabidopsis and potato [4,11] (unpublished data). Logically, the occurrence of derivatisation will depend on the chemical nature of the introduced compound, but derivatisation can also be expected to be different between plant species, depending on the activity of endogenous enzymes. If the plant already produces the target metabolite or its derivative, it should be expected that not only an increase in production would occur but also formation of "new" metabolites. In addition, when altering the subcellular location of a given enzyme, new unexpected substrates might be available for the introduced enzyme. By analysing the metabolome of wild-type plants, particularly the part related to the target metabolite, and the derivatisation of other similar metabolites in a given plant species, a large set of possible newly formed metabolites could be predicted.

Finally, it should be taken into account that new derivatives formed might have similar or opposite influences on the environment. We should therefor realise that predicting the metabolic fate of an introduced metabolite is one thing, but to predict what the biological effect of new derivatives may be (and how that will relate to the biological effect we had meant to obtain) is still out of reach.

References

[1] Aziz, M.A., Singh, S., Kumar, P.A. and Bhatnagar, R. (2002). Expression of protective antigen in transgenic plants: a step towards edible vaccine against anthrax. Biochem. Biophys. Res. Commun. 299, 345-351.

[2] Daniell, H., Lee, S.B., Panchal, T. and Wiebe, P.O. (2001). Expression of the native cholera toxin B subunit gene and assembly as functional oligomers in transgenic tobacco chloroplasts. J. Mol. Biol. 311, 1001-1009.

[3] Ray, H., Yu, M., Auser, P., Blahut Beatty, L., McKersie, B., Bowley, S. et al. (2003). Expression of anthocyanins and proanthocyanidins after transformation of alfalfa with maize Lc. Plant Physiol. 132, 1448-1463.

[4] Aharoni, A., Giri, A.P., Deuerlein, S., Griepink, F., de Kogel, W.J., Verstappen, F.W.A. et al. (2003). Terpenoid metabolism in wild-type and transgenic Arabidopsis plants. Plant Cell 15, 2866-2884.

[5] Arctander, S. (1969). Perfume and Flavor Chemicals (Aroma Chemicals). Vol. II, Las Vegas, USA: Steffen Arctander's Publications.

[6] Raguso, R.A. and Pichersky, E. (1995). Floral volatiles from *Clarkia breweri* and *C. concinna* (Onagraceae): Recent evolution of floral scent and moth pollination. Plant Syst. Evol. 194, 55-67.

[7] Rodriguez Concepcion, M. and Boronat, A. (2002). Elucidation of the methylerythritol phosphate pathway for isoprenoid biosynthesis in bacteria and plastids. A metabolic milestone achieved through genomics. Plant Physiol. 130, 1079-1089.

[8] Lange, M.B., Wildung, M.R., Stauber, E.J., Sanchez, C., Pouchnik, D. and Croteau, R. (2000). Probing essential oil biosynthesis and secretion by functional evaluation of expressed sequence tags from mint glandular trichomes. Proc. Natl Acad. Sci. USA 97, 2934-2939.

[9] McCaskill, D. and Croteau, R. (1998). Some caveats for bioengineering terpenoid metabolism in plants. Trends Biotechnol. 16, 349-355.

[10] Adam, K.P. and Zapp, J. (1998). Biosynthesis of the isoprene units of chamomile sesquiterpenes. Phytochemistry 48, 953-959.

[11] Lücker, J., Bouwmeester, H.J., Schwab, W., Blaas, J., van der Plas, L.H.W. and Verhoeven, H.A. (2001). Expression of *Clarkia S*-linalool synthase in transgenic petunia plants results in the accumulation of *S*-linalyl-beta-*D*-glucopyranoside. Plant J. 27, 315-324.

[12] Lewinsohn, E., Schalechet, F., Wilkinson, J., Matsui, K., Tadmor, Y., Nam, K.H. et al. (2001). Enhanced levels of the aroma and flavor compound *S*-linalool by metabolic engineering of the terpenoid pathway in tomato fruits. Plant Physiol. 127, 1256-1265.

[13] Lavy, M., Zuker, A., Lewinsohn, E., Larkov, O., Ravid, U., Vainstein, A. et al. (2002). Linalool and linalool oxide production in transgenic carnation flowers expressing the *Clarkia breweri* linalool synthase gene. Mol. Breed. 9, 103-111.

[14] Lücker, J., Schwab, W., Van Hautum, B., Blaas, J., van der Plas, L.H.W., Bouwmeester, H.J. et al. (2004). Increased and altered fragrance of tobacco plants after metabolic engineering using three monoterpene synthases from lemon. Plant Physiol. 134, 510-9.

[15] Ohara, K., Ujihara, T., Endo, T., Sato, F. and Yazaki, K. (2003). Limonene production in tobacco with *Perilla* limonene synthase cDNA. J. Exp. Bot. 54, 2635-2642.

[16] Hohn, T.M. and Ohlrogge, J.B. (1991). Expression of a fungal sesquiterpene cyclase gene in transgenic tobacco. Plant Physiol. 97, 460-462.

[17] Wallaart, T.E., Bouwmeester, H.J., Hille, J., Poppinga, L. and Maijers, N.C.A. (2001). Amorpha-4,11-diene synthase: cloning and functional expression of a key enzyme in the biosynthetic pathway of the novel antimalarial drug artemisinin. Planta 212, 460-465.

[18] Zook, M., Hohn, T., Bonnen, A., Tsuji, J. and Hammerschmidt, R. (1996). Characterization of novel sesquiterpenoid biosynthesis in tobacco expressing a fungal sesquiterpene synthase. Plant Physiol. 112, 311-318.

[19] Aharoni, A. and O'Connell, A.P. (2002). Gene expression analysis of strawberry achene and receptacle maturation using DNA microarrays. J. Exp. Bot. 53, 2073-2087.

[20] Verhoeven, H., Beuerle, T. and Schwab, W. (1997). Solid-phase microextraction: artefact formation and its avoidance. Chromatographia 46, 63-66.

[21] Hollander M., and Wolfe D.A. (1973). Nonparametric Statistical Methods, New York: Wiley.

[22] Chen, F., Tholl, D., D' Auria J.C., Farooq, A., Pichersky, E. and Gershenzon, J. (2003). Biosynthesis and emission of terpenoid volatiles from Arabidopsis flowers. Plant Cell 15, 481-494.

[23] Fray R.G., Wallace, A., Fraser P.D., Valero, D., Hedden, P., Bramley P.M. et al. (1995). Constitutive expression of a fruit phytoene synthase gene in transgenic tomatoes causes dwarfism by redirecting metabolites from the gibberellin pathway. Plant J. 8, 693-701.

[24] Busch, M., Seuter, A. and Hain, R. (2002). Functional analysis of the early steps of carotenoid biosynthesis in tobacco. Plant Physiol. 128, 439-453.

[25] Muller, P., Lorber P. and Haley B. (1968). Volatile growth inhibitors produced by *Salvia leucophylla*: effect on seedling growth and respiration. Bull. Torrey Bot. Club 95, 415-422.
[26] Weidenhamer, J. D., Macias, F. A., Fischer, N. H. and Williamson, G. B. (1993). Just how insoluble are monoterpenes? J. Chem. Ecol. 19, 1799-1807.
[27] Phillips, M.A. and Croteau, R.B. (1999). Resin-based defenses in conifers. Trends Plant Sci. 4, 184-190.
[28] Degenhardt, J., Gershenzon, J., Baldwin, I. T. and Kessler, A. (2003). Attracting friends to feast on foes: engineering terpene emission to make crop plants more attractive to herbivore enemies. Curr. Opin. Biotechnol. 14, 169-176.
[29] Gols, R., Posthumus, M.A. and Dicke, M. (1999). Jasmonic acid induces the production of gerbera volatiles that attract the biological control agent *Phytoseiulus persimilis*. Entomol. Exp. Appl. 93, 77-86.
[30] Hori, M. (1998). Repellency of rosemary oil against *Myzus persicae* in a laboratory and in a screenhouse. J. Chem. Ecol. 24, 1425-1432.
[31] Arabidopsis Genome Initiative (2000). Analysis of the genome sequence of the flowering plant *Arabidopsis thaliana*. Nature 408, 796-813.

Genomics for Biosafety in Plant Biotechnology
J.P.H. Nap, A. Atanassov and W.J. Stiekema (Eds.)
IOS Press, 2004

Metabolite Profiling of GM Plants: the Importance of Robust Experimental Design and Execution

Jennie LEWIS, John M. BAKER, Michael H. BEALE and Jane L. WARD*
The National Centre for Plant and Microbial Metabolomics,
Rothamsted Research, West Common, Harpenden, Hertfordshire. AL5 2JQ,
*United Kingdom; *email: jane.ward@bbsrc.ac.uk*

Abstract. The ability to simultaneously track changes in the levels of many metabolites in plants will have great potential in both system biology and biosafety applications. We have used complementary spectroscopic techniques (^{1}H-NMR and GC-MS) and multivariate statistics to examine the metabolite fingerprints of Arabidopsis aerial tissue and wheat seed extracts. Procedures for plant growth, harvesting, preparation of extracts and the collection of data have been optimised to minimise analytical and biological variation within the data set. Data is presented on the analysis of multiple Arabidopsis and wheat extracts by ^{1}H-NMR and the problems encountered during evaluation and optimisation of these protocols are discussed.

Introduction

In recent years, the 'omics' technologies have come to the forefront of biological research. Metabolomics is concerned with the analysis of the total metabolite content of a sample [1]. Formally, metabolomics has been defined as the analysis of the total population of metabolites in a given sample, cell or tissue and the integration of the data in the context of functional genomics. Such large-scale analysis is also relevant to the safety testing of genetically modified (GM) crops. This is a priority for regulatory authorities where there is a requirement for techniques that are able to detect any unintended effects of the genetic modification. Although there are many analytical techniques available, there are challenges for their use in the emerging field of metabolomics.

The cellular metabolome is complex, involving numerous classes of small molecules that vary in subunit, functional groups, concentration, size, structure and polarity. In order to release the metabolites from the plant it is necessary to prepare an extract from the plant tissue. The choice of the initial solvent used in this extraction process immediately limits the number of metabolites extracted, due to differences in chemical properties of the metabolites. Furthermore, no single analytical technology can identify and quantify, with sufficient sensitivity and precision, the diverse range of metabolites and their dynamic fluctuations in cells. Differences due to volatility and solubility mean that multiple methods need to be employed. To generate a comprehensive quantitative metabolite profile it is necessary to integrate data from several analytical methods. Such methods usually depend on partial purification of compound classes and analysis by techniques such as GC-MS and LC-MS. This kind of comprehensive quantitative analysis is not readily applicable to high throughput screening.

High-resolution proton nuclear magnetic resonance spectroscopy (^{1}H-NMR) of unpurified extracts can be used as a fast initial screen to generate fingerprints, made up of quantitative contributions from all (hydrogen-containing) metabolites in the sample. Such fingerprint data can be compared computationally by multivariate methods such as principal component analysis (PCA). This method can be used to highlight differences between samples [2,3], which can be used to guide a more targeted, chemical class approach on selected samples using GC-MS or LC-MS. Targeted and fingerprint data can then be further examined by multivariate statistical analysis. In addition the NMR technique allows easy generation of a library of spectra of standard compounds, known to exist in plants. Use of the library NMR spectra in conjunction with 'difference' spectra generated by multivariate comparison of sample spectra allows identification of metabolites responsible for qualitative and quantitative differences between lines.

In the context of GM tomato plants, metabolite profiling using high-resolution ^{1}H-NMR has already been used to detect potential unintended effects [2]. In this paper we report on our studies using ^{1}H-NMR and multivariate statistics to examine metabolite fingerprints of *Arabidopsis thaliana* (Arabidopsis) and wheat grain. Arabidopsis was selected for work on method assessment since it is well known as a model system in plant genome research due to its relatively small genome, rapid life cycle, easy cultivation and high seed production. In this work, procedures for plant growth, harvesting, preparation of extracts and the collection of data have been assessed and refined to minimise analytical and biological variation within the data sets. On the other hand wheat is one of the major crops in world agriculture. It is used to make bread, other baked goods, pasta and noodles. We also describe how the methods established using Arabidopsis, have been modified and applied to compare flour derived from GM and non-GM wheat lines

1. Metabolite Profiling by ^{1}H-NMR Spectroscopy

^{1}H-NMR of unpurified solvent extracts provides an unbiased snapshot or fingerprint of the metabolites present. There are many advantages in using this technique. The extraction method is quick and straightforward, requiring only a minimal sample preparation, thereby making it suitable for high sample throughput. The use of deuterated solvents in the extraction procedure means that no evaporation steps are required. In addition, no sample clean up, fractionation, chromatography or derivatisation is required and thus there is no loss of sample or metabolites.

The method is inherently and absolutely quantitative. The peak area is directly proportional to the amount of metabolite(s) present. This inexpensive, non-destructive technique can simultaneously detect all proton-bearing compounds in a sample, such as carbohydrates, amino acids, organic acids, esters and ethers, all of which can be found in plant tissues.

Figure 1 shows a typical ^{1}H-NMR spectrum of the Landsberg (LER) ecotype of Arabidopsis. There are many peaks in the spectrum. By far the most intense signals are observed in the "carbohydrate region" between δ5.4 and δ3.3. Whilst there are many overlapping peaks, characteristic signals of certain metabolites can be observed; for example, those due to the anomeric protons of sucrose (δ5.4) and glucose (δ5.19 and δ4.63). The aliphatic region is also well populated predominantly with signals due to amino acids and organic acids.

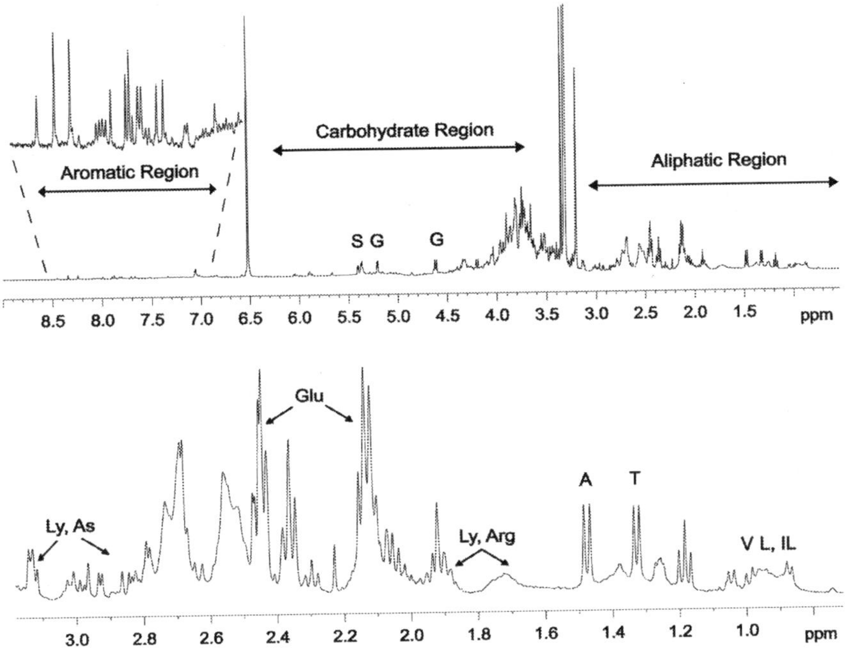

Figure 1. ^{1}H-NMR Spectrum of *Arabidopsis thaliana* ecotype Landsberg. Abbreviations: S = Sucrose, G = Glucose, V = Valine, IL = Isoleucine, L = Leucine, Ly = Lysine, A = Alanine, T = Threonine, Arg = Arginine, As = Asparagine, Glu = glutamine.

2. Development of a Robust Experimental Protocol Using Arabidopsis

Having examined preliminary ^{1}H-NMR spectra of green tissue from *Arabidopsis thaliana,* protocols were developed to enable examination and comparison of polar metabolite levels in multiple samples of plants. In order to confidently report changes in metabolite levels in plant extracts, it is imperative that differences detected are genuine and arise from the differences due to species, line etc. and are not due to environmental variation, or experimental or instrumental error. In assessing the safety of GM crops, it is paramount that the analytical procedure is robust. Preliminary experiments were carried out to assess instrument stability, extraction repeatability and consistency of plant growth and harvesting conditions.

2.1 Assessment of Instrument Stability

In order to eliminate the possibility of introducing variability due to poor instrument performance, the stability of the NMR spectrometer and the reproducibility of the data were examined over a 24-hour period. A polar extract was prepared using freeze-dried Arabidopsis aerial tissue. The same sample was run multiple times under the same conditions on the same instrument. The NMR spectra produced were found to be perfectly superimposable. The same sample was then run at intervals over the next 14 days to assess variation in the instrument performance over time and stability of the analyte. Superimposing all spectra, they overlay extremely well. Variance due to instrument variation and sample stability was calculated from this experiment and was found to be less than 1%.

2.2 Assessment of Extraction Reproducibility

Having determined that variance due to the instrumentation was less than 1%, the extraction protocol was then assessed. Multiple extracts of batches of 12 samples from the same stock of freeze-dried Arabidopsis tissue were prepared over different days. The NMR spectra generated were inspected visually and appeared to be very reproducible. As ^{1}H-NMR spectra of such mixtures contain many overlapping peaks, visual comparison and quantification of some metabolites in many samples is difficult. A solution to this problem is to utilise multivariate statistical analysis to compare spectra. PCA reduces the dimensionality of the data set by combining correlated variables or peaks in the NMR spectrum, to give a smaller set of independent variables called principal component axes or PC's. The PCs are ordered according to their ability to explain the variance contained in the original data. The PCs can be displayed in a graphical fashion as a "scores" plot. Samples having similar NMR spectra cluster together and are easily identifiable. Coefficients by which the original variables must be multiplied to obtain the PC are called "loadings." Examination of the line-plot representation of the "loadings" helps to describe which metabolites are responsible for the formation of different clusters within the scores plot.

Figure 2 shows the scores plot from the PCA analysis of multiple extracts of aliquots of a stock sample of Arabidopsis (Landsberg). The data clustered together, and there were no outliers. This result indicated the robustness in the extraction protocol, regardless of the day the samples were prepared. Variance due to experimental error was calculated from this experiment and was found to be less than 10%.

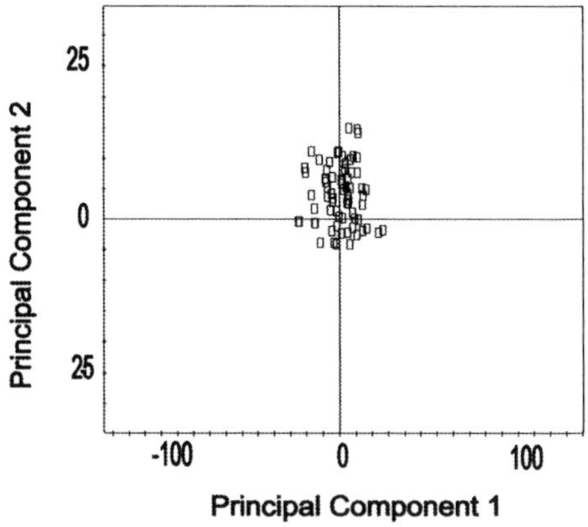

Figure 2. Scores plot from PCA analysis of multiple extracts of identical *Arabidopsis* tissue

2.3 Assessment of Growth and Harvesting Conditions

The reproducibility of the plant growth and harvesting conditions is extremely important in all 'omics' experiments One cannot hope to achieve good experimental reproducibility with poor quality plant material. Whilst it is accepted that there will be biological variation between plants, we need to ensure that errors introduced in the growing regime are not simply classed as being due to "biological variability". Initial experiments were carried out to compare the use of material from single plants to that of material produced by batching plants and sampling from the pool of material. NMR spectra of extracts from freeze-dried aerial tissue of 24 single Landsberg plants, prepared by the method above, indicated that plant-to-plant variability was quite large. The mean relative standard deviation was found to be 52% ± 7%. As a comparison, separate extracts of aliquots from material produced by pooling 24 plants (replicated x3) were also prepared and run in the same experiment. In this case the mean relative standard deviation was calculated to be 28%±3%. The plant material for the complete experiment was grown at the same time in controlled environment conditions. The data was analysed by PCA and the scores plot can be seen in Figure 3.

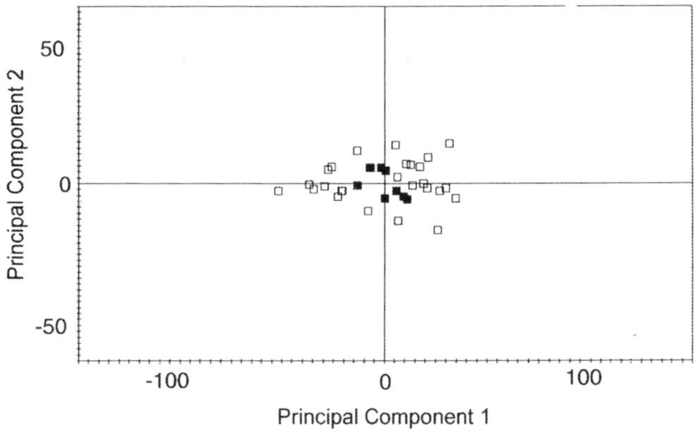

Figure 3. Scores plot from PCA analysis of extracts of identical Arabidopsis tissue from single plants (□) and from batched material (■).

Although there is still one main cluster, the area of the cluster of points generated from the individual plants (□) is much greater than that generated from the batched material (■). These results indicate that pooling of plants grown together can reduce differences in data sets due to biological variability.

Having established that the biological variability can be reduced by pooling individual plants, the next stage in the assessment procedure was to determine whether extracts prepared from the "same" plants grown on different occasions would be reproducible. This stage of the experiment would therefore assess the growth and harvesting protocols. Three cellular trays (24 plants each) of two different Arabidopsis ecotypes (LER and COL) were grown under identical long-day-controlled environment conditions. Plants were harvested at growth stage 6.1-6.5 [4]. Enzyme activity was stopped by immediately immersing the harvested material in liquid nitrogen. The tissue was lyophilised before extract preparation. Plants from three separate experiments were analysed in order to assess variability due to time plants were grown.

Figure 4A shows the scores plot from the PCA of these results. There is a clear separation between Columbia plants (■) and Landsberg Plants (▒). For the purposes of this study we were interested in the effect of growing plants at different times. It was encouraging to see that the data separated according to ecotype and not due to when the plants were grown or harvested. Figure 4B indicates the separation, within the ecotype cluster, of the separate sets of material from each growth experiment. Although there was a fairly large spread in the data, the data from the growth sets overlapped. The data indicated that the plant growth conditions used were robust and that plants from the same seed stock, grown under the developed conditions, could reproducibly generate plants with equivalent metabolite profiles.

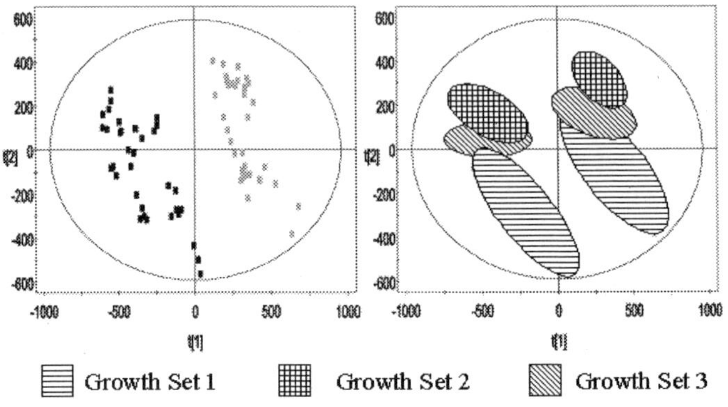

Figure 4. Scores plot from PCA analysis of extracts of Columbia (■) and Landsberg (▒) ecotypes grown on three separate occasions

3. Arabidopsis Ecotype Analysis

Gas chromatography-mass spectrometry (GC-MS) has already been used in the metabolite profiling of Arabidopsis ecotypes. This approach identified 326 distinct compounds from Arabidopsis leaf extracts [5]. The same study also showed that the metabolic profiles of two Arabidopsis ecotypes showed significant differences. Ecotypes are defined as populations that have developed in a specific environment and that have maintained their identity through isolation or selection [6]. The analysis of metabolite profiles in ecotypes of *Arabidopsis thaliana* is therefore useful in understanding the interaction of plants with their natural environment and, for this work, provided a good model system for method development before the analysis of the more important (and less available) GM wheat samples.

The protocols developed above were applied to a small set of seven different Arabidopsis ecotypes. Three analytical replicates were taken from pooled plant material from each of three replicate trays of each ecotype. As can be seen from the scores plot in Figure 5A, the NMR data sets from each ecotype form a distinct cluster in the PCA analysis. There is excellent analytical reproducibility and tray-to-tray replication is also very good. The loadings plots describe the contributing variables in the data responsible for the position of a cluster in the scores plot and therefore explain the differences between the clusters. When the loadings plots, are plotted as line plots, these resemble the original data and can therefore be considered as NMR spectra of the metabolites that contribute to the differences within the sample set. Figure 5B shows an example of one of the loadings plots from the analysis of the group of Arabidopsis ecotypes. We can see that many peaks are responsible for the separation in ecotypes and include metabolites possessing peaks at various chemical shifts. We can compare the loadings plots to our generated database of standards, run under identical conditions, in order to put assignments to the peaks from this 'difference' spectrum.

For the purposes of this study, the identity of the metabolites responsible for the differences between the ecotypes was less important. Moreover, the experiment demonstrated that by use of the methods developed we were able to reproducibly grow, process and collect NMR data for a set of different plant lines and confidently use the data to compare the metabolite content of different plant lines.

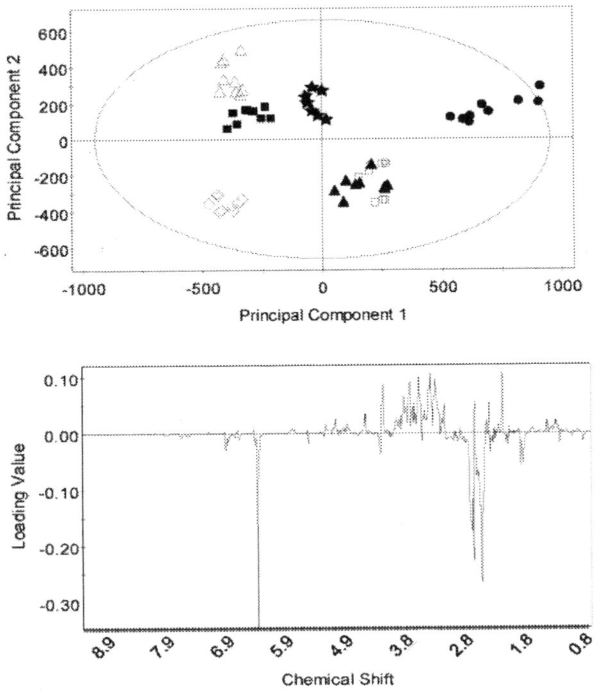

Figure 5. Scores Plot (A) and Loadings Plot (B) from PCA of Arabidopsis ecotypes. (◊) Columbia, (■) Enkheim-T, (□) Landsberg LER, (●) Enkheim-D, (▲) Ws-0, (Δ) Landsberg La-0 , (▪) An-1.

4. Refinement of Methods for the Analysis of GM Wheat

The methodologies described above for *Arabidopsis thaliana* were used to explore the suitability of NMR fingerprinting as a method to assess the substantial equivalence of GM and non-GM wheat. The project, funded by the UK Food Standards Agency, involves the analysis of multiple field plots of wheat grown in successive years at two different sites. The sample set was made up of two control lines (L88-6 and L88-31), a null segregant (B72-8-11-a) and three transgenic lines (B73-61, B102-11 and B102-12). The genetic modification involved insertion of extra genes encoding the high molecular weight subunit of glutenin under grain specific promoters [7]. The material was grown in statistically designed field trials at two sites each year from 1999-2002.

4.1 Elimination of Residual Enzymatic Activity

It was intended to directly apply the methods developed above for Arabidopsis. However problems in the translation of the methods to wheat grain were evident from the initial experiments carried out. Batches of twelve extracts from the same sample of milled whole-grain flour were prepared at the same time. It was observed that these samples were unstable on standing. Figure 6A shows the carbohydrate portions of the NMR spectra obtained from a freshly prepared extract and a second one from the same sample run, under the same conditions on the same instrument, five days later. It was evident that there were changes in the glucose and sucrose levels in the two spectra, indicating the possibility of residual enzyme activity. This was not a problem we had observed when looking at Arabidopsis green tissue. This problem was overcome, as indicated in Figure 6B, by incorporating a 2 minute 90°C heat shock into the extraction protocol. The lower picture shows the effect of the heat shock treatment of the NMR samples. It can now be observed that the sucrose and glucose levels remain stable. In addition, comparison of all regions of the spectra, indicated that introduction of this short heating step did not cause changes in any other regions of the spectrum.

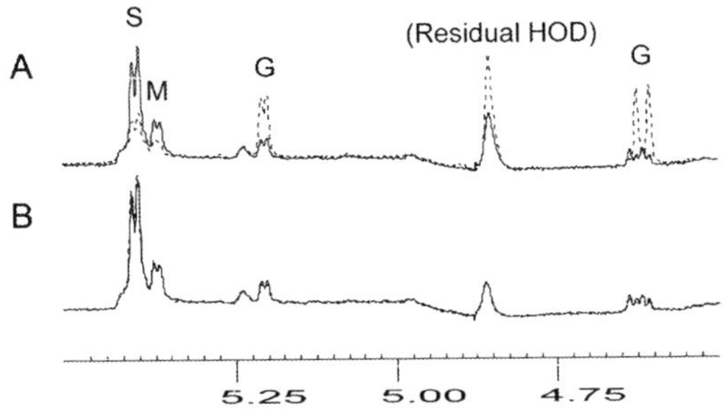

Figure 6. Expansion of the carbohydrate region of the NMR spectrum. A. Comparison of freshly prepared extract (—) vs. spectrum collected from same sample 5 days later (---) without heat shocking the samples. B. Comparison of freshly prepared extract (—) vs. spectrum collected from same sample 5 days later (---) with a 90°C heat shock incorporated into the extraction protocol.

4.2 Assessment of Experimental Reproducibility

Separate batches of twelve extracts from separate milled hole-grain flour portions were prepared at the same time. Comparison of the data by PCA analysis (Figure 7) showed that separate batches of samples prepared from the same grain clustered in different areas of the scores plot. Control samples, which in this case were prepared from commercial flour, however formed a tight cluster irrespective of when the samples were prepared, indicating that the difference observed between different batches can not be due to the extraction protocol. Therefore the variability had to be caused either by the grinding process or as a result of the natural variability of the wheat grain itself.

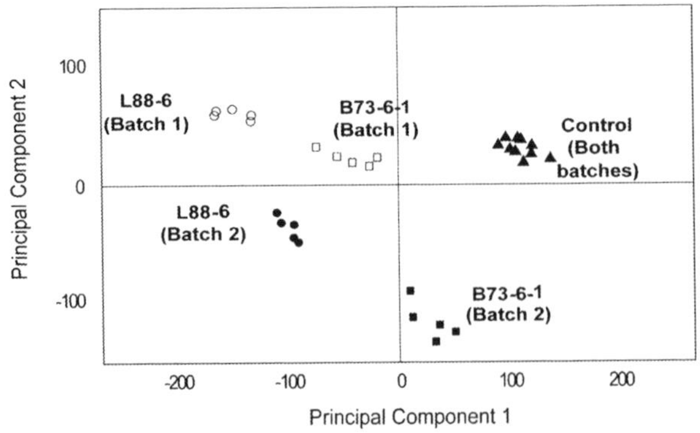

Figure 7. Scores plot from PCA of two batches of extracts of 2 wheat lines and a control sample to assess experimental variability. Control samples (▲), L88-6 batch 1 (○), L88-6 batch 2 (●), B73-6-1 batch 1 (□) and B73-6-1 batch 2 (■).

In order to determine the source of this variability an experiment was designed to distinguish between milling variability and natural variability. Two separate 1g portions of the same wheat grain were taken (X and Y). The individual grains from each portion were cut in half lengthwise to yield four portions (X_1, X_2, Y_1 and Y_2). It was assumed that the grains of wheat were symmetrical and therefore that X_1 and X_2 should be identical. Each portion was crushed under liquid nitrogen and then extracted for ^{1}H-NMR analysis.

There were three possible results from this experiment and this would determine the source of variability in the samples. If all four groups clustered together in the scores plot then it would indicate that there were no problems with the grinding or variability of the grain. This result was unlikely since we had already detected variability in the method. If all four sets of samples clustered separately, it would imply that the source of variability was introduced in the milling of the grain. Finally, if X_1 and X_2 formed one cluster and Y_1 and Y_2 formed a separate cluster, it would indicate the variability was caused by the natural heterogeneity of the seed. The results are shown in Figure 8 and we can see that both halves of the X-grains and both halves of the Y-grains cluster together. This therefore implied that the variability was not introduced in the milling process but that the variability was due to the natural variation of the grain.

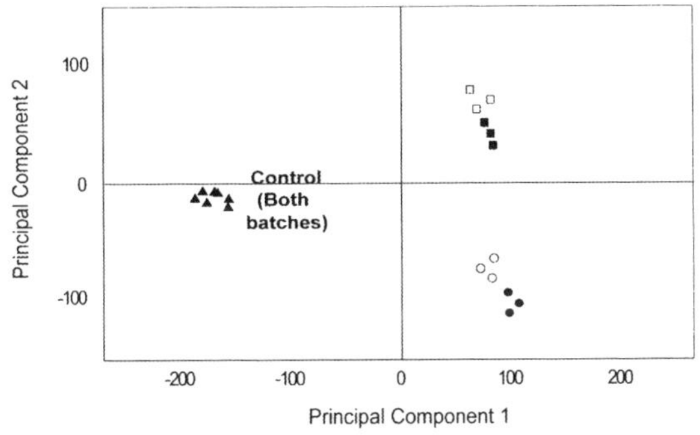

Figure 8. Scores plot from PCA of data from extracts two milled portions of half seeds and a control sample to assess source of experimental variability. Control samples (▲), L88-6 sample 1 X_1 (○), L88-6 sample 1 X_2 (●), L88-6 sample 2 Y_1 (□) L88-6 sample 2 Y_2 (■).

To generate reliable data in the analysis of the GM wheat lines the variability due to sample heterogeneity needed to be reduced. Analysis of single wheat grains would remove any issues as to whether or not the sample represented the population. Unfortunately this approach was unfeasible since single grains would not produce enough material for us to produce replicated extracts for ^{1}H-NMR analysis. Increasing the sample size, i.e. increasing the amount of grain milled, would provide a larger flour sample that may be more representative of the population. This approach was tried with portions of 5g – 20g of grain but analysis of the NMR data still indicated heterogeneity of the sample. This problem was eventually solved by batching and milling the total grain from whole field plots, each containing some 300 plants. Biological variability was assessed by examining different plots of the same line.

5. Experimental Procedures

5.1 Plant Material

Arabidopsis thaliana seeds were obtained from Nottingham Arabidopsis Seed Centre (NASC) and germinated on Murashige Skoog media with 3% sucrose at 22°C in light conditions. Plants were transferred to soil at the 2-4 leaf stage and grown in a controlled environment under long day (16 hrs), at a temperature of 23°C and 75% humidity during the day and 18°C and 80% humidity at night. Plants were harvested at growth stage 6.1-6.5 [4], immediately plunged into liquid nitrogen before freeze drying and grinding to a fine powder in a pestle and mortar. Samples were stored until required at -80°C.

5.2 Preparation of Plant Extracts for ^{1}H-NMR

Freeze-dried Arabidopsis material (15mg) or wheat flour (30mg) was weighed into an autoclaved 2ml Eppendorf tube. D_2O:CD_3OD (1ml 80:20) containing 0.05% w/v d_4-TSP (sodium salt of trimethyl-silylpropionic acid) was added to each sample. The contents of the tube were mixed thoroughly and then heated at 50°C in a water bath for 10 min. After cooling, the supernatant was transferred to a clean Eppendorf tube that was heated at 90 °C for 2 minutes. The samples were cooled in a refrigerator for 45 min. and spun down for 5 minutes. 750µl of the supernatant was added to a 5mm NMR tube.

5.3 ^{1}H-NMR Spectroscopy of Plant Extracts

^{1}H-NMR spectra were acquired under automation at a temperature of 300K on a Bruker Avance spectrometer operating at 399.752 MHz ^{1}H observation frequency using the multinuclear broadband BBO 5mm probe, and a water suppression pulse sequence with a relaxation delay of 5s. Each spectrum consisted of 2048 scans of 32k data points with a spectral width of 4845Hz. The spectra were automatically Fourier transformed using an exponential window with a line broadening value of 0.5Hz, phased and baseline corrected within the automation programme. ^{1}H-NMR chemical shifts in the spectra were referenced to d_4-TSP at δ 0.00.

5.4 Data Reduction of the NMR Spectra and Multivariate Analyses

The ^{1}H-NMR spectra were automatically reduced to ASCII files using AMIX (Analysis of MIXtures software v.3.0; Bruker Biospin). Spectral intensities were scaled to d_4 TSP and reduced to integrated regions or "buckets" of equal width (0.01 ppm) corresponding to the region of δ 9.0 to δ -0.5. The regions between δ 4.90 and δ 4.76 were removed prior to statistical analyses thus eliminating any variability in suppression of the water sample. The signals corresponding to methanol (δ3.365- δ 3.285) and TSP (δ 0.00) were also removed at this stage. The generated ASCII file was imported into Microsoft EXCEL for the addition of labels and then imported into SIMCA-P 9.0 (Umetrics, Umea, Sweden) for PCA analysis.

6. Conclusions and Future Outlook

In this paper, using NMR fingerprints, we have explored the factors that can introduce poor experimental reproducibility, thus decreasing the robustness of the overall method and therefore the reliability of the data obtained. Methods developed with *Arabidopsis thaliana* showed a robustness that was suitable for highlighting differences between different ecotypes. However, when these methods were employed in the analysis of a different type of plant material (wheat flour) initial experiments demonstrated that further development needed to be carried out before the method was able to produce robust data from the sample set. The methods were successfully modified to prevent post-extraction enzymatic degradation, and to minimise problems with heterogeneity in small samples of grain. These methods are now being used in a large-scale NMR analysis of the GM and non-GM grown in field trials over 4 years at two geographical locations. In these studies a control sample will also be prepared and run with each batch of samples. The control samples should cluster together by PCA and will serve as a good indicator of good experimental execution and stability of the samples during the course of the experiment. The results of this study will be reported elsewhere.

Acknowledgements. We thank the BBSRC-GARNet project (http://www.york.ac.uk/res/ garnet/garnet.html) (JL) and the UK Food Standards Agency (JB) for support. Rothamsted Research (MHB, JLW) receives grant-aided support from the Biotechnology and Biological Sciences Research Council of the United Kingdom.

References

[1] Tweendale, H., Notley-McRobb, L. and Ferenci, T. (1998). Effect of slow growth on metabolism of *Escherichia coli*, as revealed by global metabolite pool ("metabolome") analysis. J. Bact. 180, 5109-5116.

[2] Le Gall, G., Colquhoun, I.J., Davis, A.L., Collins, G.J. and Verhoeyen, M.E. (2003). Metabolite profiling of tomato (*Lycopersicon esculentum*) using H-1 NMR spectroscopy as a tool to detect potential unintended effects following a genetic modification. J. Agric. Food Chem. 51, 2447-2456.

[3] Ward, J.L., Harris, C., Lewis, J. and Beale, M.H. (2003). Assessment of ^{1}H-NMR spectroscopy and multivariate analysis as a technique for metabolite fingerprinting of *Arabidopsis thaliana*. Phytochemistry 62, 949-957.

[4] Boyes, D.C., Zayed, A.M., Ascenzi, R., McCaskill, A.J., Hoffman, N.E., Davis, K.R. et al. (2001). Growth stage-based phenotypic analysis of Arabidopsis: A model for high throughput functional genomics in plants. Plant Cell 13, 1499-1510.

[5] Fiehn, O., Altmann, T., Trethewey, R.N. and Willmitzer, L. (2000). Metabolite profiling for plant functional genomics. Nat. Biotechnol. 18, 1157-1161.

[6] King, G., Nienhuis, J. and Hussey, C. (1993). Genetic similarity among ecotypes of *Arabidopsis thaliana* estimated by analysis of restriction fragment length polymorphisms. Theor. Appl. Genet. 86, 1028-1032.

[7] Barro, F., Rooke, L., Békés, F., Gras, P., Tatham, A.S., Fido, R. et al. (1997). Transformation of wheat with high molecular weight subunit genes results in improved functional properties. Nat. Biotechnol. 15, 1295-1299.

2. Bioinformatics and Related Methodology for Biosafety

Genomics for Biosafety in Plant Biotechnology
J.P.H. Nap, A. Atanassov and W.J. Stiekema (Eds.)
IOS Press, 2004

Bioinformatics, Plant Genomes and Biosafety: Can Genomics Help?

Stephen RUDD
Institute for Bioinformatics, GSF-National Research Center for Environment and Health, Ingolstädter Landstraße 1, 85764 Neuherberg, Germany; current address: Centre for Biotechnology, Tykistökatu 6, FIN-20521, Turku, Finland; email: stephen.rudd@btk.utu.fi.

Abstract. Considering the power of the data that is currently hidden within the complete genome scaffolds, or even within the partial transcriptomic data available for many more plant species, it is easy to imagine that there must be information that may have direct relevance in the biosafety evaluation of how a plant may be affected by the introduction of a foreign gene. Bioinformatics has firmly established itself as one of the key disciplines within genomics. It is making in-roads in the analysis of complete genomes and the description, annotation and structuring of all data associated with the genome scaffold. Many more bioinformatics research areas extend beyond the genome sequence and aim to provide accurate methodologies for the prediction of structure and function, *e.g.* the evaluation of how proteins may interact together to form functional complexes. Here I provide a view of how contemporary bioinformatics is working to understand the genome. By considering how a raw genome sequence is annotated and described, and by integrating the data obtained within functional genomics projects, a genome information-web is woven that integrates biosafety with the genomics revolution. This may be used to predict effects of an introgressed transgene.

Introduction

With the publication of the completed *Arabidopsis thaliana* genome sequence [1] and draft sequence for two rice genomes [2,3], the plant community has been thrown into the genomics era. From within the multitude of applications that we have found for either a complete or almost complete genome there is currently little direct impact on the science of biosafety. However, once we move beyond the naïve philosophy of just considering the genome sequence and its raw annotation, we can start integrating the rich rewards from sub-systems biology, integrative biology and the large scale systematic functional genomics projects. With this accumulation of diverse data we will enter an era of genomic understanding. With this understanding we will be in a position to model how subtle changes in gene expression levels, aberrant protein localisation or the introduction of xeno-proteins will influence other genes, impact protein pathways and ultimately affect metabolite pools within any given tissue. To reach this nirvana will be a vast scientific undertaking and many aspects will undoubtedly be reliant on bioinformatics.

Bioinformatics has been brought to the forefront of genomics research with the completion of a multitude of different complete genome sequences. A genome contains one or more chromosomes that can be represented as long linear strings (or closed loops in the case of prokaryotic genomes) of four different nucleotides. The co-ordinate information on genes, gene structures and even operon layout is all encoded within this information, but is in cases highly cryptic and difficult to decipher, let alone to cleanly annotate and to unequivocally describe. Bioinformatics aims to make sense of this information and has pivotal roles within genomic research areas as diverse as the annotation of genomes, the prediction of protein-protein interactions and the basic computational representation and dissemination of accrued data.

Within this chapter a vision for a role of bioinformatics within the sciences of plant biosafety is introduced. Before we can reach a vision of how genomics and bioinformatics may touch upon biosafety we need to consider where contemporary bioinformatics interfaces with the study of complete, draft and unsequenced plant genomes. It is also essential to understand that a single eukaryotic genome may not be a useful reference within any phylogenomic context [4] beyond its nearest cladistic relatives.

1. Comparing Complete Genome Sequences

The scientific community has sequenced and accumulated over 145 prokaryotic and 21 eukaryotic genomes (http://wit.integratedgenomics.com/GOLD/). With these genomes a highly relevant observation becomes apparent – even closely related genomes contain widely different populations of proteins – genomes are not homogeneous entities when we consider their protein complement.

This can be illustrated very simply by considering the annotated protein complement of *Arabidopsis thaliana* and *Oryza sativa* [1-3]. These genomes represent the first sequenced and amongst the smallest plant genomes. These species are not closely related but might be expected to share a bulk of common genes. Investigation of the common proteins using a variety of thresholds and parameters suggested that more than 80% of the ~26,000 Arabidopsis genes have a clear homologue in rice [2,3]. The reciprocal comparison suggests that perhaps as few as 50% of rice proteins have a match in Arabidopsis [3]. Equivalent findings are seen in non-plant genomes. The yeasts represent the most studied eukaryotic group in terms of comparative genomics and profound differences have been described [5,6]. Utilising the information from three completely sequenced fungal genomes and 13 partially sequenced yeast species, insight into the evolution of yeasts has been obtained. While the numbers of genes that are shared between any two given species are not shown, the number of genes unique to a given species (not found elsewhere within the fungal protein collections) varies from very few (< 1%) to in excess of 15% of the genome total [5].

This observation has arguably focussed the genomics community. From the original aims of sequencing core and representative genomes the rationale behind genomics has been redirected to sampling groups of closely related and moderately related genomes, thus bringing the comparative aspects of genomics to the forefront [7]. The model invertebrate genome of the nematode worm *Caenorhabditis elegans* [8] has been joined by its relative *Caenorhabditis briggsae* [9]. Our model plant, Arabidopsis, has a 0.5x shotgun draft of its close taxonomic relative *Brassica oleracea* (see http://www.tigr.org/tdb/e2k1/bog1/) and a whole genome project is underway for corn (*Zea mays*), a relative of rice. In addition, several other plant genomes that are evolutionarily a little more distant, but nevertheless represent the more agronomically important crop species, are in the sequencing and preparation stages. The strong desire to compare and contrast the often unwieldy and large plant genomes has also spawned a plethora of EST sequencing projects – the rewards of which will be discussed towards the end of this chapter.

With the accumulation of more and more genomes we not only understand that there are a lot of different proteins "out there". We are also faced with the rather uncomfortable observation that we have either no or little idea of what the bulk of these proteins do. Genomics is arguably creating many more problems than it is solving. Sub-systems biology, functional genomics, and a variety of novel bioinformatics methods are working towards tidying up this ever-growing collection of loose ends. This will result in the scientific community being able to place a couple of model genomes in the enviable position of being mostly understood rather than remaining largely enigmatic.

Amongst relatively few others, *Escherichia coli* [10] (see http://www.genome.wisc.edu/functional.htm) and *Saccharomyces cerevisiae* [11] stand out as the genomes that are being exhaustively investigated and have fewest genes with unknown anchors to metabolism, development, signalling or stress response. Nevertheless, if we consider *Saccharomyces cerevisiae*, at least 1,500 of the 6,500 annotated ORFs are either similar only to predicted and experimentally uncharacterised proteins or are unique to *S. cerevisiae* and are still functionally uncharacterised [7]. When 1/6 of proteins are functional orphans within a well characterised system, many more such orphans will pollute our understanding of the dynamics and function of the more "exotic" plant genomes.

Within the science of plant biosafety the understanding of how an introgressed gene may influence the genetic neighbourhood and ultimately lead to metabolite flux or to the production of novel peptides or metabolites becomes more important. When we consider that in Arabidopsis only 67% of genes have been characterised to some extent (experimentally or on the basis of either complementary DNA (cDNA) or expressed sequence tag (EST) match) [12], and that only 3,600 genes (~13%) have clear homology to an experimentally characterised gene from Arabidopsis or another genome (http://www.tigr.org/tigr-scripts/arab_gene_phys_ev_classification.cgi) [13] the current expectations for modelling such changes seems rather bleak. How can we estimate the effect of a single gene change when the background knowledge is so incomplete?

With the genome sequences of Arabidopsis and rice in hand, we are in a position where we can at least start designing experiments to address how a change within the genome could affect the whole plant. The NSF 2010 vision for Arabidopsis (http://www.nsf.gov/pubs/2001/bio011/start.htm) is one such ambitious framework where by the year 2010 every gene within Arabidopsis should be experimentally characterised for function, role, localisation *et cetera*. While the results from this programme will have immeasurable benefit to the plant biology and genomics community, this data will become a little esoteric when we consider the plant species phylogenomically distant to Arabidopsis and the evaluation of biosafety.

Arabidopsis is of little agronomic importance beyond the fabled salads of post-doctoral researchers, and its gene content is certainly out of context for the most agronomically essential crops that have come to the forefront within biosafety evaluation. Nonetheless Arabidopsis is our model plant and is the plant of choice within functional genomics and systems biology – it therefore makes the optimal test-bed for evaluating how bioinformatics, genomics and biosafety may interface in the near future.

2. Genome-Oriented Bioinformatics for the Annotation of a Plant Genome

We have already entered the post-genomics era. When a world-renowned scientist has the ability to complete a draft sequence of his pet dog's genome [14], the probable genomic prognosis for our relevant plants are "when they will be sequenced" rather than "if they will be sequenced". The preparation of all but the smaller prokaryotic genomes remains a time-consuming and costly exercise but novel technologies will continue to emerge and eukaryotic genome sequencing will undoubtedly become routine. Currently there are initiatives world wide to sequence the genomes of *Zea mays*, *Medicago truncatula*, *Populus balsamifera* and *Brassica oleracea.* Complete genome sequencing of several other plant genomes including *Lycopersicon esculentum* have been proposed. The methodologies utilised during the analysis and annotation of the Arabidopsis genome has set a *de facto* annotation standard within the plant community, and the resulting data is certainly of relevance to the genome oriented biosafety community.

The annotation of a genome is the process of structuring and identifying data that is associated with the genome scaffold [12,13,15,16]. The annotation adds critical and explanatory notes to the genome, but is also an executive decision about the relevance, accuracy and quality of the evidence, reflecting the annotator's points-of-view [17]. The most basic level of the genome annotation involves the identification of the genes within their chromosomal context and the finer resolution of the intron and exon boundaries. When we consider the Arabidopsis genome the complexity of this task becomes apparent. Five linear strings of nucleotides representing the five chromosomes form the starting material. Within this unstructured sea of over 120 million nucleotides there are a handful of genes.

The first crucial step in describing the genome is to identify the genes. The challenge is to identify the genes without simultaneously collecting large amounts of useless information [17]. The difficulties can clearly be seen using the *Drosophila melanogaster* genome as an example. Using either "standard" or more "permissive" thresholds for the modelling of a gene wildly different numbers of genes can be identified (over 50% more genes with the permissive parameters) [18-20]. Careful examination of the predictions show that there are certainly false positives, but indeed there are also many true positives within the more permissive set including genes that are essential to cell culture viability [20]. It seems that our definition of "conservative" parameters for gene modelling is indeed over-conservative. If we consider the practical process of annotating a new genome sequence we may use two main methods for gene identification [21]: *ab initio* methods and sequence homology-based methods.

The *ab initio* gene modelling methods consider large stretches of DNA and identify gene structures on the basis of a collection of rules and syntax that has been built into the application. There are two main types of *ab initio* gene-modelling algorithms. There are algorithms that have been built on the basis of prior knowledge of coding and non-protein coding sequences. They require a substantial amount of data upon which training can be performed and usually the more training data that can be provided, the more robust the subsequent predictions will be. Such algorithms are based upon complex statistical methods based on Markov models and biases within the populations of triplet words found within coding and non-coding sequences [22]. Heuristic methods can also be applied [23], and differ only in that minimal training sets are required to perform predictions.

Homology-based methods for gene prediction relies upon having substantially related pairs of genomes (see e.g. [9]). If one genome has been well annotated for protein-coding genes, then a closely related genome will share a majority of the genes. It is then a matter of mapping gene sequences from the reference genome to the genome that is being annotated.

This method has not been used widely within the plant genomes because there has been only restricted availability of either gene or protein sequence from other plant genomes [13]. Certainly within the annotation of the Arabidopsis genome, BLAST homology [13,15] to proteins from either Arabidopsis or other plants was used as strong evidence for the existence of a gene within a neighbourhood and to assist within the delineation of possible gene structures.

Complementary DNAs (cDNAs) and expressed sequence tags (ESTs) have great relevance within gene identification and annotation [12,13,16,24]. They are potent indicators of where transcribed sequences exist and are invaluable within the homology-based characterisation of genes [25]. The ease with which these cDNA-based sequences may be produced also allows for the preparation of very large sequence collections. In addition to providing basic evidence on the existence of a gene within a neighbourhood, there is additional information on the alternative splice forms that may be present [24,26]. Crude information on the relative expression levels for the various splice forms and the ratios between the different genes themselves may also be available [27,28].

To obtain as robust a gene model as is possible, genome annotators will integrate and use annotation evidence from as many lines of evidence as is possible [15,29,30]. An example of a computer-interface to a gene-modelling pipeline is shown in Figure 1. In addition to the *ab initio* gene models, extrinsic features including splice-site predictions and homology to EST sequences are available as additional lines of evidence that can assist in the selection of an optimal gene model.

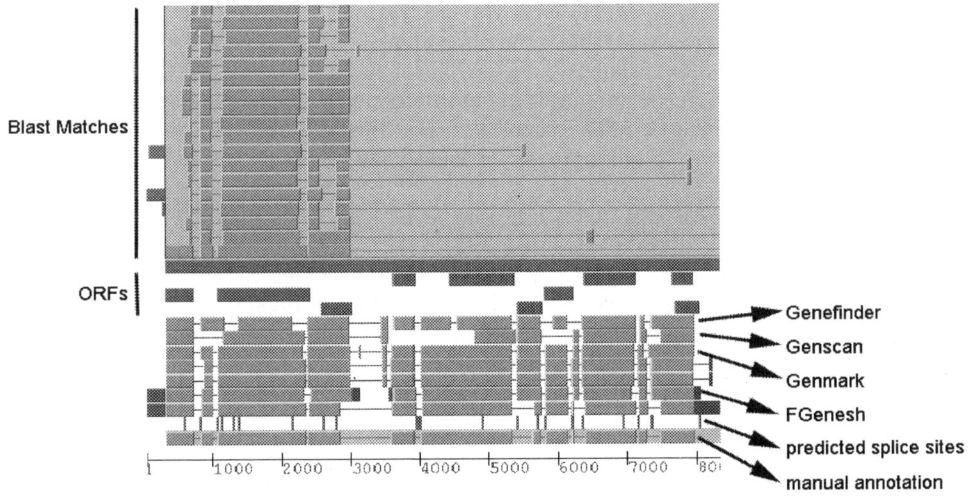

Figure 1. An example of the gene-modelling interface used for annotation of the *Arabidopsis thaliana* and *Neurospora crassa* genomes. The example shown [31] reveals an 8.5 kbp segment of the *Neurospora crassa* genome. In the lower half of the figure are a variety of gene-models that stem from different gene-modelling algorithms. Most algorithms suggest that there is a single gene in this neighbourhood. By evaluating the BLAST matches at the top of the figure, extrinsic homology based evidence suggests that there are in fact two genes, only one of which has been described in the sequence databases. This is in agreement with a gene prediction by the FGenesh algorithm. On the basis of this information the annotator creates the manual annotation.

In addition to the mapping and annotation of the protein coding genes within the genome, there is a wide range of other genetic elements that can be comprehensively described. Such elements may range from other transcribed genetic elements such as miRNAs [32,33], tRNAs [34,35] or structural elements, such as nuclear scaffold/matrix attachment regions (S/MARs) [36]. While the function of S/MARs, and to a lesser extent the miRNAs, may be currently poorly characterised and defined, within the broadest scope of biosafety and genomics the detailed map of the genome will allow us to decide as to whether a transgene has been introgressed into an apparently "inert" genomic neighbourhood or if falls within protein coding, protein level regulating or other genetic elements that may influence the underlying pools of metabolites. The spectre of transcriptional read through has been apparent since the first anecdotal observations of plant gene silencing (reviewed in [37]). Such detailed gene annotation concomitantly allows for the identification of neighbouring genes that may be partially and erroneously expressed as either sense or antisense constructs by read-through mechanisms.

3. Plant Proteomes and Functional Annotation

With the annotation of a plant genome we are presented with a long list of proteins. These proteins belong to different classes of biological "understanding". These range from the previously known and experimentally characterised and published proteins (currently a minority of plant proteins) through to proteins that have been annotated to the genome scaffold on the basis of gene-modelling predictions alone. In-between these extremes, there is a wide continuum of other classes, including clear orthologues of known proteins and gene models validated by cDNA evidence. In Arabidopsis, for example, 26,000 proteins have been predicted and or identified on the chromosomal sequence. These can be split into classes as described in [12, 13]. This proteome information needs curation, structuring and classification before it can be meaningfully interpreted or interrogated. This structuring is achieved through annotation and the structure, function and other protein attributes may be placed within biological context [38].

Before comprehensive annotation and analysis efforts can be undertaken, it is necessary to consider what data will be yielded from such an undertaking and how it can be best represented and disseminated. While plain-text description of each entity within a genome allows for the propagation of additional information, the information remains unstructured. Ideally information should be both human- and machine-readable. This requires a common and controlled vocabulary with which all attributes for a specific feature can be labelled.

Gene ontology [39,40] and the MIPS Functional catalogue (Funcat) [41] are among the widely known and accepted functional catalogues. In the MIPS Funcat a hierarchical classification schema is used. Such a catalogue can be used for high-resolution identification of proteins having clearly defined roles, e.g. proteins involved in mRNA degradation. The data can also be viewed using just the top levels of the hierarchy. This presents a basic overview of the distribution of protein functions across the genome, and the data is summarised in the now ubiquitous pie charts that appear in many genomics publications. Figure 2 shows an example of the distribution of functional classes across an EST collection from sugarbeet.

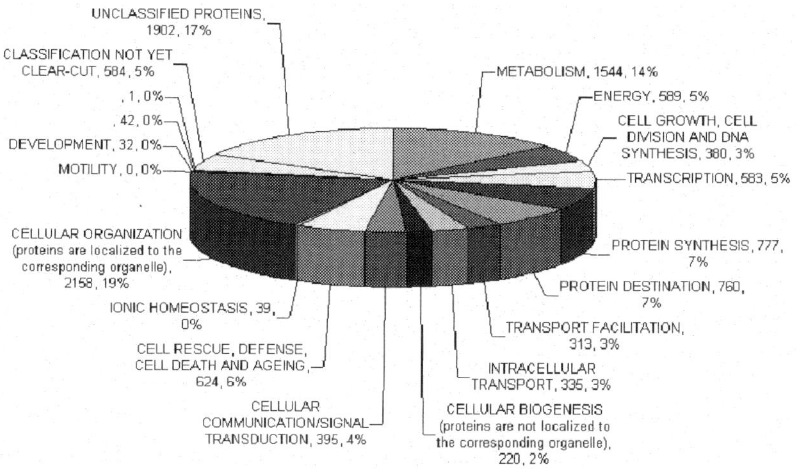

Figure 2. A pie chart showing the top-level functional assignments from over 9,500 peptides derived from a sugarbeet EST collection. The functional distribution when compared to equivalent charts from other genomes allows assessing whether any given functional class is over- or under-represented. We can see that the proteins assigned roles within metabolism and cellular organisation are dominant in this collection. Summarising top-level information stored in ontologies allows for fast surveys of large data sets.

Annotation of proteins can be achieved in two main ways. Manual curation involves an experienced researcher evaluating a protein sequence and placing the sequence within contemporary biological context on the basis of sequence pointers, prior knowledge and intuition. This requires research into the published data that describes function, role and specifics of the protein when available. If a protein has not been described or published before, then likely orthologues can also be used as annotative reference sequences. By definition an orthologous sequence will perform a similar role and annotative attributes can be safely inherited. However, in many cases true orthology cannot be established. Curation then requires the careful assignment of function and role on the basis of moderately conserved homologous sequences. In many cases groups of proteins may be homologous only to themselves and no member may have been experimentally characterised – such groups can only be described as "currently unclassifiable". The plant community is facing many such ambiguous protein families.

In addition, and in many cases as an alternative to manual curation, automated annotation can be used (see Figure 3). Assuming that we have access to a collection of already described and curated data from a different genome we can use homology based methods and stringent expectation values to transfer annotation between genomes and to rapidly perform crude classification on large numbers of proteins. In addition to merely using homologous reference sequences automated annotation uses the fruits of state-of-the-art bioinformatics methods to further add structured information to the protein object.

Annotations typically include an analysis of the protein domain content. The reference method for domain analysis is currently InterPro [42]. By combining several extensive protein domain databases under a single computational umbrella this tool has become indispensable in genome oriented bioinformatics and identified domains in many cases can be linked directly to functional or structural attributes within the Gene Ontology [39]. In addition to protein domains other valuable information may be added. TMHMM predicts the number and location

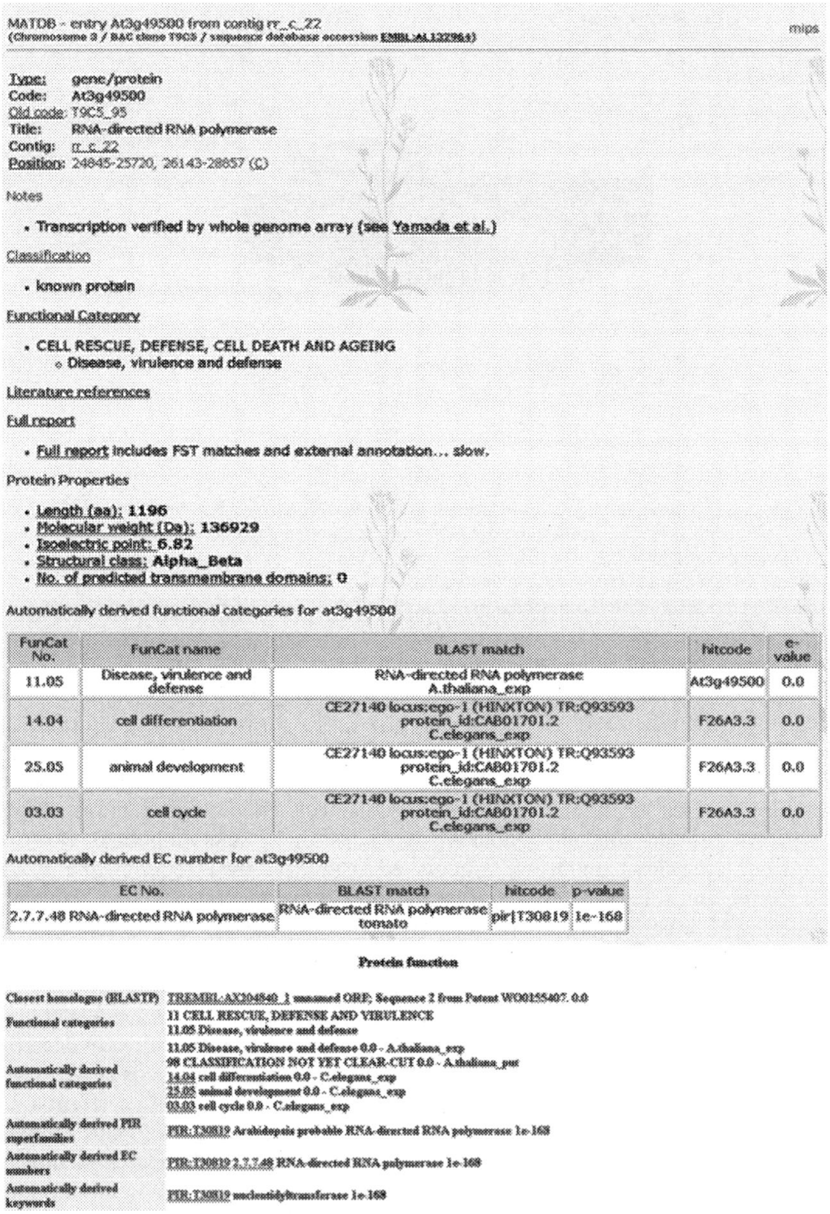

MATDB - entry At3g49500 from contig rr_c_22 mips
(Chromosome 3 / BAC clone T9C5 / sequence database accession EMBL:AL132964)

Type: gene/protein
Code: At3g49500
Old code: T9C5_95
Title: RNA-directed RNA polymerase
Contig: rr_c_22
Position: 24845-25720, 26143-28857 (C)

Notes

- Transcription verified by whole genome array (see Yamada et al.)

Classification

- known protein

Functional Category

- CELL RESCUE, DEFENSE, CELL DEATH AND AGEING
  - Disease, virulence and defense

Literature references

Full report

- Full report includes FST matches and external annotation... slow.

Protein Properties

- Length (aa): 1196
- Molecular weight (Da): 136929
- Isoelectric point: 6.82
- Structural class: Alpha_Beta
- No. of predicted transmembrane domains: 0

Automatically derived functional categories for at3g49500

FunCat No.	FunCat name	BLAST match	hitcode	e-value
11.05	Disease, virulence and defense	RNA-directed RNA polymerase A.thaliana_exp	At3g49500	0.0
14.04	cell differentiation	CE27140 locus:ego-1 (HINXTON) TR:Q93593 protein_id:CAB01701.2 C.elegans_exp	F26A3.3	0.0
25.05	animal development	CE27140 locus:ego-1 (HINXTON) TR:Q93593 protein_id:CAB01701.2 C.elegans_exp	F26A3.3	0.0
03.03	cell cycle	CE27140 locus:ego-1 (HINXTON) TR:Q93593 protein_id:CAB01701.2 C.elegans_exp	F26A3.3	0.0

Automatically derived EC number for at3g49500

EC No.	BLAST match	hitcode	p-value
2.7.7.48 RNA-directed RNA polymerase	RNA-directed RNA polymerase tomato	pir|T30819	1e-168

Protein function

Closest homologue (BLASTP)	TREMBL:AX204840_1 unnamed ORF; Sequence 2 from Patent WO0155407. 0.0
Functional categories	11 CELL RESCUE, DEFENSE AND VIRULENCE 11.05 Disease, virulence and defense
Automatically derived functional categories	11.05 Disease, virulence and defense 0.0 - A.thaliana_exp 98 CLASSIFICATION NOT YET CLEAR-CUT 0.0 - A.thaliana_put 14.04 cell differentiation 0.0 - C.elegans_exp 25.05 animal development 0.0 - C.elegans_exp 03.03 cell cycle 0.0 - C.elegans_exp
Automatically derived PIR superfamilies	PIR:T30819 Arabidopsis probable RNA-directed RNA polymerase 1e-168
Automatically derived EC numbers	PIR:T30819 2.7.7.48 RNA-directed RNA polymerase 1e-168
Automatically derived keywords	PIR:T30819 nucleotidyltransferase 1e-168

Figure 3. A screen-shot of a feature report for the *Arabidopsis thaliana* Sde-1 protein from within the MAtDB database [45]. The annotation includes functional information extracted from the Sde-1 publication [46] in addition to functional assignments based on a homologue from C. elegans. This report includes both manually annotated information and automatically generated annotation such as assignment to PIR superfamilies and derived EC numbers.

of potential transmembrane domains [43] and TargetP can be used to predict where a peptide may specifically targeted to within the cell [44]. Genomics platforms as PEDANT [41] and Sputnik [47] utilise such automated annotation pipelines to organise, describe and interpret large collections of data from multiple genomes. The annotation of an individual protein has little impact on plant biosafety. However, the annotated proteome from a plant forms a robust foundation that is the core of a bioinformatics-oriented vision for biosafety evaluations.

4. Expression Profiling and Pools of Proteins

Microarrays have become a wide spread and predominant method for the parallel profiling of relative gene expression level across either whole genomes or subsets of genes from the genome. The technology behind DNA microarrays involves the fixation of a certain number of probe molecules (either cDNA sequences or oligonucleotides) (from hundreds to hundreds of thousands in number) to a solid substrate. Complementary target molecules (DNA or RNA) can subsequently be bound to the array and with suitable labelling the abundance of target molecules can be estimated [48]. Expression data however provide only indirect and noisy information about regulatory relationships between genes and the experiments themselves reflect distinct physiologies as directed by research foci [49]. With the accumulation of masses of microarray information, it is easy to find ways to exploit this data within genomics and to at least perceive the relevance within biosafety.

So far within this chapter we have been working towards the identification of the unknown set of genes that make up the active transcriptome within a candidate transgenic plant. The difficulties encountered in determining whether a predicted gene-structure is biologically meaningful are clear. Microarray technology has been successfully applied in the search for genes. By using permissive gene prediction models in the Drosophila genome, over 22,000 putative genes were identified. With a whole-transcriptome array on the basis of these predictions and performing different hybridisations using nine different stages of the Drosophila life-cycle, 67% of the predictions could be confirmed as being at least transcribed [20]. This is a clear illustration that microarray analyses are a rapid and reliable method for validating the occurrence of transcription. An analogous experiment has been performed in Arabidopsis using chromosome two as target [50].

Microarray data are more typically used to identify and dissect the varying patterns of expression that occur between genotypes, tissues, developmental stages and between normal and stress conditions [51]. The linked data is invaluable in ascertaining the functional role of genes. However, to get a complete picture of the expression of genes, it is desirable to have all of the genes of an organism represented in the microarray. The value of such whole genome arrays has been shown by examining the expression of genes across the Arabidopsis root [52]. Digital *in situ* data was obtained by generation of high-resolution spatial and temporal expression profiles through a developmental gradient along the root. The analysis revealed that at least 10,500 Arabidopsis genes are expressed within the root, and of these 5,700 are apparently differentially regulated across root sub-zones [52]. The value of such data within genome annotation is invaluable. The picture of when and where a transcript is expressed allows us to perform detailed data mining on the genome information. The integration of other data attributes such as function allows us to propose pathways and interaction networks, especially when we consider such genes as transcription factors.

genome size are not well correlated. Amongst the larger genomes there are several of our agronomically desirable crops *e.g.* bread wheat (*Triticum aestivum*) with a genome size of 17,000 Mbp. Such genomes have such large sizes because of massive accumulation of repetitive elements, this unfortunately (using today's sequencing technologies) renders these genomes unsequenceable since bioinformatics methods cannot make unequivocal sense from millions of related and similar sequence reads [61,62].

The desire to understand and describe the gene sets even within these largest genomes has driven other sequencing strategies. We do not really need the complete genome sequence, so just sequence the currently interesting protein encoding component. Several technologies have been employed to sample the gene space of plant genomes. Methyl-filtration of the gene space is based upon the observation that the repetitive (and non-protein coding) component of the genome tends to be highly methylated [63,64]. By selecting for the under-methylated regions of the genome a significant enrichment of the protein coding features can be achieved. Another sampling method showing promise is the Rescue-Mu method that uses the observation that certain transposing retroelements integrate specifically into or close to genes [65].

The most commonly used method for sampling a component of the gene space, or the transcriptome, is cDNA sequencing. The ease with which cDNA libraries can be prepared and high-throughput sample handling and sequencing pipelines has led to the production of over 4.3 million plant ESTs (EMBL release 77, Nov. 2003). The plant species with the largest EST collections are listed in table 1. It can clearly be seen that there are very large EST collections for some of the agronomically important and unsequenceable genomes (*e.g. Triticum aestivum, Glycine max* and *Hordeum vulgare*).

Table 1. Plant species that have more than 25,000 ESTs within the public domain. This collection contains model genetic systems such as *Physcomitrella patens* and *Arabidopsis thaliana* and species of agricultural interest such as *Vitis vinifera, Lactuca sativa* and *Saccharum officinarum.* The biodiversity within these species makes ESTs suitable for comparative genomics. The number of unigenes for some of these collections (see http://mips.gsf.de/proj/sputnik) comes close to the number of genes that we might expect from the genome. This satisfies the argument that large EST collections form a usable alternative to whole genome scaffolds.

Plant species	# ESTs
Triticum aestivum (bread wheat)	511732
Zea mays	384391
Hordeum vulgare	372432
Glycine max (soybean)	344524
Oryza sativa	260901
Saccharum officinarum	246301
Arabidopsis thaliana (thale cress)	190741
Medicago truncatula (barrel medic)	187763
Sorghum bicolor (sorghum)	161766
Chlamydomonas reinhardtii	154600
Lycopersicon esculentum (tomato)	150228
Vitis vinifera	135712
Pinus taeda (loblolly pine)	110622
Physcomitrella patens subsp. patens	102219
Solanum tuberosum (potato)	94525
Lactuca sativa	68188
Helianthus annuus (common sunflower)	59841
Populus tremula x Populus tremuloides	56013
Gossypium arboreum	38915
Brassica napus (rape)	37159
Lotus corniculatus var. japonicus	36311
Phytophthora sojae	31847
Populus balsamifera subsp. trichocarpa x Populus deltoides	30296
Ipomoea nil (Japanese morning glory)	25899
Mesembryanthemum crystallinum (common iceplant)	25803

The previous sections within this chapter have shown that a plant genome with high quality annotation and the integration of masses of extrinsic data are required before we can start to model how a particular cell will behave when a gene is introgressed – there is clearly no substitute for the fullest collection of data. However, this is impractical and we now need to utilise the bulk of the available information for the evaluation of biosafety. These EST collections represent part of a genome and a functioning intermediate can than be used to establish a handle on the unsequenceable genomes [25].

A variety of bioinformatics platforms have been established that provide some basic groundwork on these EST collections. The more notable resources include Sputnik [47], The TIGR Gene Indices [66], PlantGDB [67] and the NCBI Unigenes [68]. In each of these database subsets of the available EST collections have been collated and much of the underlying redundancy has been stripped from the collections, providing non-redundant sequence assemblies or unigenes [25]. Large EST data sets also have additional applications that have relevance within the general scope of biosafety evaluation. In addition to providing evidence for alternative splice forms [12], the underlying populations of non-synonymous sequence polymorphisms also become apparent [69]. In the absence of microarray data, EST evidence may also be used for *in silico* assessment of expression profiles [27].

Within the Sputnik resource for 'reconstructomics' derived peptide sequences for each of the unigene assemblies allows these unigene features to be annotated and analysed in a manner analogous to the peptides derived from a complete genome sequence scaffold [47]. Certainly within the context of the Sputnik framework, a large collection of ESTs may be considered as a partial substitute for a complete genome sequence. The partial reconstructome sequences clearly have applications within comparative genomics [70]. The foundation peptide sequences form a diverse core that can be fed into our proposed biosafety evaluation pipeline, namely the structured annotation and description and evaluation for likely interaction and pathway partners. While a large EST collection will not be a substitute for the complete genome scaffold with associated high-quality annotation we are afforded at least a glimpse of the genome and can place the sequences that we can view within some form of meaningful biological context.

Even a very modest EST collection of only thousands of sequences is enough to shed light on a genomic system. The collection of 4000 EST sequences from *Cycas rumphii* has been used in conjunction with biochemical knowledge to predict pathways for the production of BMAA ((S)+beta-methyl-alpha, beta-diaminopropionic acid), a potent antagonist of animal glutamate receptors that has been associated with Guam's dementia [71]. This link of plant biochemistry with clinical physiology is certainly a robust demonstration that the application of plant "reconstructomics" has a place within contemporary plant biosafety – both within the context of the biological implications of introgression of foreign genes and the assessment of the metabolite content of existing plants.

8. Future Outlook on Plant Genomics and Biosafety

In the field of plant biosafety we are faced with a continuum of risks and hazards. The only definitive genome within which we should hypothesise the effects of an introgressed or introduced gene should be the genome within which the gene has been inserted. This is clearly not currently practical for the bulk of genomes. While one alternative is to invest effort into making Arabidopsis a consumer-friendly crop plant, it is more sensible to view the biosafety concerns from a pragmatic point-of-view. Certainly we are now in a position where genomics is in a sound position to let us know if an introgressed gene may have an untoward effect in a local genetic neighbourhood. Once we need to start hypothesising if the product of a transgene is likely to have any effects elsewhere within the plant the bioinformatics community really is not yet in a position to help.

There is an enormous number of bioinformatics groups that are working alone to solve crucial questions within genomics, transcriptomics, proteomics and even metabolomics. With each release of a tool, a software package or even a database the quality and applicability of the underlying data become more robust and usable. It is undoubtedly only a matter of time before multiple bioinformatics based resources will be in a position to start answering questions on our core concerns. There are certainly a variety of informatics projects and networks that strive to achieve interoperability and data dissemination between diverse and mutually incompatible data resources e.g. [72]. These are resources that we must watch as data mountains continue to swell. The groundwork for predictive biosafety evaluation has been established and the foundations are strong. Genomics and bioinformatics continue to make small steps towards complete characterisation of a genome. While the mid-term outlook is certainly bright, Arabidopsis may not be is a suitable model for predictive biosafety evaluation in other plant species.

References

[1] AGI (2000). Analysis of the genome sequence of the flowering plant *Arabidopsis thaliana*. Nature 408, 796-815.
[2] Goff, S.A., Ricke, D., Lan, T.H., Presting, G., Wang, R., Dunn, M. et al. (2002). A draft sequence of the rice genome (*Oryza sativa* L. ssp. *japonica*). Science 296, 92-100.
[3] Yu, J., Hu, S., Wang, J., Wong, G.K., Li, S., Liu, B. et al. (2002). A draft sequence of the rice genome (*Oryza sativa* L. ssp. *indica*). Science 296, 79-92.
[4] Eisen, J.A. and Fraser, C.M. (2003). Phylogenomics: intersection of evolution and genomics. Science 300, 1706-1707.
[5] Herrero, E., de la Torre, M.A. and Valentin E. (2003). Comparative genomics of yeast species: new insights into their biology. Int. Microbiol. 6, 183-190.
[6] Kellis, M., Patterson, N., Endrizzi, M., Birren, B. and Lander, E.S. (2003). Sequencing and comparison of yeast species to identify genes and regulatory elements. Nature 423, 241-254.
[7] Hardison, R.C. (2003). Comparative Genomics. PLoS Biol. 1, 156-160.
[8] The *C. elegans* Sequencing Consortium (1998). Genome sequence of the nematode *C. elegans*: a platform for investigating biology. Science 282, 2012-2018.
[9] Stein, L.D., Bao, Z., Blasiar, D., Blumenthal, T., Brent, M.R., Chen, N. et al. (2003). The genome sequence of *Caenorhabditis briggsae*: a platform for comparative genomics. PLoS Biol. 1, 166+.
[10] Serres, M.H., Gopal, S., Nahum, L.A., Liang, P., Gaasterland, T. and Riley, M. (2001). A functional update of the *Escherichia coli* K-12 genome. Genome Biol. 2, 35.
[11] Bader, G.D., Heilbut, A., Andrews, B., Tyers, M., Hughes, T. and Boone, C. (2003). Functional genomics and proteomics: charting a multidimensional map of the yeast cell. Trends Cell Biol. 13, 344-356.
[12] Haas, B.J., Delcher, A.L., Mount, S.M., Wortman, J.R., Smith, R.K., Jr, Hannick, L.I. et al. (2003). Improving the Arabidopsis genome annotation using maximal transcript alignment assemblies. Nucl. Acids Res. 31, 5654-5666.
[13] Wortman, J.R., Haas, B.J., Hannick, L.I., Smith, R.K., Jr., Maiti, R., Ronning, C.M. et al. (2003). Annotation of the Arabidopsis genome. Plant Physiol. 132, 461-468.
[14] Kirkness, E.F., Bafna, V., Halpern, A.L., Levy, S., Remington, K., Rusch, D.B. (2003). The dog genome: survey sequencing and comparative analysis. Science 301, 1898-1903.
[15] Mayer, K.F., Lemcke, K., Schuller, C.M., Rudd, S. and Zaccaria, P. (2000). Arabidopsis genome analysis as exemplified by analysis of chromosome 4. Brief Bioinform. 1, 389-397.
[16] Haas, B.J., Volfovsky, N., Town, C.D., Troukhan, M., Alexandrov, N., Feldmann, K.A. et al. (2002). Full-length messenger RNA sequences greatly improve genome annotation. Genome Biol. 3, 29.
[17] Oliver, B. and Leblanc, B., (2003). How many genes in a genome? Genome Biol. 5, 204.

[18] Adams, M.D., Celniker, S.E., Holt, R.A., Evans, C.A., Gocayne, J.D., Amanatides, P.G. et al. (2000). The genome sequence of *Drosophila melanogaster.* Science 287, 2185-2195.
[19] Misra, S., Crosby, M.A., Mungall, C.J., Matthews, B.B., Campbell, K.S., Hradecky, P. et al. (2002). Annotation of the Drosophila melanogaster euchromatic genome: a systematic review. Genome Biol. 3, 83.
[20] Hild, M., Beckmann, B., Haas, S., Koch, B., Solovyev, V., Busold, C. et al. (2003). An integrated gene annotation and transcriptional profiling approach towards the full gene content of the Drosophila genome. Genome Biol. 5, R3.
[21] Mathe, C., Sagot, M.F., Schiex, T. and Rouze, P. (2002). Current methods of gene prediction, their strengths and weaknesses. Nucl. Acids Res. 30, 4103-4117.
[22] Salzberg, S.L., Delcher, A.L., Kasif, S. and White, O. (1998). Microbial gene identification using interpolated Markov models. Nucl. Acids Res. 26, 544-548.
[23] Besemer, J. and Borodovsky, M. (1999). Heuristic approach to deriving models for gene finding. Nucl. Acids Res. 27, 3911-3920.
[24] Zhu, W., Schueter, S. and Brendel, V. (2003). Refined annotation of the *Arabidopsis thaliana* genome by complete EST mapping. Plant Physiol. 132, 469-484.
[25] Rudd, S. (2003). Expressed sequence tags: alternative or complement to whole genome sequences? Trends Plant Sci. 8, 321-329.
[26] Volfovsky, N., Haas, B.J. and Salzberg, S.L. (2003). Computational discovery of internal micro-exons. Genome Res. 13, 1216-1221.
[27] Ewing, R.M., Kahla, A.B., Poirot, O., Lopez, F., Audic, S. and Claverie, J.M. (1999). Large-scale statistical analyses of rice ESTs reveal correlated patterns of gene expression. Genome Res. 9, 950-959.
[28] Ewing, R.M. and Claverie, J.M. (2000). EST databases as multi-conditional gene expression datasets. Pac. Symp. Biocomput. pp. 430-442.
[29] Allen, J.E., Pertea, M. and Salzberg, S.L. (2004). Computational gene prediction using multiple sources of evidence. Genome Res. 14, 142-148.
[30] Shah, S.P., McVicker, G.P., Mackworth, A.K., Rogic, S. and Ouellette, B.F.F. (2003). GeneComber: combining outputs of gene prediction programs for improved results. Bioinformatics 19, 1296-1297.
[31] Galagan, J.E., Calvo, S.E., Borkovich, K.A., Selker, E.U., Read, N.D., Jaffe, D. (2003). The genome sequence of the filamentous fungus *Neurospora crassa.* Nature 422, 859-868.
[32] Ambros, V., Bartel, B., Bartel, D.P., Burge, C.B., Carrington, J.C. and Chen, X. (2003). A uniform system for microRNA annotation. RNA 9, 277-279.
[33] Griffiths-Jones, S. (2004). The microRNA Registry. Nucl. Acids Res. 32, D109-111.
[34] Lowe, T.M. and Eddy, S.R. (1997). tRNAscan-SE: a program for improved detection of transfer RNA genes in genomic sequence. Nucl. Acids Res. 25, 955-964.
[35] Tsui, V., Macke, T. and Case, D.A. (2003). A novel method for finding tRNA genes. RNA 9, 507-517.
[36] Frisch, M., Frech, K., Klingenhoff, A., Cartharius, K., Liebich, I. and Werner, T. (2002). *In silico* prediction of scaffold/matrix attachment regions in large genomic sequences. Genome Res. 12, 349-354.
[37] Cerutti, H. (2003). RNA interference: traveling in the cell and gaining functions? Trends Genet. 19, 39-46.
[38] Mewes, H.W., Amid, C., Arnold, R., Frishman, D., Guldener, U., Mannhaupt, G. et al. (2004). MIPS: analysis and annotation of proteins from whole genomes. Nucl. Acids Res. 32, D41-44.
[39] Harris, M.A., Clark, J., Ireland, A., Lomax, J., Ashburner, M., Foulger, R. et al. (2004). The Gene Ontology (GO) database and informatics resource. Nucl. Acids Res. 32, D258-261.
[40] Hill, D.P., Blake, J.A., Richardson, J.E. and Ringwald, M. (2002). Extension and integration of the gene ontology (GO): combining GO vocabularies with external vocabularies. Genome Res. 12, 1982-1991.
[41] Frishman, D., Albermann, K., Hani, J., Heumann, K., Metanomski, A., Zollner, A. et al. (2001). Functional and structural genomics using PEDANT. Bioinformatics 17, 44-57.
[42] Apweiler, R., Attwood, T.K., Bairoch, A., Bateman, A., Birney, E., Biswas, M. et al. (2000). InterPro - an integrated documentation resource for protein families, domains and functional sites. Bioinformatics 16, 1145-1150.
[43] Sonnhammer, E.L., von Heijne, G. and Krogh, A. (1998). A hidden Markov model for predicting transmembrane helices in protein sequences. Proc. Int. Conf. Intell. Syst. Mol. Biol. 6, 175-182.
[44] Emanuelsson, O., Nielsen, H., Brunak, S. and von Heijne, G. (2000). Predicting subcellular localization of proteins based on their N-terminal amino acid sequence. J. Mol. Biol. 300, 1005-1016.
[45] Schoof, H., Ernst, R., Nazarov, V., Pfeifer, L., Mewes, H.W. and Mayer, K.F. (2004). MIPS *Arabidopsis thaliana* Database (M*At*DB): an integrated biological knowledge resource for plant genomics. Nucl. Acids Res. 32, D373-376.

[46] Dalmay, T., Hamilton, A., Rudd, S., Angell, S. and Baulcombe, D.C. (2000). An RNA-dependent RNA polymerase gene in Arabidopsis is required for posttranscriptional gene silencing mediated by a transgene but not by a virus. Cell 101, 543-553.
[47] Rudd, S., Mewes, H.W. and Mayer, K.F. (2003). Sputnik: a database platform for comparative plant genomics. Nucl. Acids Res. 31, 128-132.
[48] Zhu, T. (2003). Global analysis of gene expression using GeneChip microarrays. Curr. Opin. Plant Biol. 6, 418-425.
[49] Bergmann, S., Ihmels, J. and Barkai, N. (2004). Similarities and differences in genome-wide expression data of six organisms. PLoS Biol. 2, E9.
[50] Kim, H., Snesrud, E.C., Haas, B., Cheung, F., Town, C.D. and Quackenbush, J. (2003). Gene expression analyses of Arabidopsis chromosome 2 using a genomic DNA amplicon microarray. Genome Res. 13, 327-340.
[51] Rishi, A.S., Nelson, N.D. and Goyal, A. (2002). DNA microarrays: gene expression profiling in plants. In Reviews in Plant Biochemistry and Biotechnology (Goyal, A., Mehta, S. and Lodha, M. Eds) pp. 81-100.
[52] Birnbaum, K., Shasha, D.E., Wang, J.Y., Jung, J.W., Lambert, G.M., Galbraith, D.W. et al. (2003). A gene expression map of the Arabidopsis root. Science 302, 1956-1960.
[53] Brun, C., Chevenet, F., Martin, D., Wojcik, J., Guenoche, A. and Jacq, B. (2003). Functional classification of proteins for the prediction of cellular function from a protein-protein interaction network. Genome Biol. 5, R6.
[54] Kumar, A. and Snyder, M. (2002). Protein complexes take the bait. Nature 415, 123-124.
[55] Vazquez, A., Flammini, A., Maritan, A. and Vespignani, A. (2003). Global protein function prediction from protein-protein interaction networks. Nat. Biotechnol. 21, 697-700.
[56] Valencia, A. and Pazos, F. (2002). Computational methods for the prediction of protein interactions. Curr. Opin. Struct. Biol. 12, 368-373.
[57] Walhout, A.J. and Vidal, M. (2001). Protein interaction maps for model organisms. Nat. Rev. Mol. Cell. Biol. 2, 55-62.
[58] Kanehisa, M., Goto, S., Kawashima, S. and Nakaya, A. (2002). The KEGG databases at GenomeNet. Nucl. Acids Res. 30, 42-46.
[59] Kanehisa, M., Goto, S., Kawashima, S., Okuno, Y. and Hattori, M. (2004). The KEGG resource for deciphering the genome. Nucl. Acids Res. 32, D277-280.
[60] Mueller, L.A., Zhang, P. and Rhee, S.Y. (2003). AraCyc: a biochemical pathway database for Arabidopsis. Plant Physiol. 132, 453-460.
[61] Hoskins, R.A., Smith, C.D., Carlson, J.W., Carvalho, A.B., Halpern, A., Kaminker, J.S. et al. (2002). Heterochromatic sequences in a Drosophila whole-genome shotgun assembly. Genome Biol. 3, 85.
[62] Kent, W.J. and Haussler, D. (2001). Assembly of the working draft of the human genome with GigAssembler. Genome Res. 11. 1541-1548.
[63] Whitelaw, C.A., Barbazuk, W.B., Pertea, G., Chan, A.P., Cheung, F., Lee, Y. et al. (2003). Enrichment of gene-coding sequences in maize by genome filtration. Science 302, 2118-2120.
[64] Palmer, L.E., Rabinowicz, P.D., O'Shaughnessy, A.L., Balija, V.S., Nascimento, L.U., Dike, S. et al. (2003). Maize genome sequencing by methylation filtration. Science 302, 2115-2117.
[65] Raizada, M.N., Nan, G.L. and Walbot, V. (2001). Somatic and germinal mobility of the RescueMu transposon in transgenic maize. Plant Cell 13, 1587-1608.
[66] Quackenbush, J., Cho, J., Lee, D., Liang, F., Holt, I., Karamycheva, S. et al. (2001). The TIGR Gene Indices: analysis of gene transcript sequences in highly sampled eukaryotic species. Nucl. Acids Res. 29, 159-164.
[67] Dong, Q., Schlueter, S.D. and Brendel, V. (2004). PlantGDB, plant genome database and analysis tools. Nucl. Acids Res. 32. D354-359.
[68] Wheeler, D.L., Church, D.M., Federhen, S., Lash, A.E., Madden, T.L., Pontius, J.U. et al. (2003). Database resources of the National Center for Biotechnology. Nucl. Acids Res. 31. 28-33.
[69] Kota, R., Rudd, S., Facius, A., Kolesov, G., Thiel, T., Zhang, H. et al. (2003). Snipping polymorphisms from large EST collections in barley (*Hordeum vulgare* L.). Mol Genet Gen. 270, 24-33.
[70] Rudd, S. (2003). Aktuelle Probleme der vergleichenden Pflanzengenomforschung. GSF Jahresbericht 2003, München: GSF, pp. 33-38.
[71] Brenner, E.D., Stevenson, D.W., McCombie, R.W., Katari, M.S., Rudd, S.A., Mayer, K.F. et al. (2003). Expressed sequence tag analysis in Cycas, the most primitive living seed plant. Genome Biol. 4, R78.
[72] Wilkinson, M.D. and Links, M. (2002). BioMOBY: an open source biological web services proposal. Brief. Bioinform. 3, 331-341.

Genomics for Biosafety in Plant Biotechnology
J.P.H. Nap, A. Atanassov and W.J. Stiekema (Eds.)
IOS Press, 2004

New Tools and Technologies for Controlled Plant Transgene Expression and Transgene Flow: Safety by Design

Yuri GLEBA
Icon Genetics AG, Biozentrum Halle, Weinbergweg 22,
D-06120 Halle/Saale, Germany;
email: gleba@icongenetics.de

Abstract. Safety and control can be derived from safe technology rather than from legal regulations. This is the approach of the 'engineer', *i.e.* a technologist, who is undertaking the task to develop a process that satisfies the requirements and specifications of a producer with respect to technical efficiency, compliance with business requirements and compatibility with existing or predicted infrastructure. It complies with the requirements of a governmental regulator in view of regulatory compliance and sustainability, and it meets the wishes and desires of an end user. This applies to new plant biotechnology for controlled hybridization systems or plant gene expression. In the same way, it can apply to any new genomics approach to biosafety.

Introduction

Plant biotechnology as a commercial process is a reality. Data released by USDA and ISAAA show that in 2003, the global area of genetically modified (GM) crops continued to grow for the seventh consecutive year at a sustained double-digit rate and reached 67.7 million hectares (167.2 million acres). Such growth reflects benefits enjoyed by the various participants in the enterprise, including about 6 million farmers. GM plants grown at present were modified by incorporating input traits to facilitate crop production. GM plants with output traits are also being developed, for example for the production of pharmaceutical proteins. Although there are currently no such plant-made drugs on the market, several clinical trials of 'biogenerics', antibodies and enzymes are in progress or announced. It seems only a matter of time before such products are approved and marketed.

Fears of, and resistance against, plant biotechnology-based activities are a reality too. Publications dealing with safety and regulatory issues relating to GM plants make up a significant portion of publications in the popular press and major science journals. The dialogue is active, often emotionally charged, and involves diverse members of society. Governmental and epigovernmental regulators all over the world are busy defining the legislative framework for safety rules and restrictions that would satisfy both their electorate and economy, while remaining scientifically sound.

The bottom-line of any new technological development, let it be plant biotechnology or genomics, is that safety and control can be derived from safe technology rather than from legal regulations. This is, or should be, the approach of the 'engineer', that is, a technologist, who is undertaking the task to develop a process that satisfies the requirements and specifications of different customers. In the first place, the producer will have requirements with respect to technical efficiency, compliance with business requirements and compatibility with existing or predicted infrastructure. The process should also comply with the requirements of a governmental regulator who looks for regulatory compliance and sustainability. Finally, the process, or resulting product, should meet the wishes and desires of an end user. Such a process has to satisfy currently legislated safety thresholds and, ideally, be ready to satisfy future safety concerns. Being unable to predict future legal requirements, the engineer should rely on safety that is generated by underlying technology. Safety by design can be achieved with appropriate technical components and protocols using those components.

1. Classification of Plants for Biosafety

Plant biotechnology, including genomics, represents a potentially powerful new technology platform for multiple applications and economic activities, including agriculture, food, feed, specialty chemicals, human and animal health and more. The broad applicability and the nascent nature of this technology platform are complicating the analysis. Obviously, safety issues relating to GM plants engineered to be insect-resistant will be different from those relating to plants engineered to make pharmaceutical proteins. It therefore would seem to make sense to define classes of GM plants, that could be produced under broadly similar production processes and would require similar safety locks and circuits. For the purpose of this review, the following four classes are proposed:

- (1). Plants with improved 'input traits' such as pest control (disease resistance, insect resistance or herbicide tolerance), yield enhancement through abiotic stress tolerance, improved plant nutrition and water use. Examples are corn, cotton, soybean and canola varieties that are resistant to herbicides or insects.

- (2). Plants intended for food or feed with improved 'output traits' present in 'physiologically safe' amounts, such as improved nutrients (protein, amino acid, oil, carbohydrates, polysaccharides composition and quality; phytochemicals, e.g. antioxi-dants, isoflavones; micronutrients, e.g. vitamins, iron, etc.), reduced allergenicity, reduced waste, extended shelf life, improved processing and consistency and similar applications. Examples are plants with elevated vitamin content ('golden rice') or tomato varieties with extended shelf life.

- (3). Plants for food/feed with output traits that are present in high ('non-physiological') amounts which have to be processed and adjusted for the amount of active ingredient before use as a food/feed. Examples are canola varieties that are engineered for high expression levels of omega-3 fatty acids, high level of phytase, or plants expressing high amounts of sweet proteins such as thaumatin.

- (4). Plants that are not intended for food or feed, but are used as production hosts for pharmaceutical proteins or industrial enzymes.

GM plants of the classes III and IV will require much higher level of biological separation and control than GM plants that are safe as food.

Although it is often declared that 'safety cannot be compromised', in the real world, we (have to) settle for a calculated risk and accept numerous benefit versus risk bargains. One of the more telling phenomena of our political reality is that, despite more than one million annual deaths in automobile-related accidents worldwide, including over 100000 in the US and EU alone, and the devastating effect of transportation on the environment, there is little serious political opposition to the industry. The general public seems to regard the price worth the benefits enjoyed.

In case of plant biotechnology, the same public is much more negative, and can afford to be so negative. The benefits for the individual other than the about 6 million farmers who grow GM crops are still largely hypothetical. Rather than awaiting future benefits, a very negative attitude is effectively created and exploited by non-governmental organizations that thrive on selling technophobe's fears. A business professional will only regard a plant biotechnology-based business as worthwhile enterprise if the costs of safety-related modifications of production and distribution allow for measurable and reasonable profit margins. As outlined above, only processes and scenarios that satisfy all parties involved will have a chance to become a commercially feasible technology platform. Therefore, plant biotechnology has to present solutions that combine seemingly conflicting requirements.

2. Appropriate Control Systems

Plants are 'self-replicating entities'. Therefore, the plant biotech industry is faced with problems that are strikingly similar to those experienced today by industries whose manufactured products can be easily copied, such as digital music and movies. Plant breeding companies have always tried to develop technologies that would limit the farmer's ability to replant part of harvested seed. The commercial success of corn as a crop is to a great extent the result of development of hybrid seed varieties, which limit the yield benefits of the purchased seed to a single generation, thus providing better control and better profit margins for the developer of the germplasm and the seeds.

With the rise of plant genetic engineering, agrochemical companies such as Monsanto, have developed a novel business model that is no longer built around patented, difficult-to-manufacture chemical compounds as pesticides. Instead, the model is based on the sales of germplasm as a 'self-replicating product'. To successfully exploit this new technology and to recoup the high costs invested in research and development, plant biotechnology companies have to decide how to prevent unauthorized and/or illicit replication and exploitation of their products.

The very high pace of the new technology as well as the development of radically different new plants and products threatens to contaminate existing traditional productions processes and products, potentially leading to health risks, ecological hazards and genetic pollution. Plant biotechnology engineer is therefore faced with the situation where powerful new products generated by a new technology are produced using the 'old' infrastructure of open fields. This infrastructure has so far only slightly been adjusted to suit the 'new plants'.

Another remarkable peculiarity of today's plant biotechnology industry is the absence of potential 'blockbuster' products in research and development pipeline of the 'output traits' category. Except perhaps for omega-3 fatty acids, all other traits either have too small potential markets to appeal to large companies, or their development faces serious technical challenges.

Examples of the latter type include phytase, modified oils, collagen, silk protein, polyhydroxy-butyrates (biodegradable plastics) or human serum albumin. In most cases, the level of expression of the trait of interest achieved is not high enough for an economically feasible production process. Moreover, in those cases where a sufficiently high level of expression can be achieved, agronomic performance and yield of the crop is severely affected.

The apparent conflict between high level of expression of a gene of interest, on one side, and growth and development of the expressing this gene, on the other side, is a well-known phenomenon in biotechnology. The problem can be overcome by separating the 'growth phase' from the 'production phase', as is also often done in industries that use microorganisms. In plants, some endogenous or GM proteins and molecules are being produced at very high levels, with up to 35% of total protein in a leaf (large subunit of ribulose-1,5-bisphosphate carboxylase/oxygenase), and up to 35% of total protein in a seed. There are reasons to believe that the same should be possible with commercially useful products without compromising agronomic performance. Achieving such high productivity levels will, however, require highly sophisticated expression controls. Just using a strong constitutive promoter, as is the case with GM crops today, will not be enough. A more controlled process of expression will not only bring safety benefits for consumers and the environment, but it will also provide for a more efficient protocol for production.

Uncontrolled spread of transgenes to non-transgenic plants and to wild relatives may have harmful consequences both for humans and the environment. Consequently, the design of a GM plant has to provide for a reasonable and controlled containment. Transgene flow should be controlled to prevent escape of the transgene to non-modified plants of the same or related species. There are multiple barriers that the engineer may consider to use, including mechanical, physical, ecological, geographic, temporal, regulatory, and, last but not least, biological. In real world situations, multiple barriers are likely to be used in a multi-layer safety system. We will here limit ourselves to the biological barriers. Several biological processes could be targeted to design safety barriers. These include spread due to vegetative propagation, sexual reproduction, or interspecific gene transfer, and expression of a transgene after spread. In Table 1, a summary of biological control mechanisms is presented.

Table 1. Possible biological components of biosafety systems

Level of biosafety control	Applications/examples
Sexual transmission	Designated crop species Vegetative propagation, apomixis Cytoplasmic control, male sterility Prevention of hybridization
Transgene expression	Encryption Multicomponents
Transgene presence	Tracking Fitness control Negative selection

3. Controlled Hybridization in Plants

One of the potentially most useful technologies that may help to prevent the spread of transgenes to other varieties or to related wild species is the control of sexual hybridization. Since a plant is a 'self-replicating product' and reproduction of most crop plants is by sexual hybridization and seed formation. Therefore, plant breeders have always been interested in methods that would allow them to control sexual reproduction, either to stabilize and exploit an interesting individual genotype by asexual reproduction, or to limit use of the desired trait(s) of interest to one generation only by hybrid seed systems.

3.1 Asexual Reproduction: Vegetative Multiplication and Apomixis

Stabilization of an interesting genotype can be achieved for perennial tree species by vegetative multiplication. In such a case, sexual hybridization would be used only for breeding. Mature plants obtained by this technology are fertile and capable of sexual hybridization, and therefore, this sort of control is not especially useful for our purpose. One particular version of *de facto* vegetative multiplication is apomixis [1-4]. Genetic control of apomixis is however far from being understood. It is probably polygenic. Several genes involved in the control of fertilization-independent seed development have already been identified [5,6]. In some species, apomictic propagation is accompanied by the development of viable pollen that can fertilize non-apomictic plants [7], thus limiting usefulness of this system for transgene control.

3.2 Hybridization Systems

Most domesticated plants were originally capable of cross-pollination and had genetic mechanisms of self-incompatibility that favored cross-pollination over selfing. During decades of breeding, which aimed at generating genetically uniform and high-performing populations, those self-incompatibility genes were bred out, for example in tomato [8]. Lines were selected that were predominantly self-pollinating, thus giving rise to the current breeding material that is often strictly self-pollinating. The inability to cross-pollinate is an excellent genetic barrier that should be used by a genetic engineer. Transgenic varieties of soybean, for example, are much more contained than transgenic canola lines, because soybean, unlike canola varieties, is a facultative self-pollinator [9]. Such a 'free ride' is however not possible for many crops that retain a high level of cross-fertilization, such as corn, canola, sugar beet, barley or rice.

For the purpose of safety, an engineer can extract much more efficacy and control from a system that is using controlled hybridization. Crossing is by far the easiest and best 'process switch' that we currently have available for use in genetic engineering. There are many systems in use today for production of hybrid seed. They include systems based on mechanical castration, nuclear or cytoplasmic male sterility developed by breeders, and transgenic male sterility.

Genetic systems based on nuclear or cytoplasmic male sterility have been used for decades, but they are relatively complex and sometimes problematic. The hybrid seed systems developed by plant breeders were designed to ensure consistent and improved yield for one generation only. The seeds that are harvested by the farmer are fully capable of further sexual reproduction. The notable exception is sterile triploid sugar beets that are obtained by hybridization between tetraploid and diploid parents.

Male sterile phenotypes were discovered as a result of random selection or by interspecific hybridization. For their propagation, one needs a so-called 'maintainer line'. Also, in most crops, the sterility phenotype has to be fully reversed in the resultant hybrid, bringing up the need for 'fertility restorer genes'. All genetic methods for male sterility in use today (for corn, rice, tomato and more.) are based on two- or three-component genetic systems that are difficult to introgress and maintain.

Several companies have had projects aiming at developing transgenic systems for hybrid seed production. To our knowledge, only one has been commercialized so far. This is the canola hybrid seed technology developed by Plant Genetic Systems (now Bayer) [10]. The male sterility is achieved by expression of the toxic enzyme barnase under control of a highly specific and tightly controlled tapetum-specific promoter. Maintenance of sterile lines requires expression of an additional gene, barstar, to reverse the toxic effect of barnase.

A refined approach may represent an additional and useful system for generation and maintenance of male sterile plants and for hybrid seed production [11,12]. The process relies on co-expression of two pairs of genes coding for non-functional barnase fragments (or any other toxic protein) that cause male sterility in anther tissues when both protein fragments are present. Assembly of a functional protein is achieved by intein-based *trans*-splicing (see below) but functionality can also be achieved by simple co-expression of separate polypeptides. For maintenance of the male sterile parent, the genes expressing the barnase fragments are linked to a second pair of genes expressing fragments of a herbicide resistance gene (acetolactate synthase). The two pairs of complementary genes are placed in isoallelic position on homologous chromosomes, thus allowing for easy selection of male sterile 'heterozygotes' as plants that are resistant to the herbicide [12].

These hybrid seed systems can be used in biosafety approaches for two different reasons. When one of the parents is male sterile, spread of the transgene via pollen can be avoided, provided the transgene of interest is only present in the genome of the male sterile parent. Alternatively, and possibly more importantly, the hybrid seed system can b used as an effective switch to turn on the synthesis of the desired product and simultaneously render the resulting progeny sterile, i.e. incapable of subsequent sexual reproduction.

3.3 Hybridization Control Relying on Maternal Transgene Inheritance

Another example of a hybrid seed system is cytoplasmic male sterility. In a plant cell, two organelles in addition to the nucleus, have their own genetic system: plastids and mitochondria) The nucleus contains about 99% of the genetic material. In many important crops, plastid genes are transmitted strictly maternally. Therefore, introducing transgenes into organelle DNA rather than in the nucleus offers a biological level of safety by preventing unwanted spread of transgene through pollen. Plastid-encoded transgenes also offer the advantage of high levels of gene expression [13].

At present, however, there are several limitations to the use of extranuclear transgenics. Despite more than a decade of efforts by numerous groups, plastids of only one species, tobacco, can be transformed routinely. There are still no reliable methods for the genetic transformation of mitochondria. Proteins expressed from organelle DNA are confined to the organelle and are therefore not glycosylated. For the application in biosafety, the most important limitation is that, like in most other genetic systems, the paternal exclusion is not absolute. Transfer of organelle genes via pollen can occur and is detected at low frequencies. Moreover, the organelle-encoded transgene could 'escape' from the plastid to the nucleus.

4. Control of Transgene Expression in Plants

In addition to the prevention of sexual hybridization, plant biotechnology focuses on new tools and technologies to control transgene expression as means to develop safe technology. Many of these tools have left the designer table and were shown applicable in model crops. The challenge now is to apply them in the agronomically important crops.

In designing any system that gives reliable control over transgene transmission ('germplasm control') and/or expression ('expression control'), we inevitably come to a scenario that requires at least two or more components. In its most abstract formulation, it is a two-factor concept consisting of a 'blocking sequence' linked to the gene(s)-of-interest and a 'recovering sequence', all assembled within one construct [14]. Action of the blocking sequence leads to cell death or to impairment of sexual reproduction, while the action of the recovering sequence is induced by externally applied chemical or physical factors, leading to the recovery of the normal plant phenotype. An example of blocking and recovery sequences are the barnase and barstar genes under control of tissue-specific and inducible promoters, respectively.

Depending on the configuration of the components, such a system acquires the possibility to be in two states: 'on' or 'off'. These configurations are designed and implemented with molecular genetics tools. Their interaction should be implemented as a part of normal plant development or life cycle. Switching between components of a multicomponent system can be achieved in several ways. Hybridization of two plants brings the two genetic components together in the hybrid cell. Delivery of a message/ component from outside the cell can be a small molecule that acts as a chemical switch [15] or a virus or bacterium that is a biological switch. Biological switches tend to be more complex. At present, the most reliable 'switching step' is hybridization and several approaches that involve sexual hybridization as switch will be discussed below.

4.1 Two-Component Systems for Transcriptional Control

The general design of a two component system that uses chemical switches for transcriptional control of a gene-of-interest requires a gene of interest under control of a promoter that is regulated by a suppressor or an activator. The suppressor/activator is expressed from a separate gene and its interaction with the promoter of the gene of interest requires presence of a small molecule, such as tetracycline or a steroid [15]. Consequently, the gene can be selectively activated by application of the chemical inducer. Among the limitations of these systems are: (i) a rather low expression level provided by regulated promoters in the 'on' state, (ii) leakiness of the promoters in the 'off' state, (iii) high costs and (iv) safety concerns associated with use of most small chemical activators.

Another two-component system aims to control seed germination [16]. The system relies on a seed-lethal gene, in this particular example, the *Agrobacterium* genes *iaa*M and *iaa*H that cause auxin overproduction in embryos and result in seed lethality. The system also requires a transcriptional repressor that is preferably located on another chromosome and that is required to prevent expression of the seed-lethality gene. The seed-specific phaseolin promoter used to drive the first lethality gene was engineered to contain a binding site for the bacterial Tet repressor. The system requires the presence of both the seed lethality gene linked to the gene-of-interest and the lethality suppressor. Any illicit cross that leads to transmission of only one component will deactivate repression and lead to elimination of all plants with the gene-of-interest from the population. When the two components are positioned at the same locus on two homologous parental chromosomes ("linked in repulsion"), the configuration will lead to complete elimination of all progenies carrying the gene of interest upon cross-pollination [16].

4.2 Control with Site-Specific Recombination

Other two-component systems were used to design a system that would allow restricting expression of a transgene to a particular stage of plant development, to a particular tissue, to particular environmental conditions or to a particular seed generation. Such systems include the use of repressible promoters and a repressor, as well as the use of a gene that can be activated by site-specific recombination [17]. This system aimed to develop a technology that would limit transgene expression or the formation of a viable plant to the first hybrid seed generation only. It would thus be impossible to use part of the harvest for replanting. Several variations in this theme now exist [18].

Although the technical feasibility of such approaches is only scarcely documented, the mere idea of the design caused quite some uproar. Opponents of plant biotechnology dubbed the approach as 'Terminator technology' and argued that its implementation would threaten farmers notably in the developing world by making these farmers even more dependent on large seed and agrochemical companies. The regretful result of this political campaign was that Monsanto officially, and several other companies unofficially, took the decision to not pursue the development of this or similar technologies. In our opinion, however, the principles described in this invention are useful technical concepts that, if implemented, would also allow for control of transgene movement [18], which is essential for transgenic plants of the classes 3 or 4 as defined above.

A number of site-specific recombination systems have been tested and shown to work in plants with high fidelity [19]. Site-specific recombination results in insertions/ excisions, inversions or translocations, depending on the relative orientation of the recombination sites on a chromosome, or their position on different chromosomes. It is therefore possible to design non-functional constructs by addition of disruptive sequences flanked by recombination sites, or by "flipping" part of the construct between two recombination sites. Such configurations will reactivate upon expression and activity of the appropriate recombinase.

Recombination can be induced either by hybridization with a plant expressing recombinase or by transcriptional activation of a recombinase gene using a regulated promoter. For example, an inactive GUS reporter gene in rice plants could be reactivated in the hybrid progeny by expression of Cre recombinase and removal of a blocking DNA insert [20]. So far, a large proportion of the work done in plants has used site-specific recombination to delete parts of an integrated vector, either to remove a selectable marker [20-23], or to resolve complex integration patterns [24].

4.3 Control with Encryption

A different type of approach is also based on multi-component systems, but is not relying on transcriptional control of the gene-of-interest. Such systems use genes-of-interest or constructs with altered or "scrambled" coding instructions. As a result, any escaping genetic material would be completely non-functional if incorporated into an organism that does not possess the appropriate decoding or 'message descrambling' capabilities. Employing existing biological processes, one can design "decryption" mechanisms at the level of transcription (site-specific recombinases, transposases), RNA processing (ribozymes, maturase), translation (tRNA, ribozymes, acetyl synthases), or post-translational modification (inteins, specific targeting to subcellular compartments).

4.3.1 Ribozymes

Intermolecular RNA recombination catalyzed by engineered *trans*-splicing ribozymes has been actively explored. Ribozymes have been engineered mainly from Group I and Group II introns [25-28]. RNA *trans*-splicing was tested in plant and animal systems, primarily to mediate repair of defective mRNAs or to confer toxicity and eliminate cells with undesired RNAs, such as viral RNAs [29-32]. This specific use was dictated by the relatively low efficiency of *trans*-splicing. The application of ribozyme-mediated *trans*-splicing is limited to applications where a very low amount of *trans*-spliced RNA is sufficient to obtain the desired phenotype. The process is however highly specific and can be tailored to target almost any RNA. As such, ribozymes represent a very interesting tool for the genetic engineer's toolbox.

4.3.2 Trans-Splicing Inteins

Inteins are protein fragments that are embedded within precursor proteins, and that are able to excise from the precursor by a post-translational processing event called protein splicing [33-35]. Inteins are able to catalyze not only *cis*-splicing reactions, but also *trans*-splicing events between separate protein fragments [36-38]. Therefore, inteins can reconstitute a functional protein from inactive fragments. Reconstitution of a functional protein by intein-mediated *trans*-splicing in plants was demonstrated recently [39].

The ability of intein fragments to catalyze *trans*-splicing events can be used for building multi-component systems characterized by a highly reduced possibility of transgene escape. For example, a transgenic plant containing both fragments of an intein-split gene will express a fully functional protein. Only part of the progeny inheriting both transgenes will contain the functional protein. If the gene fragments are inserted at the same locus, but on a separate (homologous) chromosome, no progeny of an "illicit" cross with a non-transformed plant will inherit both copies and express the transgene.

In an alternative configuration, the two coding fragments can be transformed into different cell compartments, such as the nucleus and the chloroplast [40]. Since plastid genes are inherited uniparentally (maternally) in most species, the frequency of escape of a functional trait by outcrossing is lowered by several orders of magnitude. By additionally equipping the system with a male sterility gene, one could very effectively suppress the ability of the transgene to escape. One limitation of such a system is the current lack of plastid transformation methods for species other than tobacco.

Intein-assisted *trans*-splicing is a potentially very versatile tool since many different inteins have been described, and because several sites can be used within one protein. Therefore, inteins can be used by the genetic engineer to design complex systems with multiple independent components. We have used intein-based *trans*-splicing to split and reconstitute several proteins, including non-selectable (GUS) and selectable (acetolactate synthase) markers, barnase (a toxic RNAse), T7 polymerase and a recombinase (integrase). Reconstitution of protein function can sometimes also be achieved by alternative means. In some cases, co-expression of non-overlapping peptide fragments can be sufficient for function [41,42]. In other cases, reconstitution of protein function was obtained by introducing affinities (*e.g.* with leucine zippers and alike) that do not involve peptide bond formation [43-45]. Unassisted assembly of barnase from non-overlapping fragments [41,42] is part of a hybrid seed system to confer male sterility [11].

4.3.3 Alternative Coding

In theory, the 'ideal' solution for a safe 'transgene operating system' would be to express the transgene using coding instructions different from those operated by nature. The system would require a message that is using new coding instructions, and a tool that would allow the message to be decoded. The advantage of this approach would be that the transgene would not be expressed in a plant lacking the new coding instructions. In practice, however, a modified coding system in an otherwise normal cell would lead to chaos. However, such a system could still be implemented by a temporary 'violation' of normal coding instructions, for example to activate a genetic switch or to trigger a cascade.

Similar systems, although with fewer modifications than proposed here) have in fact been known in genetics for decades. Most of them are based on nonsense mutation suppressor tRNAs (suppressor tRNAs) that lead to incorporation of amino acids instead of terminating protein translation, when reading stop codons [46]. In its simplest form, the system requires a transgene containing one or more non-sense mutations, and a corresponding suppressor tRNA gene that can be activated when needed. Similar schemes exist for *trans*-activation in plants [47-49]. A mutant GUS gene with three stop codons could be transiently expressed in tobacco protoplasts after co-electroporation with suppressor tRNA genes [48]. After supplying suppressor tRNA genes by hybridization, the mutant gene was also expressed in *Arabidopsis thaliana* [47] and tobacco [49].

4.4 Control of Chromosomal Position

An important issue in the design of proper technology is the relative position of multiple components in the genome [16,50]. Chromosomal position will affect the segregation patterns of transgenes in the progenies after either intentional or illicit crosses. For example, if a transgene or trait of interest is combined with a seed lethality gene to prevent escape of the transgene by outcrossing [16], transgene and repressor have to be present on the same chromosome, on the same locus. Engineering such a linkage is usually not a problem and is achieved by introducing both genes as one DNA construct. Correct positioning and introduction of the other component of the lethality system, such as a repressor of seed lethality is much less straightforward. It can be designed in a number of ways requiring different genetic engineering steps. One of the most powerful solutions is when both components of the system (transgene of interest/lethality gene and lethality repressor) are located at the same locus, but on different homologous chromosomes. The two components, alleles, are then 'linked in repulsion'. In the process of meiosis, each gamete will inherit either one or the other component, but not both [50].

5. Conclusions and Future Outlook

As illustrated in this overview, the GM toolbox to establish safe technology with proper control of transgene expression and transgene flow is filled with promising options and interesting possibilities. Most of these tools still have to prove themselves in the real-world market for large-scale stability and efficacy in agronomically important crops. In the meantime, plant biotechnology research is shifting towards genomics and its omics variants, such as transcriptomics, proteomics and others. The genome-wide and high-throughput analysis of plant genomes and plant gene expression will deepen the understanding of plant life and development. The detailed scrutiny of transgenic events will help to unravel regulatory mechanisms of control of gene expression in plants. In the future, such understanding may result in new and other sophisticated tools for developing safe technology. Such genomics-based tools will face the same requirements for proper design as outlined above: satisfaction of producer, regulator as well as end user. The detailed analyses of any engineered plant material that are now possible in the genomics and post-genomics era, rather than result in legal regulations, should also contribute to the definition, design and use of safe technology for appropriate safety and control.

Acknowledgements. I wish to express my gratitude to Dr. Sylvestre Marillonnet and Dr. Victor Klimyuk at Icon Genetics (Munich, Germany) for valuable discussions during preparation of this review.

References

[1] Merkle , S.A and Dean, J.F. (2000). Forest tree biotechnology. Curr. Opin. Biotechnol. 11, 298-302.
[2] Koltunow, A.M. (1993). Apomixis: embryo sacs and embryos formed without meiosis or fertilization in ovules. Plant Cell 5, 1425-1437.
[3] Koltunow, A.M., Bicknell, R.A. and Chaudhury, A.M. (1995). Molecular strategies for the generation of genetically identical seeds without fertilization. Plant Physiol. 108, 1345-1352.
[4] van Dijk, P. and van Damme, J. (2000). Apomixis technology and the paradox of sex. Trends Plant Sci. 5, 81-84.
[5] Chaudhury, A.M., Luo, M., Miller, C., Craig, S., Dennis, E.S. and Peacock, W.J. (1997). Fertilisa-tion-independent seed development in *Arabidopsis thaliana*. Proc. Natl Acad. Sci. USA 94, 4223-4228.
[6] Luo, M., Bilodeau, P., Koltunow, A., Dennis, E., Peacock, W.J. and Chaudhury, A.M. (1999). Genes controlling fertilization-independent seed development in *Arabidopsis thaliana*. Proc. Natl Acad. Sci. USA 96, 296-301.
[7] Van Dijk, P.J. (2003). Ecological and evolutionary opportunities of apomixes: insights from *Taraxacum* and *Chondrilla*. Phil. Trans. Royal Soc. London. Series B. Biol. Sci. 358, 1113-1121.
[8] Kondo, K., Yamamoto, M., Itahashi, R., Sato, T., Egashira, H., Hattori, T. et al. (2002). Insights into the evolution of self-incompatibility in *Lycopersicon* from a study of stylar factors. Plant J. 30, 143-153.
[9] Rieger, M.A., Lamond, M., Preston, C., Powles, S.B. and Roush, R.T. (2002). Pollen-mediated movement of herbicide resistance between commercial canola fields. Science 296, 2386-2389.
[10] Williams, M. and Leemans J. (1998). Maintenance of male-sterile plants. U.S. Patent 5,750,867.
[11] Burgess, D.G., Ralston, E.J., Hanson, W.G., Heckert, M., Ho., M., Jenq, T. et al. (2002). A novel, two-component system for cell lethality and its use in engineering nuclear male sterility in plants. Plant J. 31, 113-122.
[12] Klimyuk, V.I., Ivanov, P., Marillonnet, S., Zenk, M. and Gleba, Y. (2002). Amplification vectors based on trans-splicing. PCT Patent WO 02/097080.
[13] Maliga, P. (2003). Progress towards commercialization of plastid transformation technology. Trends Biotechnol. 21, 20-28.
[14] Kuvshinov, V.V., Koivu, K., Kanerva, A. and Pehu, E. (2001). Molecular control of transgene escape from genetically modified plants. Plant Sci. 160, 517-522.
[15] Gatz, C. (1997). Chemical control of gene expression. Ann. Rev. Plant Physiol. Mol. Biol. 48, 89-108.

[16] Schernthaner, J.P., Fabijanski, S.F., Arnison, P.G., Racicot, M. and Robert, L.S. (2003). Control of seed germination in transgenic plants based on the segregation of a two-component system. Proc. Natl Acad. Sci. USA 100, 6855-6859.
[17] Oliver, M.J., Quisenberry, J.E., Trolinder, N.L.G. and Keim, D.L. (1998). Control of plant gene expression. U.S. Patent 5,977,441.
[18] Visser, B., van der Meer, I., Louwaars, N., Beekwilder, J. and Eaton, D. (2001). The impact of 'terminator' technology. Biotechnol. Devel. Mon. 48, 9-12.
[19] Ow, D.W. (2002). Recombinase-directed plant transformation for the post-genomic era. Plant Mol. Biol. 48, 183-200.
[20] Hoa, T.T., Bong, B.B., Huq, E. and Hodges, T.K. (2002). Cre/lox site-specific recombination controls the excision of a transgene from the rice genome. Theor. Appl. Genet. 104, 518-525.
[21] Sugita, K. Kasahara, T., Matsunaga, E. and Ebinuma, H. (2000). A transformation vector for the pro-duction of marker-free transgenic plants containing a single copy transgene at high frequency. Plant J. 22, 461-469.
[22] Zuo, J., Niu, Q.W., Moller, S.G. and Chua, N.-H. (2001). Chemical-regulated, site-specific DNA excision in transgenic plants. Nat. Biotechnol. 19, 157-161.
[23] Cornelie, S., Lutz, K., Svab, Z. and Maliga, P. (2001). Efficient elimination of selectable marker genes from the plastid genome by the Cre-lox site-specific recombination system. Plant J. 27, 171-178.
[24] Srivastava, V., Anderson, O.D. and Ow, D.W. (1999). Single-copy transgenic wheat generated through the resolution of complex integration patterns. Proc. Natl Acad. Sci. USA 96, 11117-11121.
[25] Mikheeva, S. and Jarrel, K.A. (1996). Use of engineered ribozymes to catalyze chimeric gene assembly. Proc. Natl Acad. Sci. USA 93, 7486-7490.
[26] Ayre, B.G., Koehler, U., Turgeon, R. and Haseloff, J. (2002). Optimization of trans-splicing ribozyme efficiency and specificity by *in vivo* genetic selection. Nucl. Acids Res. 30, E141.
[27] Saldanha, R., Mohr, G., Belfort, M. and Lambowitz, A.M. (1993). Group I and group II introns. FASEB J. 7, 15-24.
[28] Cech, T. (1995). Group I introns: new molecular mechanisms for mRNA repair. Biotechnol. 13, 323-326.
[29] Ayre, B.G., Koehler, U., Goodman, H.M. and Haseloff, J. (1999). Design of highly specific cytotoxins by using trans-splicing ribozymes. Proc. Natl Acad. Sci. USA 96, 3507-3512.
[30] Lan, N., Howrey, R.P., Lee, S.W., Smith, C.A. and Sullenger, B.A. (1998). Ribozyme-mediated repair of sickle beta-globin mRNAs in erythrocyte precursors. Science 280, 1593-1596.
[31] Perriman, R., Bruening, G., Dennis, E.S. and Peacock, W.J. (1995). Effective ribozyme delivery in plant cells. Proc. Natl Acad. Sci. USA 92, 6175-6179.
[32] Sullenger, B.A. and Cech, T.R. (1994). Ribozyme-mediated repair of defective mRNA by targeted trans-splicing. Nature 371, 619-622.
[33] Perler, F.B. (1998). Protein splicing of inteins and hedgehog autoproteolysis: structure, function and evolution. Cell 92, 1-4.
[34] Perler, F.B. and Adam, E. (2000). Protein splicing and its applications. Curr. Opin. Biotechnol.11, 377-383.
[35] Liu, X.Q. (2000). Protein-splicing intein: Genetic mobility, origin and evolution. Ann. Rev. Genet. 34, 61-76.
[36] Wu., H., Hu, Z. and Liu, X.Q. (1998). Protein *trans*-splicing by a split intein encoded in a split *dnaE* gene of *Synechocystis sp.* PCC6803. Proc. Natl Acad. Sci. USA 95, 9226-9231.
[37] Wu, H., Xu, M.Q. and Liu, X.Q. (1998). Protein *trans*-splicing and functional mini-inteins of a cyanobacterial *dnaB* intein. Biochim. Biophys. Acta 1387, 422-432.
[38] Liu, X.Q. and Yang, J. (2003). Split *dnaE* genes encoding multiple novel inteins in *Trichodesmium erythraeum*. J. Biol. Chem. 278, 26315-26318.
[39] Yang, J., Fox, G.C. and Henry-Smith, T.V. (2003). Intein-mediated assembly of a functional β-glucuronidase in transgenic plants. Proc. Natl Acad. Sci. USA 100, 3513-3518.
[40] Chin, H.G., Kim, G.-D., Marin, I., Mersha, F., Evans, T.C., Chen, L. et al. (2003). Protein *trans*-splicing in transgenic plant chloroplast: reconstitution of herbicide resistance from split genes. Proc. Natl Acad. Sci. USA 100, 4510-4515.
[41] Sancho, J. and Fersht, A.R. (1992). Dissection of an enzyme by protein engineering. The N- and C-terminal fragments of barnase form a native-like complex with restored enzymatic activity. J. Mol. Virol. 224, 741-747.
[42] van der Ent, F.M.I., Vos, A. and Plasterk, R.H.A. (1999). Dissecting the role of the N-terminal domain of Human Immunodeficiency Virus integrase by *trans*-complementation analysis. J. Virol. 73, 3176-3186.
[43] Karimova, G., Pidoux, J., Ullmannn, A. and Ladant, D. (1998). A bacterial two-hybrid system based on a reconstituted signal transduction pathway. Proc. Natl Acad. Sci. USA 95, 5752-5756.

[44] Pelletier, J.N., Campbell-Valois, F.-X. and Michnick, S.W. (1998). Oligomerization domain-directed reassembly of active dihydrofolate reductase from rationally designed fragments. Proc. Natl Acad. Sci. USA 95, 12141-12146.

[45] Pelletier, J.N., Arndt, K.M., Plueckthun, A. and Michnick, S.W. (1999). An *in vivo* library-versus-library selection of optimized protein-protein interactions. Nat. Biotechnol.17, 683-690.

[46] Beier, H. and Grimm, M. (2001). Misreading of termination codons in eukaryotes by natural nonsense suppressor tRNAs. Nucl. Acids Res. 29, 4767-4782.

[47] Betzner, A.S., Oakes, M.P. and Huttner, E. (1997). Transfer RNA-mediated suppression of amber stop codons in transgenic *Arabidopsis thaliana.* Plant J. 11, 587-595.

[48] Carneiro, V.T., Pelletier, G. and Small, I. (1993). Transfer RNA-mediated suppression of stop codons in protoplasts and transgenic plants. Plant Mol. Biol. 22, 681-690.

[49] Choisne, N., Martin-Candell, A. and Small, I. (1997). Transactivation of a target gene using a suppressor tRNA in transgenic tobacco plants. Plant J. 11, 597-604.

[50] Fabijanski, S.F. and Arnison, P.G (1999). Methods and genetic compositions to limit outcrossing and undesired gene flow in crop plants. PCT Patent WO 00/37660.

Genomics for Biosafety in Plant Biotechnology
J.P.H. Nap, A. Atanassov and W.J. Stiekema (Eds.)
IOS Press, 2004

Biosafety Assessment of GM Crops and Their Wild Relatives with Molecular Markers for Biodiversity

Alexander N. IGNATOV*, Dinara Sh. MALLABAEVA, Dmitry B. DOROKHOV and Konstantin G. SKRYABIN
Centre for Bioengineering, Russian Academy of Science,
Pr. 60-letiya Oktyabrya, 7/1, 117312, Moscow, Russian Federation;
**email: ignatov@biengi.ac.ru*

Abstract. DNA fingerprinting present a reliable way to detect even minor genetic differences between genotypes, for example between genetically modified (GM) and non-GM crops. Different PCR-based DNA fingerprint systems were applied for research into genomic variability, management of germplasm and genome stability. The major problem in DNA fingerprinting is the absence of criteria for evaluation of the value, use and reproducibility of the data obtained. An ' ideal' DNA fingerprint should result in a profile that reveals genetic distances on three levels: (1) intra- and inter-accession level, (2) inter-accession and inter-subspecies level, and (3) inter-subspecies and inter-species level. In a model study of crop plants and their wild relatives, several PCR methods were evaluated. Statistical analysis of genetic distances calculated from those fingerprints demonstrated significant differences in information capacity and usefulness of the various methods. It is concluded that a combination of methods is most consistent and informative.

Introduction

There are several major applications for DNA fingerprinting of plants: survey, characterisation and management of genetic resources as breeding material and genetically modified (GM) crops. They include the investigation of genetic variability among cultivated and wild plants for the management of germplasm collections and monitoring of biodiversity, for phylogenetic analysis and tagging and mapping of desired trait/genes, for the identification of genotypes and for the evaluation of plant genome stability. Whole-genome sequencing enables to evaluate the utility of each of many different DNA fingerprinting methods regarding genome coverage, reliability and reproducibility.

1. Plant Genome Organisation

At least three plant genomes have been or are being fully sequenced: *Arabidopsis thaliana*, *Medicago truncatula* and *Oryza sativa*. Many more are physically and genetically mapped. The range of plant genome size is large; extending from approximately the size of the genome of animals to over five times the human genome. The genome of *Arabidopsis* contains approximately 25,498 genes encoding proteins from 11,000 families [1]. Analysis of the rice genome indicates that it may have 40,000-50,000 functional genes with about 15,000 gene families [2]. Almost 70,000 ESTs are mapped and cover about 80% of the coding rice genome but most of them were placed in distal parts of chromosomes [3]. Similar to all plant genomes, duplications have occurred randomly in most regions of the chromosomes, resulting in a chimeric nature of the genome [4]. Numerous disease resistance genes belonging to protein kinase families are clustered on certain chromosomes [5]. Up to 50% of the rice genome, including some parts of protein-coding genes, are made up of repetitive sequences [6]. A large proportion of the moderately repeated sequences consists of transposon- and other mobile DNA-related sequences [7]. Distribution and transposition of endogenous retrotransposon sequences may contribute to the organisation of genome structure and evolution, but this remains purely speculative.

The stability of the structure of genomes is challenged by recombination events. Major rearrangements such as inversions are thought to operate by homologous recombination between inverted repeats. Therefore, both the presence and the distribution of such repeats in genomes are related to the conservation of chromosomal structure. Strong under-representation of inverted repeats is found in bacterial genomes and this may contribute to their stability [8]. This indicates the challenges imposed on genome structure by the presence of inverted repeats. It is unknown whether there is a mechanism for gene order control and location on the chromosomes. Such a mechanism would play an important role in both chromosome configuration and functional organisation of the genome.

2. Sources of Variations in Plant Genomes

Differences in genome sequences can be caused by various mechanisms, each of which has its own preferred genotyping method. These mechanisms include:

- Indels (insertion/deletion) and nucleotide replacement in a coding region
  - single nucleotide polymorphisms (SNPs);
- Indels and nucleotide replacement in a non-coding and regulatory region
  - SNPs;
- Indels and nucleotide replacement in a non-coding and non-regulatory region
  - SNPs;
- Short tandem repeats change
  - SSR, ISSR;
- Genomic mobile elements insertion/deletion
  - ISTR; TE-display; ITE-PCR;
- Genome segments rearrangement (recombination)
  - RFLP, AFLP, ISSR, RAPD, AP-PCR;
- Genome segments multiplication or deletion
  - cytological analysis;
- Whole genome duplications
  - chromosome counting.

Study of several orthologous regions in barley and wheat revealed unequal recombination after the ancestors of barley and wheat diverged, and significant variations among plants in non-synonymous substitution rates. Intergenic retrotransposon blocks, many of them organised as nested insertions, mostly account for the lower gene density of the barley and wheat regions. All but two of the retrotransposons were found in the regions between genes, while all but 2 of the 51 inverted repeat transposable elements were found as insertions in gene regions and outside the retrotransposon blocks [9].

3. Molecular Genotyping

Molecular genetics approaches have enriched the resolution of plant genome analysis. The ability to clone and sequence specific genome regions has added sequence-based information to our understanding of plant genomes derived from cytogenetics. While the database of DNA sequences is exponentially growing, other methods are needed to investigate plant genomes at a level of complexity above the primary sequence, but below the cytogenetic arrangement. Historically, molecular typing started from sequencing of nuclear ribosomal RNA genes. Ribosomal DNA analysis identifies interspecies variation, straightforward & reproducible, but not sensitive enough to reveal intraspecies differences. Southern blot-based detection of restriction fragment length polymorphisms (RFLPs) was the first DNA fingerprinting technology, but it has now generally been replaced by polymerase chain reaction (PCR)-based protocols. RFLP analysis was conceptually simple and easy in the laboratory, but in comparison to the new PCR methods, it was very costly, labour intensive and time consuming.

Analysis of single nucleotide polymorphism (SNP), simple sequence repeats (SSR) and simple sequence repeat polymorphism (SSLP) analysis are very sensitive, co-dominant, robust and reproducible [10]. They allow identification of small sequence variations, but require preliminary sequence knowledge and investigation of possible allelic variants for all tested loci. PCR amplification and detection of intrinsic genetic elements in a plant genome can make use of 'random' sequences flanked by short inverted repeats (AP-PCR, RAPD, DAF, etc.) or use short palindromes/restriction sites (AFLP). Other methods are based on repetitive sequences, such as inter single sequence repeat (ISSR) PCR or randomly amplified microsatellite (RAM) PCR, or utilise transposable elements (ISTR, TE-display, ITE) or gene-specific elements, for example for analysing resistance gene families based on PCR primers directed towards conserved sequence domains.

Single, arbitrary primer-based DNA amplification techniques [11,12] have extended the utility of PCR to general genome analysis. Several protocols based on random (arbitrary) amplification of DNA to generate marker for DNA fingerprinting have been reported in the literature. Polymorphisms could be caused by differences in nucleotide sequences at the priming sites (such as point mutations), or by structural rearrangements within the amplified sequence, (e.g., insertions, deletions that facilitate genetic analysis). Such techniques include digital amplification fingerprinting (DAF), multiple arbitrary amplicon profiling (MAAP), inter-repeat amplification (IRA) [13], and amplified fragment length polymorphism (AFLP) analysis [14].

AFLP involves arbitrarily amplification of restriction fragments. A major advantage of AFLP is that a high number of markers can be detected per PCR reaction. All arbitrary PCR fingerprints (including AFLP) are rather simple and cost-effective, probably reveals variation related to different genomic elements, but needs optimisation of the PCR [11]. The detection of genetic fragments flanked by short random sequences (like RAPD, DAF or AFLP) in analysis of the whole genome provides a skeleton of polymorphic sites without knowledge of exact genomic positions before whole-plant genome data became available.

The vast majority of sequenced PCR loci deposited to Genebank have no similarity to any known plant genes or genomic regions except for some retrotransposon-like fragments (unpublished data). Therefore, there is little chance for their localisation. Mobile genetic elements such as transposable elements (TE), retrotransposons and retroviruses integrate into the plant genome to alter band patterns produced by RAPD or AFLP. Such events can be traced by inter-TE PCR, inverted sequences terminal repeats (ISTR), TE-display or other methodologies.

4. Primer Design and Power of PCR-based DNA Fingerprints

The sum of the effective number of alleles (SENA), the polymorphic information content (PIC, also indicated with H_{av}) and the marker index (MI) are used to evaluate the information contents of PCR results applied in genetics studies [15]. Exclusion of identical or highly similar bands from analysis improves the performance of DNA profiling by two-to-three fold [16]. The same study showed that about two-third of all polymorphic RAPD fragments are tightly linked one to another.

Factor analysis (FA) was applied to determine the high informative RAPD and ISSR bands and to remove ' noisy' markers [17]. The ratio between ISSR and RAPD markers among informative loci was approximately equal. In percentage, SENA for information content of loci was 68%, when SENA for the total marker set was only 56%. The average PIC and MI for markers selected by FA in populations of *A. tauschii* and several other *Aegilops* species were higher than the average PIC values calculated for total data [17]. The markers selected by FA reflected better the differences between the species, but were less effective in revealing intraspecies polymorphisms [17].

It is believed that the information content of PCR markers could be increased by application of primers complementary to DNA repeats abundant in genome or to DNA fragments participating in structural events. The distribution of small (8 bp) DNA repeats located 150-2000 bp from each other on plus or minus strands of DNA, and thus capable to serve as PCR primers alone or in pair with another repeat, was analysed in the complete sequenced genome of cereal organelles. Repeats flanking 10 to 30 PCR-able fragments of wheat chloroplast DNA were identified and used for octamer-based genome scanning (OBGS). The OBGS PCR fingerprints obtained for chloroplasts of wheat and barley lines were compared to the fingerprints that were predicted on the basis of the sequence. Compared to RAPD analysis, OBGS gave significantly higher PIC, MI and Genetic Diversity (GD) values [18].

Computer analyses of DNA sequences tightly associated with the synaptonemal complex allow identification of short conservative sequences of about 15 bp in length that occur frequently in genomes of species in different taxonomic groups. These have been called short conservative sequences (SCS). SCS are found at a very high frequency in structural DNA that is associated with the nuclear lamina, nuclear matrix and other structural components [19]. Application of SCS as primers for arbitrary PCR fingerprinting was highly effective [unpublished data]. RAPD and AP-PCR primers were evaluated for linguistic complexity [20] and for information content obtained for *Aegilops* species. Primers of higher linguistic complexity were significantly more informative [A. Serova, personal communication]. Thus, DNA fingerprinting with PCR primers selected for homology to direct and inverted repeat sequences can be used for the survey of plant genomes.

5. Evaluation of Different DNA Fingerprinting Methods

Different genes and genomic regions provide different levels of evolutionary information for determining genome relationships, depending on whether they are highly variable or are prone to intragenomic recombination and rapid evolution. Comparative analyses of different gene and sequence region classes can provide detailed information about plant relationships. However, there are no clear criteria to evaluate the use and validity of PCR-based DNA fingerprints.

The 'ideal' DNA fingerprint profile should produce genetic distances at different levels (1) between intra-accession and accession level, (2) between accession and accession group or subspecies level, and (3) between subspecies and species levels.

Interspecies relationships between plants can be revealed correctly in some particular cases, and it depends on genetic process of species formation. For example, the distributions of genetic distances within Aegilops sp. accessions (Figure 1) indicate clear differences between RAPD, ISSR and ITE-PCR methods. The data show that RAPD fingerprinting reveals intra-species variation better in comparison to ISSR and ITE, and ISSRs are better for revealing interspecies relations than RAPDs and ITEs.

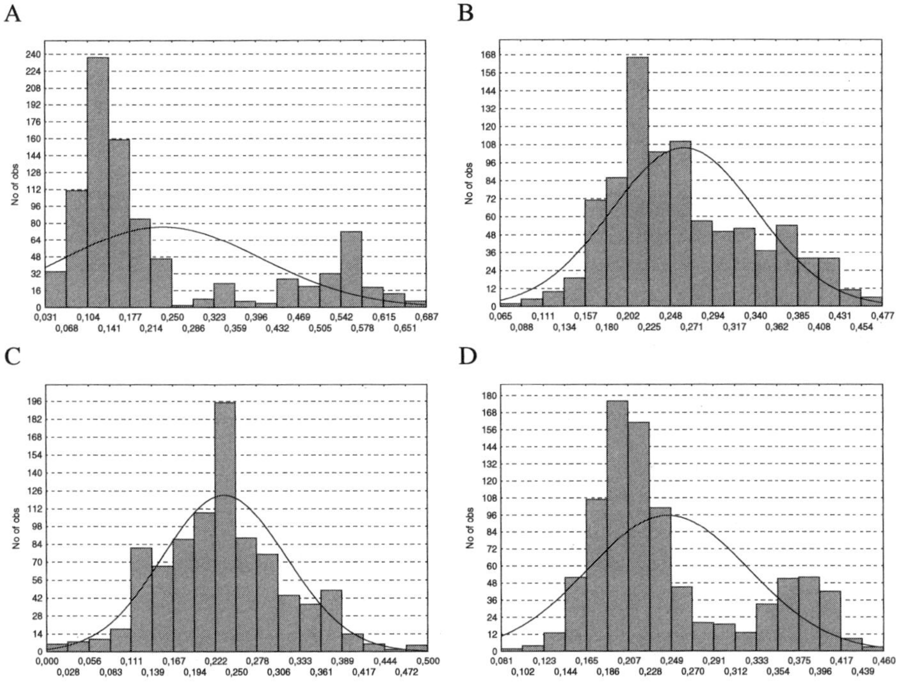

Figure 1. Genetic distance (Jaccard's) between 43 accessions of *Aegilops* sp. calculated from about 100 loci. A. RAPD; B. ISSR, C. ITE-PCR D. total data. Data are plotted against the corresponding Gaussian distribution curve. Two peaks correspond to intra-species and inter-species pairs.

Previously, the correspondence between distance matrices was evaluated by Mantel's Z test [21], co-phenetic correlation between dendrograms [22], and Pearson's product-moment correlation [23,24]. Prince et al. [25] reported a significant correlation (0,44) between RFLP analyses and isozymes estimates of genetic variation among accessions of *Capsicum* sp. In contrast, Beer et al. [26] found a highly variable correlation (0,02-0,76) between isozyme and RFLP-based matrices obtained for 177 accessions of *Avena sterilis*. Heun et al. [27] reported 0,36 between RAPD and isozyme analysis of *A. sterilis*.

Comparison of AFLP and RAPD markers on *Pyrus* species [28] demonstrated a very high correlation between the two methods (0.96). Correspondence between the distance matrices strongly depends on the number and genetic variation of the accessions that are compared. Despite a high correlation between co-phenetic values obtained between RAPD and RFLP matrices at the intra-specific level (0.932), only a low correspondence between the methods was found at the inter-species level (0.363) [29].

Other studies have demonstrated that an increase of loci number above 50 for DNA fingerprinting would not change the co-phenetic relationships significantly [30]. It was reported that 100-150 dominant DNA markers could be necessary to obtain a robust cluster with high co-phenetic similarity for genetically highly related plant germplasm of *Vitis vinifera* (with Jaccard's distance ranging from 0.22 to 0.44) [16]. However, the same study also demonstrated that a matrix obtained with only 52 RAPD loci revealed a reliable correlation (0.53, p<0.01) with the matrix obtained for the total data set (387 loci) [16].

Among different statistical approaches, the confirmatory factor analysis (CFA) [31] can test a hypothesis about the factor structure of object characteristics across several data sets. Such data are obtained by different genotyping methods including PCR fingerprinting. If the stability of obtained genetic distance structure is considered the criterion for evaluation of genotyping methods, CFA can classify the correlation matrixes obtained by the various methods. With this purpose, we have evaluated data obtained by RAPD, ISSR and inter-transposable elements (ITE) PCR analysis of *Aegilops* sp. and *Zea mays,* [17,32], and the public data on morphological, isozyme or DNA variation of soybean [33], wheat [34], barley [35], and brassica [29] (Table 1).

Table 1. The structure of major factors calculated for different genotyping data by CFA.

Plant	Factor %	Method and factor loadings	Jaccard's distance range	Reference
Aegilops sp.	F1=68.7 F2=33.3	RAPD 0.88/0.91 ISSR 0.82/0.82 ITE 0.03/0.81	0.08-0.46	[17]
Zea mays	F1=52.3 F2=28.8 F3=17.2	RAPD 0.06/0.06/0.98 ISSR 0.96/-0.14/0.07 ITE -0.11/0.98/0.09	0.05-0.17	[32]
Glycine max *G. soja*	F1=55.9 F2=38.63	Morphology 0.96/-0.23 Isozymes-0.23/ 0.94 RFLPs 0.44/ 0.84 All 0.86/ 0.48	0.32-0.56	[33]
Hordeum vulgare	F1=72.3	SSR -0.56 AFLP-0.91 RAPD 0.99 RFLP -0.86	0.02-0.4	[35]
Brassica sp.	F1=52.2 F2=34.3	RAPD 0.67/ -0.59 GRFLP 0.94/ 0.09 CRFLP 0.04/ 0.95	0.1-0.9	[29]
Triticum aestivum	F1=99.7	RAPD 0.99 ISSR 0.99 ALL 0.99	0.07-0.79	[34]

The results demonstrate stable differences in the structure of the analysed matrices. Application of another multivariate analysis technique, discriminant analysis (DA) [36], indicated that none of the PCR fingerprinting methods could correctly classify every analysed genotype, but all methods together are successful [V. Zotov, personal communication]. Stable differences in components of genetic variability reflected by different PCR analyses were confirmed by analysis of over 700 markers sequenced and submitted to GeneBank. The majority (~60%) of theses RAPD fragments represent sequences without any significant similarity according to BLAST. Only ~30% has homology to some parts of retro-transposable elements. The majority of ISSR markers were rich in simple repeats of different structure, when a few sequenced ITEs have been similar to gene and inter-gene regions with TE insertions [unpublished data].

6. Concluding Remarks

Several techniques for DNA fingerprinting have been developed. These show varying applicability. Primers for arbitrary PCR analysis may be designed for abundant genome sequences, based on linguistic complexity, and functional role in genome structure. PCR fingerprinting requires optimisation and selection of informative methods, primers and PCR bands for analysis. RAPD-PCR was applicable for identification of genetic variability at the intra-accession level. Therefore, it is probably best in recognising differences between GM and corresponding non-GM crops. ITE-PCR fingerprints showed the highest correlation with morphological differences and the best information content at the levels of inter-variety and sub-species comparison. ISSR-PCR is most useful at the inter-specific level. The combination of the RAPD-ISSR-ITE methods produced the DNA fingerprints with information parameters that were most consistent at each level.

Acknowledgements. We are grateful to the NATO-ARW program for financial support of our participation in the Workshop.

References

[1] Arabidopsis Genome Initiative (AGI) (2000). Analysis of the genome sequence of the flowering plant *Arabidopsis thaliana*. Nature 408, 796-815 (http://www.arabidopsis.org/info/agi.jsp).

[2] Goff S.A. et al. (2002). A draft sequence of the rice genome (*Oryza sativa* L. ssp. *japonica*) Science 296, 92-100.

[3] Wu, J. et al. (1998). Physical mapping of duplicated genomic regions of two chromosome ends in rice. Genetics 150, 1595-1603.

[4] Vision, T.J., Brown, D.G., Shmoys, D.B., Durrett, R.T. and Tanksley, S.D. (2000). Selective mapping: a strategy for optimizing the construction of high-density linkage maps. Genetics 155, 407-420.

[5] Wang, Z. et al. (2001). Rice ESTs with disease resistance gene or defense response gene-like sequences mapped to regions containing major resistance genes or QTLs. Mol. Gen. Genet. 265, 302-310.

[6] Kurata, N. et al. (1994). A 300 kilobase interval genetic map of rice including 883 expressed sequences. Nat. Genetics 8: 365-372.

[7] Mao, L. et al. 2000. Rice transposable elements: a survey of 73,000 sequence-tagged-connectors. Genome Res. 10, 982-990.

[8] Achaz, G., Rocha, E.P.C., Netter P and Coissac E. (2002). Origin and fate of repeats in bacteria. Nucl. Acids Res. 30, 2987-2994.

[9] Ramakrishna, W. et al. (2002). Different types and rates of genome evolution detected by comparative sequence analysis of orthologous segments from four cereal genomes. Genetics 162, 1389–1400.

[10] Tautz, D. (1989). Hypervariability of simple sequences as a general source for polymorphic markers. Nucl. Acids Res. 17, 6463-6471.

[11] Welsh, J. and McClelland, M. (1990). Fingerprinting genomes using PCR with arbitrary primers. Nucl. Acids Res. 18, 7213-7218.

[12] Williams, J.G.K., Kubelic, A.R. Livak, K.J.Rafalski, J.A. and Tingey, S.V. (1990). DNA polymorphisms amplified by arbitrary primers are useful as genetic markers. Nucl. Acids Res. 18, 6531 -6535.

[13] Caetano-Anollés, G. and Gresshoff, P.M. (Eds) (1997). DNA Markers: Protocols, Applications and Overviews. New York: Wiley.

[14] Vos, P. et al. (1995). AFLP: a new technique for DNA fingerprinting. Nucl. Acids Res. 23, 4407-4414.

[15] Powell, W. et al. (1996). The comparison of RFLP, RAPD, AFLP and SSR (microsatellite) markers for germplasm analysis. Mol. Breed. 2, 225-238.

[16] Fanizza, G., Colonna, G., Resta, P. and Ferrara, G. (1999). The effect of the number of RAPD markers on the evaluation of genotypic distances in *Vitis vinifera*. Euphytica 107, 45-50.

[17] Serova, A.S., Ignatov, A.N. and Dorokhov, D.B. (2002). Complex molecular marker system applied for estimation of genetic polymorphism within genera Aegilops and Triticum. Bull. State Nikitsky Bot. Gardens 85, 14-16.

[18] Ignatov, A.N. et al. (2002). Distribution of short inverted repeats-flanked DNA fragments in cereal chloroplast genomes and their application for PCR-fingerprinting. Abstract 3rd International Conference On Bioinformatics Of Genome Regulation And Structure. Novosibirsk, Russia, July 14 - 20, p. 121.

[19] Dadashev, S. (2003). Short conservative sequences of DNA - characteristics, localisation, functions Computer Applications in Fundamental and Applied Chemistry, Biology, Pharmaceutics and Medicine. IVTN Online Conferences. 2003. (http://www.ivtn.ru.)
[20] Troyanskaya, O.G., Arbell, O., Koren, Y., Landau, G.M. and Bolshoy, A. (2002). Sequence complexity profiles of prokaryotic genomic sequences: a fast algorithm for calculating linguistic complexity. Bioinformatics 18, 679-688.
[21] Mantel, N. (1967). The detection of disease clustering and a generalized regression approach. Cancer Res. 27, 209-220.
[22] Sneath, P.H.A. and Sokal, R.R. (1973). Numerical Taxonomy. The principles and practice of numerical classification. San Francisco: WH Freeman and Co.
[23] Pearson, K. (1900). On the criterion that a given system of deviations from the probable in the case of a correlated system of variables is such that it can be reasonably supposed to have arisen from random sampling. Phil. Mag. 50, 157-175.
[24] Smouse, P.E., Long, J.C. and Sokal, R.R. (1986). Multiple regression and correlation extensions of the Mantel test of matrix correspondence. Syst. Zool. 35, 627-632.
[25] Prince, J.P., Loaiza-Figueroa, F. and Traksley, S.D. (1992). RFLP and genetic distance among Mexican accessions of *Capsicum*. Genome 35, 726-732.
[26] Beer, S.C., Goffreda, J., Phillips, T.D. Murphy, J.P. and Sorrells, M.E. (1993). Assessment of genetic variation in Avena sterilis using morphological traits, isozymes, and RFLPs. Crop Sci. 33, 1386-1393.
[27] Heun, M., Murphy, J.P. and Phillips, T.D. (1994). A comparison of RAPD and isozyme analysies for determining the genetic relationships among *Avena sterilis* L. accessions. Theor. Appl. Genet 87, 686-696.
[28] Monte-Corvo, L., Cabrita, L., Oliveira, C. and Leitao, J. (2000). Assessment of genetic relationships among Pyrus species and cultivars using AFLP and RAPD markers. Genet. Res. Crop Evol. 47: 257-265.
[29] Thormann, C.E., Ferreira, M.E., Camargo, L.E.A., Tivang, J.G. and Osborn, T.C. (1994). Comparison of RFLP and RAPD markers to estimating genetic relationships within and among cruciferous species. Theor. Appl. Genet. 88, 973-980.
[30] Kresovich, S., Williams, J.G.K. and McFerson, J.R. (1992). Characterisation of genetic identities and relationships of *Brassica oleracea* L. via a random amplified polymorphic DNA assay. Theor. Appl. Genet. 85, 190-196.
[31] Long, J.S. (1983). Confirmatory Factor Analysis. Beverly Hills, USA: Sage Pubs.
[32] Mallabaeva, D.Sh., Ignatov, A.N., Komarova, G.E. and Dorokhov, D.B. (2003). Comparative analysis of information revealed by PCR markers from somaclons of corn. 2nd Moscow International Congress "Biotechnology: State of the Art and Prospects of Development", November 10-14, C.2.18.
[33] Gai, J.Y. et al. (2000). Studies on the evolutionary relationship among ecotypes of *G. max* and *G. soja* in China. Acta Agron. China 26, 513-520.
[34] Nagaoka, T. and Ogihara, Y. (1997). Applicability of inter-simple sequence repeats polymorphisms in wheat for use as DNA markers in comparison to RFLP and RAPD markers. Theor. Appl. Genet. 94, 597-602.
[35] Russel, J.R. et al. (1997). Direct comparison of levels of genetic variation among barley accessions detected by RFLPs, AFLPs, SSRs, and RAPDs. Theor. Appl. Genet. 95, 714-722.
[36] Huberty, C.J. (1975). Discriminant Analysis. Rev Educat. Res., 45, 543-598.

Genomics for Biosafety in Plant Biotechnology
J.P.H. Nap, A. Atanassov and W.J. Stiekema (Eds.)
IOS Press, 2004

Bioinformatics for Biosafety: Predicting the Potential Allergenicity of GM Food

Willem J. STIEKEMA[1,2,3] and Jan-Peter NAP [1,2,4*]
[1]*Centre for BioSystems Genomics, P.O. Box 98, 6700 AB Wageningen, The Netherlands;* [2]*Plant Research International, Wageningen University and Research Centre, The Netherlands;* [3]*Laboratory of Bioinformatics, Wageningen University and Research Centre, The Netherlands;* [4]*Hanzehogeschool Groningen, Groningen, The Netherlands;*
**email: janpeter.nap@wur.nl*

Abstract. The potential of increased allergenicity of genetically modified (GM) food is a major issue in the biosafety assessment of GM crops. Food allergy is a complex adverse reaction towards food that is difficult to diagnose or treat. Potential unintended effects in food and food processing related to allergenicity increase the complexity of the safety assessments. A detailed decision tree to evaluate the potential allergenicity of GM food starts with the source of the gene. Comparison to known allergens is the only known way to predict the allergenic potential of a protein without laboratory tests. Biochemical characteristics have limited predictive power. Here, an overview is given of bioinformatics approaches towards prediction of allergenicity. These are based on advanced multi-alignment, clustering and pattern recognition tools as well as on three-dimensional structural analysis, using a variety of databases. The criteria what establishes 'sufficient similarity' to identify allergens are still being discussed. Cur-rent criteria may need improvements. Ongoing developments will contribute to the future ability to predict the potential allergenicity of proteins *in silico* and may result in new approaches to circumvent the issue for future novel crops.

Introduction

A major issue in the biosafety of genetically modified (GM) food is the potential of increased allergenicity of foods derived from GM crops [1-3]. Food can cause various adverse reactions, known as food sensitivity. Food allergy is an immune response towards a proteinaceous food component mediated by immunoglobulin E (IgE) [4]. A change of food characteristics by the introduction of novel protein components could change the allergenicity of food. Initially, the allergenicity issue focussed on the protein components deliberately added to the crop by genetic engineering, such as the proteins conferring herbicide tolerance or insect resistance. More recently, it has become clear that the transformation events, also when mediated by *Agrobacterium tumefaciens*, could generate protein components that are not deliberately added [5]. This adds to the complexity of predicting future allergenicity. It would be advantageous if the potential allergenicity of any protein or protein component could be evaluated in advance. Towards this end, tools of bioinformatics and genomics are being developed. In this paper, we will introduce the issues of food allergy and protein allergenicity and relate them to GM crops in the context of the *in silico* prediction of potential allergenicity.

1. Food Allergy

An allergy is an abnormal reaction, or overreaction, of the immune system. The word allergy means literally 'altered reactivity'. Allergy can concern all kinds of substances, even water. True food allergy is an adverse reaction to food caused by immunoglobulin E (IgE) [4,6], although other classifications are proposed [7]. The agent eliciting the overreaction is called 'antigen'. IgE is a protein antibody that recognizes an allergen. It circulates in the blood and attaches to the surface of mast cells. When one or multiple IgE's bind to the allergen, chemicals are released that cause the allergic reaction. The allergenicity of a given antigen is attributable to the particular (abnormal) sensitivity of the individual consuming it. Together with other IgE-mediated responses, such as allergies to pollen or dust, food allergy is also known as type I or immediate hypersensitivity, atopic allergy or atopy [8]. IgE-mediated reactions can affect up to 25% of the population in developed countries [9].

Signs of immediate hypersensitivity are elicited within a few minutes after exposure to the antigen. The reaction may be acute and life threatening (anaphylaxis). It is estimated that about three individuals per 100.000 a year suffer from anaphylaxis [10]. Symptoms vary in extent and severity, depending on the type and amount of allergen involved. Eczema, swellings and itching are common signs involving the skin. Asthma, rhinitis and pneumonitis reflect involvement of the respiratory system. Gastrointestinal involvement is reflected by abdominal distress, vomiting and diarrhea. Also, behavioral problems may be related to the direct or indirect involvement of the nervous system [11]. The physiology and cellular mechanisms of an allergic reaction to food are highly complex and involve sensitization (induction) and provocation (actual reaction). Several seemingly independent physiological and biochemical criteria must be met before recognition and response of an antigen result in a hypersensitive immune reaction. For details, we refer to textbooks and reviews [4,6,8,9]. In addition, there are non-IgE-mediated reactions in response to food, particularly in children These involve cell-mediated reactions of lymphocytes rather than antibodies. This reaction is slower than the IgE mediated reaction. The food components that induce such non-IgE mediated reactions and the molecular mechanisms have yet to be fully defined [12]. For example, celiac disease, or gluten-sensitive enteropathy, is a form of food allergy in which the body reacts to gluten. This disease affects about one in every 300 to 3000 individuals [13].

The reason why a given individual develops a true food allergy for a particular antigen is not very well understood. The higher the IgE concentration in the blood serum, the higher is the likelihood for developing IgE-mediated allergy. Although the genetics of the individual determines the level of IgE molecules in the serum, this genetics is not well understood. Regions on chromosomes 5, 6 and 11 are proposed as areas conferring propensity towards allergy, but several others exist [14]. In addition, there is a multitude of physiological and external factors that are supposed to increase the likelihood of developing food allergy. Factors may comprise diet, infections or disease, as well as age or physical condition.

Common causes of food allergy are milk, eggs, peanuts, tree nuts (walnuts, almonds, cashews, pistachios etc.), wheat, soybean, fish and shellfish. These eight food groups are thought to account for over 90% of all allergic reactions to food [13]. Various foods contain multiple allergens. Egg white, for example, is a complex mixture of at least five or six allergenic proteins [8]. The list of protein allergens of the International Union of Immunological Societies (IUIS) now contains well over 400 allergens and over 200 isoforms (see www.allergen.org/List.htm). An extensive list of over 160 less common, but potentially allergenic foods [15], comprises mainly plant-derived compounds [16]. Allergic reactions to fresh fruits and vegetables, the so-called oral allergy syndrome, are also common. These give relatively mild symptoms that are confined to the area of the mouth.

How often true food allergy occurs in populations is not very clear [17]. The tools to come to the diagnosis of food allergy are limited [12]. Roughly 20% of all adverse reactions to food should be considered true food allergy. The diagnosis of food allergy relies on a careful history, physical examination, appropriate skin testing or *in vitro* testing with food extracts, and/or double blind, placebo-controlled food challenges [12,18]. Despite this battery of diagnostic methods comprising immunoassays, molecular approaches and *in vivo* tests (such as the well-known skin prick), current diagnostic procedures are insufficiently standardized to allow firm conclusions. Food allergy is thought to represent only a small fraction of all allergic diseases, affecting less than 2.5 % of the population in the developed world [19]. Children are more commonly affected. In the European Union, nearly 1 in 10 (10%) children have some form of food allergy [20]. In children it is usually a transient phenomenon, with the notable exception of peanut allergy [21]. The prevalence of allergy varies between countries and depends on the local diet. In the USA, peanut is the main food allergen, in Japan rice and soybean, whereas in Scandinavia it is codfish [21].

There is the feeling that allergy in general is on the rise [17]. This is certainly true for pollen allergy, but the public may overestimate the importance of true food allergy as health problem. In the Dutch adult population, self-reported food allergy was 12.4%, whereas its true prevalence confirmed by clinical trials was just under 3% [22]. The prevalence of food allergy is therefore a matter of debate [18]. It appears to rise particularly in 'westernized' societies and has been linked to the 'hygiene hypothesis' and the increased diversity of food consumption worldwide [23]. Altered dietary preferences and the introduction of new food (e.g. kiwi fruit) may have (had) considerable impact [13].

All foods are in principle capable of eliciting an allergic reaction. Almost all food allergens are proteins, although other food components may act as so-called haptens, small molecules that interact with proteins and cause these to become allergenic. Another complicating factor is the occurrence of cross-reactivity between antigens as result of common epitopes among antigens [24]. An individual with an allergy for tree pollen often also suffers from intolerance to nuts and fruits such as apples [25]. The study of allergic reactions to food is furthermore complicated by the fact that most foods are cooked or otherwise processed and consist of many different (proteinaceous) ingredients [8], so that pinpointing the allergic culprit is not straightforward. A next complication is the existence of compounds that are non-toxic and non-allergic themselves, but are able to enhance the sensitization to antigens. These are the so-called adjuvants. Various food products contain such adjuvants. Therefore, the prediction of the allergenic potential of a given protein should be evaluated in the presence of the molecules with which it is consumed [8].

2. GM Crops and Food Allergy

2.1 Issues and Approaches

In a GM crop, generally one or a few new proteins are introduced. It is possible that potentially allergenic proteins are transferred from a donor organism into a plant or food. This gives rise to valid concerns. Obviously, the transgene-encoded protein should not be allergenic. Allergens can maintain their allergenicity in heterologous hosts. This was shown for a GM soybean containing a methionine-rich, but allergenic protein from Brazil nut [26]. The agrofood industry obviously does not want to harm its customers and may even consider making hypo-allergenic food for allergic customers a niche market. Yet there may not be sufficient knowledge about food allergy to be able to predict the future allergenicity of a protein product in a new crop. The relevant issues with respect to food allergenicity are:

- is the transgene-encoded protein itself allergenic?
- can the transgene-encoded protein itself become allergenic in the new host?
- can the transgene-encoded protein alter or contribute the allergenicity of the recipient?
- is there any other undesired change of the allergenic characteristics of the GM crop?

Proteins can be transferred within varieties of the same food, i.e. from one allergenic variety of apple to another, such as in breeding. In such cases, there would be very little reason for concern for allergic consumers. Such a consumer would avoid apples anyway, or select by trial and error a variety without allergenic properties. In contrast, the transfer of proteins across species, or across food families with GM should be investigated in terms of allergenicity. Moreover, new proteins could be produced in the process of genetic engineering (see below, unintended effects). Allergic sensitization can develop after a long-term period of exposure, so after-market surveillance and review of unforeseen and undesired allergenicity should be considered. The diagnosis of true food allergy is not easy and would become even more complicated if a potentially allergenic protein were used in a wide variety of foods. This could also raise the incidence of allergy, especially when children would become sensitized at an earlier age.

There has been and still is quite a lot of discussion about the issue of labeling GM food. In the context of allergy, labeling would need to provide the necessary and sufficient information to potentially affected allergic consumers [27]. The labeling should be specific and not generic, as now in place in the EU. It could be developed as a system as used for the labeling of food additives [1]. Notably for potential allergy, the issue of tolerance and thresholds of ingredients is of importance. Any GM strategy should focus on preventing the transfer of known or potential allergens and on preventing the development of new allergens. For this, the methodology to predict and/or assess the allergenicity of proteins should be markedly improved.

A fundamental issue is therefore when to consider a protein an allergen. For academic purposes, a single IgE-mediated reaction could be enough to classify a protein as allergenic. In this context, any protein should be considered as (potentially) allergenic, but probably only a minority is able to induce the immune response that results in immediate hypersensitivity [28]. Each classification as 'non-allergenic' might simply mean that the individual with the sensitive immune system has not yet been identified. Proteins for which many patients are sensitized are considered 'major allergens'. In a food safety context, some criterion must be set for the number of people affected. The allergenicity assessment should have an objective, for example the GM crop should be as safe as its non-modified counterpart [29], similar to the starting point of the assessment of ecological safety of GM crops [30]. In the decision trees currently in use (see below), such a criterion is defined.

A different angle to the relationship between GM crops and allergenicity are approaches where the increased knowledge on allergy and allergens is used to reduce allergenicity. If food could be modified to reduce or eliminate the allergenic proteins, this could benefit the allergic population in the future. GM technology could be of help. For example, recombinant derivatives of allergens can be used to develop vaccines [31]. Antisense or other gene silencing technologies [32] [33] can be used, in addition to other breeding approaches and technological possibilities [34]. Selection schemes considering the (relative) amounts of allergens deserve more attention, assuming such a reduction in allergens does not affect the agronomic performance of the crop. Soybean contains as many as 15 proteins that should be considered allergenic. The ultimate production of a hypoallergenic soybean could therefore have high impact [32]. Other examples include the elimination of pollen allergens from ragweed or rye, and hypoallergenic rice with reduced seed allergens [33,35,36]. Production of such 'hypoallergenic' food is, however, also hampered by incomplete methods for assessing the allergenic potential of food [18].

2.2 Decision Tree Approaches

In 1996, the International Food Biotechnology Council (IFBC) and the Allergy and Immunology Institute of the International Life Science Institute (ILSI) proposed a decision tree for the evaluation of the potential allergenicity of GM foods [37] that has seen wide adoption. The approach involves (i) the source of the gene, (ii) the sequence homology to known allergens, (iii) the affinity of the introduced protein to IgE from individuals allergic to the source of the genetic material and (iv) the physicochemical properties. The endpoint of the IFBC/ILSI assessment is focussed on the decision to allow marketing the GM food.

An important entry point to consider potential allergenicity is the source of the gene. If it is obtained from a source known to be allergenic, it should be shown that the gene does not encode an allergen. The immune reactivity of the protein should be assessed with serum from a minimum of 14 patients known to be sensitive to the source [37]. If all fail to react to the protein, it is considered unlikely that the protein is a significant food allergen. Where fewer than five relevant sera or allergic individuals are available, other analyses are required. The assessment of the binding of the novel protein to IgE from the blood serum of individuals allergenic to the source of the gene would provide a high level of confidence that no major allergens were transferred.

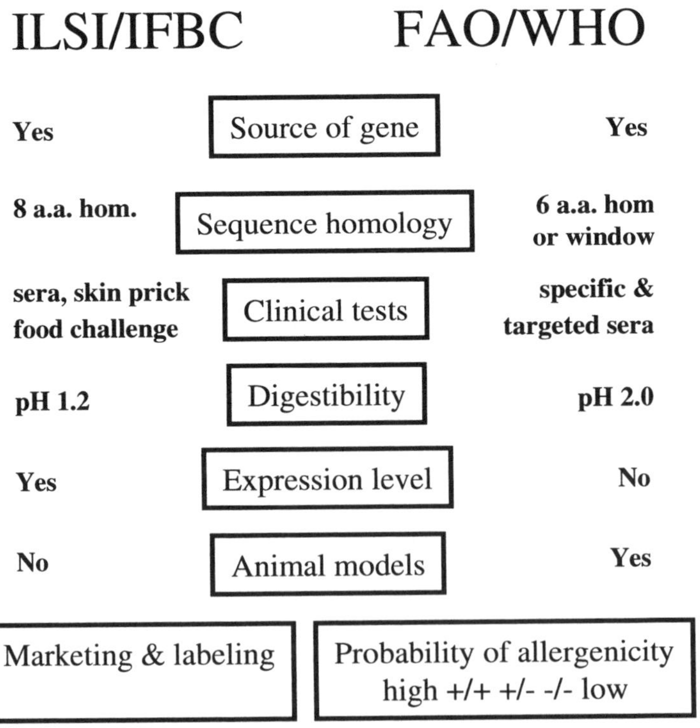

Figure 1. Comparison of the key issues in the IFBC/ILSI [37] and FAO/WHO [38] recommendations for the assessment of the allergenicity of novel foods. Adapted from [10,27].

For proteins derived from sources not associated with allergy, the assessment is more difficult, because the source of the protein and the protein itself, may have no history of being eaten. The first step is comparison. The amino acid sequence should be compared with the amino acid sequence of known allergens using bioinformatics approaches (see below). If regions of homology with known protein allergens spanning eight or more consecutive amino acids are found, this may indicate the presence of common epitopes. If no significant homology with known allergens is found, the protein is not likely to be an important allergen, although stability in simulated gastric fluids should still be tested. A protein from a non-allergenic source with no homology to known allergens, but stable to digestion and processing is considered a minor indication of potential allergenicity. Otherwise, it is not considered a potential allergen. The expression level, tissue specificity and other characteristics of the novel protein are also issues of interest for assessments [27].

In the new FAO/WHO tree [38,39], the endpoint is to give an assessment of the likelihood that the GM food is allergenic. Other adjustments are the introduction of additional serum screening and the use of animal models. Skin prick testing and oral food challenge are not included anymore. Targeted antisera screens for novel proteins from sources without a history of allergy was proposed, using sera from individuals that are taken from patients that are allergic to organisms that are distantly related to the source of the new protein. Also, the use of animal models was recommended. In addition, it was decided to lower the threshold for short identical stretches from eight to six contiguous amino acids. In case of six or more identical amino acids, the protein should be considered as 'likely allergenic' [38]. This criterion is still being debated (see below).

The decision tree approach has rather good predictive value in case of known allergens for which sera of allergic patients exist. In case of potential allergens for which no such sera are available, the predictive value is less good, although it is still considered to result in a reasonably certainty of no evidence of allergenicity [21]. The existing approaches lack the assessment of clinical relevance that should also be taken into account. A certain protein may have allergic potential, but that does not necessarily imply that consumers will be sensitised and suffer. The concentration in the food can be too low, or factors in the food processing may do away with the allergic potential. There may be a threshold dose below which allergic individuals will not react and there may be a threshold dose for sensitization to a protein [40], but more standardization and as wide a range of patients as possible is required to establish such thresholds [41].

Vice versa, an existing food may be used for setting thresholds. For example, kiwi fruit is giving allergenic reactions in a significant fraction of consumers of the population in Western Europe [32]. Yet, kiwi fruit is allowed on the market. No decision-maker has, however, concluded that this implies that for any novel food a threshold of susceptibility in the population that equals to kiwi fruit would be acceptable. In terms of risk assessments, the current decision trees and approaches are mainly focusing on hazard identification. When combined with data on expected exposure and allergenic potency relative to known allergens, it may develop into a real risk assessment. It may become possible to identify minor risks to manage and major risks to avoid [21].

3. Assessment of Allergenicity

Predicting the potential allergenicity of a protein is a complex issue and as outlined above, detailed decisions trees are in place to assess such allergenicity. In addition to testing on patients, the main approaches are the use of general biochemical characteristics, the use of bioinformatics tools and databases for sequence prediction and comparison and the use of animal models. These approaches are discussed in detail below.

3.1 General Biochemical Characteristics

The only currently known way to predict the allergenic potential of a given protein is the comparison of the characteristics of the protein to known allergens. The comparison usually involves common biochemical characteristics, but now includes the amino acid signatures of the protein (Table 1).

Table 1. Biochemical characteristics of proteins in relationship to their potential allergenicity.

Protein characteristic	Criterion for potential allergenicity
Relative abundance	Higher than 0.1-1% of total protein content
Stability	High stability in simulated gastric fluids
Size	Between 10 to 80 kD
Solubility	Required
Glycosylation	Increased
Amino acid signature	Identity

The common biochemical characteristics are abundance, stability, size, solubility and glycosylation. Many common food allergens are present as major protein components in food and can vary from 1-80%. Sensitization may require as little as 0.1-1% [41]. The amount of antigen required for symptom development varies considerably between individuals. In -exceptional- cases very small or trace amounts can be sufficient. In such cases, consumers should be on guard for often-unsuspected 'hidden' ingredients in food.

The more stable a protein is in processing and digestion, the higher may be the likelihood of encountering the immune system. Tests in simulated gastric fluids indicated that proteins not known to be an antigen are digested within 30 seconds, whereas known food allergens were stable for 2 minutes up to over an hour [8]. The relative stability is thought to reflect properties of proteins that contribute directly to their allergenicity, such as stability in serum or proteolysis [28]. Digestive processes might increase allergenicity by unmasking new epitopes or the formation of new epitopes [21]. Although digestibility may establish a well-defined parameter for assessment of potential allergenicity, it is not universally true and many exceptions to the rule exist [28]. Notably proteins associated with the oral allergy syndrome can be unstable. Another study concludes that there is no usable relationship between *in vitro* digestibility and allergenicity [42].

The food allergen must be small enough to pass the gastrointestinal membrane, and large enough to allow two IgE molecules to attach. Laboratory evidence shows a range of 10-80 kD, with most known allergens in the 14-40 kD range [8,21]. Solubility is another prerequisite to become absorbed in the gastrointestinal tract, but this does not imply that proteins with lipophilic parts cannot be antigenic. Also, the role of glycosylation is debated. Most food allergens are glycoproteins, but most glycoproteins are not allergens [21]. The covalent linkage of sugar moieties may increase antigenicity directly, or affect one of the other characteristics, such as the two- or three-dimensional structure of the protein [43].

Overall, the predictive value of the general biochemical characteristics for allergenicity is fairly limited. Although a given protein can be rated for all characteristics, it will not be possible to predict or, more importantly, exclude the potential allergenicity of a protein on the basis of general biochemical characteristics.

3.2 Sequence and Structure Prediction and Comparison

Now that the amino acid sequences of entire allergens are available, proteins can be compared on the sequence level. Bioinformatics provides tools for analyzing allergens that are largely complementary to the laboratory investigations into allergy and allergens. Bioinformatics applications include the structural analysis and prediction of allergens, the identification of epitopes, the assessment of allergenicity and cross-reactivity as well as genome analysis [44]. The impressive quantity of data produced by current high throughput genomics is likely to grow in importance in allergy research and reveal new leads for the analysis of allergy and allergenicity. The approaches of functional genomics, micro-arrays and proteomics are being exploited in allergy-related data [45]. Transcriptome analysis of specific cells playing a role in the immune response will yield more insight in that cellular response on the molecular level. Microarrays are likely to develop in the platform of choice to detect specific IgE's in patient samples [44,46]. EST/genome comparison will help to identify allergic isoforms of gene family members. Associated bioinformatics has become an essential tool for management, storage, interpretation and planning of experiments. In the framework of allergenicity prediction, however, tools should be developed that enable to predict the potential allergenicity, preferably detailing the genetics of the affected patients as well, of any given protein or peptide.

All common and more specialized bioinformatics tools for protein multi-alignment and protein structural analysis are obviously applicable to allergen research and the prediction of allergenicity [44,45]. Bioinformatics tools have been proposed for the prediction of T-cell epitopes [47] and prediction of allergenicity [48]. The subunit sequence of a protein and the three-dimensional shape of the epitope both influence its ability to ignite an allergy sufferer's immune system [43,49]. Two types of epitopes of allergen have been identified, linear and conformational. For linear epitopes, only the primary sequence of a protein is involved in antibody binding. Conformational epitopes are formed from a number of segments in the protein chain, which can be far apart in the primary amino acid sequence. Notably the helper T-cell epitopes depend on the linear amino acid sequence, whereas others, in particular the B-cell epitopes are conformational, so depend (in part) on the 3D structure of the protein.

Analysis of about 40 allergenic proteins indicated the existence of different structural families, without distinguishing structural characteristics [43]. Another analysis of the common properties of plant food allergens identified three protein superfamilies with allergenic properties: prolamins, cupins and cysteine proteases [25]. Cloning of genes encoding allergens has furthermore revealed that most, if not all, allergens are heterogeneous. The significance of this 'isoallergenic variation' is not at present fully understood, however, the implication is that several molecules exist with amino acid substitutions, up to 25% of the amino acids may be substituted. With respect to nomenclature, each isoallergen should be identified by a separate name. What is also poorly understood is the difference in allergic reaction between closely related proteins. For example, the family of 2S seed storage proteins contains some of the most potent plant allergens, while other members seem not to be allergens at all.

The molecular characterization of an allergen is not very different from the desired characterization of any protein from sequence data: sequence comparison and three-dimensional (3D) structural analysis. There is a vast suite of algorithms proposed for de novo structure prediction of protein structure [50] and expectation is that this field will see much progress in the future. Protein allergenicity is determined in part by the 3D structure of the antigenic site [51]. Structural knowledge can therefore contribute to understanding allergen function [45]. Sequence comparison using BLAST or FASTA algorithms with their refinements are essential to point to potential allergenicity and allergic cross-reactivity.

Table 2. Web resources for allergens and predictions of allergenicity and antigenicity

URL	Contents
www.allergen.org	Allergen nomenclature (IUIS)
Fermi.utmb.edu/sdap/sdapf_src.html	Food allergens (SDAP)
www.iit.edu/~sgendel/fa.htm	Food allergy (BIFS)
www.expasy.org/cgi-bin/lists?allergen.txt	Allergen index (Swiss-Prot)
www.ifr.bbsrc.ac.uk/protall/	Plant food allergens (Protall)
Allergen.csl.gov.uk	Allergen database
www.imtech.res.in/raghava/bcipep/index.html	B-cell epitopes
Www2.licr.org/cancerimmunomedb/	CIDB
www.allermatch.org	Online allergenicity prediction
www.allergenonline.com/	Allergen database (Farrp)
www.allergome.org	Official names of allergens
www.allallergy.net	General allergen database
Cooke.gsf.de/asthmagen/main.cfm	Asthma and allergy*
www.hon.ch/library/theme/allergy/glossary/allergy.html	Allergy glossary
Allergyadvisor.com	Crossreactions
Arbl.cvmbs.colostate.edu/molkit/hydropathy/index.html	Antigenicity prediction
Us.expasy.org/cgi-bin/protscale.pl	Antigenicity prediction
Bioinformatics.weizmann.eac.il/hydroph/	Antigenicity prediction

*discontinued

The Protall database contains biochemical and clinical information about plant food allergens involved in classical IgE-induced hypersensitivity reactions. It documents over 77 allergens from 48 plant species [25]. There are many foods with a case history of an allergic reaction for which the allergens responsible have not been described. These are not included in the database. The use of this database is experimental in nature. The database includes unique known and putative allergens (658 unique entries) that were identified by searching publicly available protein databases using the key word allergen, and by searching Medline. A sequence was considered unique if at least one amino acid difference was observed. Not all proteins in the database have been shown to induce clinical allergic responses in individuals. This web site includes a sequence comparison routine to evaluate whether the query protein sequence is homologous to entries in the Protall database.

The Biotechnology Information for Food Safety (BIFS) web site [63] contains a non-redundant list of protein allergens aimed at facilitating sequence comparisons. The Food Safety and Quality Directorate of the Central Science Laboratory (CSL) web site maintains a searchable database of allergens and relevant links. The Farrp Protein Allergen Database contains a list of protein allergens developed and maintained by Monsanto [60]. The web site can be used as a bioinformatics tool to compare any protein sequence to the protein allergen database using FASTA. The Swiss-Prot index lists the entries of many IUIS-designated allergens. The AllAllergy database also includes chemical allergens. The asthma and allergy database [64,65] is a sad example of the 'bioinformatics trap' of individual initiatives: this overview of linkage and mutational studies for allergy-related traits was discontinued due to lack of funding end of 2003.

Allermatch is a recent web tool for the efficient and standardized prediction of potential allergenicity of proteins and peptides according to the current recommendations in the Codex alimentarius [66]. A query amino acid sequence is compared with all known allergenic proteins retrieved from the SwissProt database, using a sliding window approach to identify stretches of 80 amino acids with more than 35% identity or identical small stretches of at least six amino acids.

3.4 Animal Models

Although sequence homology and protein stability provides valuable information as part of the safety assessment, they do not supply a definitive answer. Many factors can contribute to the (potential) allergenicity of any given protein. Some may be essential, such as the presence of epitopes with allergenic potential, others may play minor, but nevertheless critically important roles. Allergens can be detected by serological methods such as enzyme immunoassays, *in vitro* histamine release or *in vivo* skin test procedures in sensitized individuals. Double-blind-placebo controlled food challenges of individuals known to be sensitized to the food source are the best available test [28]. These tests are expensive, time-consuming and difficult to perform experimentally [25], for example with respect to allergen masking. They may require sensory testing by means of paired comparisons [67].

There is a growing interest in the application of animal models to assess allergenic potential as direct approach for determining the potential of a protein to cause allergy. Animal models can be used to investigate the mechanisms of IgE-mediated disease and to develop prophylaxis or therapies for allergic disease. With the advent of GM crops, such models become more important. Numerous variations exist. These include guinea pig intradermal sensitization, or a guinea pig intratracheal test, a mouse intranasal test, oral dosing of rats or mice, or an intraperitoneal injection of mice [28,68]. Systemic administration to Balb/c mice is a promising possibility [69]. The atopic dog and the neonatal swine models are gaining importance [70]. If animals are used to predict allergenicity, it is essential that their immune response identifies the same epitopes as humans. Limited data suggest that they do [71], although the outcome of the different approaches is varying considerably.

Many issues that should be considered, discussed, solved and validated when establishing proper animal models [71]: what is the most appropriate use of an animal model in a decision tree? What is the most appropriate endpoint? What constitutes a positive allergic response in an animal model, notably in relation to estimates of the probability of allergenicity? What design, route of exposure, and form of protein should be used? How to validate an animal model [10]? A reliable, validated and predictive animal model would prove valuable for the assessment of potential allergenicity and help in hazard identification of GM crops. At present it is not clear if this will be achieved. More studies are required for the evaluation and validation of the sensitivity, specificity and reproducibility of such models [10]. Once established, proper data management integrating the various different data sources will be at the basis of science-based decision making.

4. Unintended Effects

As a result of GM, GM crops may give rise to unintended or unexpected changes in food composition that can be relevant for the assessment of potential allergenicity. Examples have been put forward that show that genetic engineering may result in an undesired change in amino acid signature in a crop [5]. In individual cases, the detailed analysis of the transgene insertion site has given evidence for unintended expression and peptide or protein generation. In GM products new interactions may occur that may be of influence for other or known allergens [21]. This might also happen after consumption. Also, if a minor allergen is consumed more often, the occurrence of its allergy may increase. Such undesired changes have to be taken into account in the biosafety considerations. The potential for unintended proteins and peptides adds another layer of complexity to the assessment of food allergy.

Unintended effects may sometimes be predictable on the basis of detailed characterization of the insertion place(s) of the incoming DNA. Unintended effects may be due to incomplete insertion, duplication or random insertion. Also, portions of the bacterial vector DNA can be transferred. Such events could disrupt existing genes, modify gene expression or result in new proteins. The alteration of metabolism may result in a change of overall protein patterns. For example, GM rice with low glutelin content, made with antisense technology for brewing purposes, was associated with an unintended increase in amount of prolamins, a known allergen [38]. Although this change did not affect the purpose of the GM, it could affect the allergenic potential if the GM rice would enter the food chain. Therefore it is important that intended as well as unintended differences between parental line and GM offspring need to be documented and evaluated. This applies to the sequence of the newly inserted DNA, of the flanking DNA and the method of transformation, as well as the expression of the new genetic material.

The Dutch Health Council Committee on Safety Assessment of Novel Foods stresses the importance to identify differences between the DNA of the GM plant used as food and that of the parental, non-modified line [5]. Important is the DNA sequence of the newly inserted DNA and that of the flanking DNA in the GM plant. The final configuration of the incoming DNA often differs from the intended configuration [5]. Therefore, the Committee currently requires sequencing one thousand bases on each side of the insert. The purpose is to detect potential DNA rearrangements and/or possible effects on nearby plant genes. Experience is accumulating that especially around integration sites unexpected events can occur. This also occurs when using the *Agrobacterium tumefaciens* transformation system, which is generally considered the most precise of currently available methods. The example given in Figure 2 is the detection of a number of small additional DNA fragments in the genome of herbicide-tolerant maize [72] as well as in both herbicide-tolerant and insect-resistant GM maize (not shown) [73]. The sequence information should be used to assess whether the particular pattern of integration can be anticipated to have effects on the characteristics of the plant, for example by making novel fusion products. Potential promoters should be identified and all possible translation product should be compared to databases of toxins, allergens and functions. If results show that a new protein or potential allergen could be formed, additional experimental data are required. Experimental confirmation is needed to determine whether the RNA actually exists, the supposed protein is actually being synthesized and is allergenic [5]. In case the molecular information raises more questions, exposure studies in animals and testing for allergenicity become increasingly important (see below).

Targeted approaches as described above can be used to assess known or hypothetically new allergenic compounds in GM food to investigate if their abundance and/or allergenicity has not been influenced. However, changes in GM crops and food may not be reflected by changes in preselected compounds. Non-targeted profiling approaches will be essential for the identification of any unintended effects that are not predictable [74, 75] and can precede allergenicity assessments. Genomics strategies with associated bioinformatics infrastructure will help to identify changes at the level of RNA, protein or metabolism. The routine application of genomics profiling methods for the detection of unintended effects does, however, require more work in standardisation, knowledge on natural variation in crops and foods, the development of proper experimental, statistical and analytical procedures, as well as an appropriate bioinformatics infrastructure and suite of tools [74].

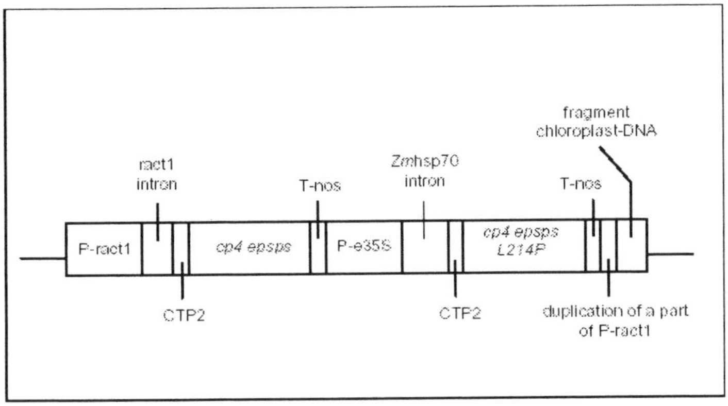

Figure 2. Detailed characterisation of a T-DNA insertion site in herbicide-tolerant GM maize (NK603) based on DNA sequence reveals unintended events at the borders of the integration. See [72] for details. Such unintended events should be analysed using bioinformatics for their potential to generate novel proteins with undesired characteristics.

5. Future Outlook

Any new and novel food has the potential to introduce new allergens. Potential allergenicity of a protein is a complex issue and on top of this comes the complexity of food and food processing. Along with the increasing world distribution of agricultural products, new allergies may be elicited among consumers. The fear of potential allergies in GM crops has had broad impact, for example influencing some African countries to reject (US) food aid. The potential of GM crops or GM crop-derived foods to induce or exacerbate allergic reactions is an easily understandable concern that has resonated with consumers and consumer organizations.

However, at present there is no evidence that current GM crops actually pose allergenic risks that are even a small fraction of the hazards encountered daily from naturally occurring plant allergens. As far as known, no foreign protein deliberately expressed in commercially allowed GM crops share identical stretches of eight of more amino acids with known allergenic proteins [56]. Conventional crops may contain uncharacterized peptides that warrant clinical testing. Many possibilities for assessing potential allergenicity exist. Individually none of the tests or characterizations gives full or sufficient reliability. The caveat remains that whatever criteria are applied to define a potential allergen, exceptions can be found within known food allergens [76]. A balanced judgment of all available data could contribute to sufficient safety of foods on the market place. Better would be a comprehensive screening of the potential allergenicity for all crops, rather than focusing on the GM component of crops.

Ongoing developments in the laboratory and *in silico* will contribute to our future ability to predict potential allergenicity of GM crops. The combination of validated laboratory methodology with bioinformatics approaches will help establish a more efficient and effective assessment of allergenicity and understanding of allergy and allergy cure. This will then result in new and better approaches to ensure the future safety of GM crops. Moreover, it may also develop into an *in silico* toolkit to apply genomics and design breeding and GM strategies in such a way that potential allergenicity should no longer be considered an issue of concern for future novel crops.

Acknowledgements. The authors would like to thank Gijs Kleter (RIKILT) and Mark Fiers (Plant Research International) for discussions and material as well as NATO for funding.

References

[1] Nap, J.P., Metz, P.L. J., Escaler, M. and Conner, A J. (2003). The release of genetically modified crops into the environment - part I. Overview of current status and regulations. Plant J. 33, 1-18.
[2] Kuiper, H.A., Kleter, G.A., Noteborn, H. and Kok E.J. (2001). Assessment of the food safety issues related to genetically modified foods. Plant J. 27, 503-528.
[3] Konig, A., Cockburn, A., Crevel, R.W.R., Debruyne, E., Grafstroem, R., Hammerling, U., et al. Assessment of the safety of foods derived from Genetically Modified (GM) crops. Food Chem. Tox., in press.
[4] Roitt, I.M. (1997). Roitt's Essential Immunology. Oxford, UK: Blackwell.
[5] Health Council of The Netherlands (2002). Safety asessment of novel foods. The Hague: Health Council of The Netherlands (www.gr.nl/pdf.php?ID=519).
[6] Burks, A.W. and Sampson, H.A. (1999). Anaphylaxis and food allergy. Clin. Rev. Allergy Immunol. 17, 339-360.
[7] Pichler, W.J. (1998). IgE-mediated food allergy: classification based on the way of sensitization. Allergologie 21, 441-450.
[8] Gilissen, L.J.W. and Nap, J.P. (1997). Assessment of allergenic potential of foods from genetically modified plants. Wageningen, The Netherlands: CPRO-DLO, 40 pp.
[9] Mekori, Y.A. (1996). Introduction to allergic diseases. Crit. Rev. Food Sci. Nutr. 36, S1-S18.
[10] Ladics, G.S., Holsapple, M.P., Astwood, J.D., Kimber, I., Knippels L.M.J., Helm R.M., et al. (2003). Workshop overview: Approaches to the assessment of the allergenic potential of food from genetically modified crops. Tox. Sci. 73, 8-16.
[11] Reddy, C.S. and Hayes, A.W. (1998). Food-borne toxicants. In Principles and Methods of Toxicology, (Hayes, A.W. Ed.) New York, USA: Raven Press, pp. 67-110.
[12] Metcalfe, D.D. (2003). Introduction: What are the issues in addressing the allergenic potential of genetically modified foods? Environ. Health Persp. 111, 1110-1113.
[13] FAO/WHO (2000). Safety aspects of genetically modified foods of plant origin. Report of a joint FAO/WHO expert consultation on foods derived from biotechnology. Geneva, Switserland: WHO/FAO (www.fao.org/es/ESN/food/pdf/gmreport.pdf).
[14] Ono, S.J. (2000). Molecular genetics of allergic diseases. Ann. Rev. Immunol. 18, 347-366.
[15] Hefle, S.L., Nordlee, J.A. and Taylor, S.L. (1996). Allergenic foods. Crit. Rev. Food Sci. Nutr. 36, S69-S89.
[16] Ebner, C. and Breiteneder, H. (2002). A survey of plant food allergens. Allergologie 25, 527-533.
[17] Fernandez-Rivas, M. and Miles, S. (2004). Food allergies: clinical and psychosocial perspectives. In Plant Food Allergens. (Mills, E.N.C. and P. R. Shewry, P.R. Eds). Oxford, UK: Blackwell, pp. 1-23.
[18] Ebo, D.G. and Stevens, W.J. (2001). IgE-mediated food allergy - extensive review of the literature. Acta Clin. Belg. 56, 234-247.
[19] Anderson, J.A. (1996). Allergic reactions to foods. Crit. Rev. Food Sci. Nutr. 36, S19-S38.
[20] Davies, P.J. (2004). Preface. Historical and cultural background to plant food allergies. In Plant Food Allergens. (Mills, E.N.C. and P. R. Shewry, P.R. Eds).Oxford, UK: Blackwell, pp. IX-XV.
[21] Penninks, A., Knippels, L. and Houben, G. (2001). Allergenicity of foods derived from genetically modified organisms. In Safety of Genetically Engineered Crops. (R. Custers, R. Ed.) Gent, Belgium: VIB, pp. 108-134.
[22] Jansen, J.J.N., Kardinaal, A.F.M., Huijbers, G., Vliegboerstra, B.J., Martens, B.P.M. and Ockhuizen, T. (1994). Prevalence of food allergy and intolerance in the adult Dutch population. J. Allergy Clin. Immunol. 93, 446-456.
[23] Helm, R.M. and Burks, A.W. (2000). Mechanisms of food allergy. Curr. Opin. Immunol. 12, 647-653.
[24] Ring, J., Brockow, K. and Behrendt, H. (2001). Adverse reactions to foods. J. Chrom. B-Anal. Techn. Biomed. Life Sci. 756, 3-10.
[25] Mills, E.N.C., Madsen, C., Shewry, P.R. and Wicher, H.J. (2003). Food allergens of plant origin - their molecular and evolutionary relationships. Trends Food Sci. Techn. 14, 145-156.
[26] Nordlee, J.A., Taylor, S.L., Townsend, J.A., Thomas, L.A. and Bush, R.K. (1996). Identification of a brazil-nut allergen in transgenic soybeans. New Engl. J. Med. 334, 688-692.
[27] Taylor, S.L. and Hefle, S.L. (2001). Will genetically modified foods be allergenic? J. Allergy Clin. Immunol. 107, 765-771.
[28] Huby, R.D.J., Dearman, R.J. and Kimber, I. (2000). Why are some proteins allergens? Tox. Sci. 55, 235-246.

[29] Kimber, I. and Dearman, R.J. (2002). Approaches to assessment of the allergenic potential of novel proteins in food from genetically modified crops. Tox. Sci. 68, 4-8.
[30] Conner, A.J., Glare, T.R. and Nap, J.P. (2003). The release of genetically modified crops into the environment - part II. Overview of ecological risk assessment. Plant J. 33, 19-46.
[31] Mahler, V., Vrtala, S., Kuss, O., Diepgen, T.L., Suck, R., Cromwell, O., et al. (2004). Vaccines for birch pollen allergy based on genetically engineered hypoallergenic derivatives of the major birch pollen allergen, Bet v 1. Clin. Exp. Allergy 34, 115-122.
[32] Herman, E.M., Helm, R.M., Jung, R. and Kinney, A.J. (2003). Genetic modification removes an immunodominant allergen from soybean. Plant Phys. 132, 36-43.
[33] Bhalla, P.L. and Singh, M.B. (2004). Knocking out expression of plant allergen genes. Methods 32, 340-345.
[34] Wichers, H., Matser, A., van Ameringen, A., Wichers, J. and Soler-Rivas, C. (2004). Monitoring of and technological effects on allergenicity of proteins in the food industry. In Plant Food Allergens. (Mills, E.N.C. and P. R. Shewry, P.R. Eds).Oxford, UK: Blackwell, pp. 196-212.
[35] Nakamura, R. and Matsuda, T. (1996). Rice allergenic protein, an molecular-genetic approach for hypoallergenic rice. Biosci. Biotechnol. Biochem. 60, 1215-1221.
[36] Tada, Y., Nakase, M., Adachi, T., Nakamura, R., Shimada, H., Takahashi, M., et al. (1996). Reduction of 14-16 kda allergenic proteins in transgenic rice plants by antisense gene. FEBS Lett. 391, 341-345.
[37] Metcalfe, D.D., Astwood, J.D., Townsend, R., Sampson, H.A., Taylor, S.L. and Fuchs, R.L. (1996). Assessment of allergenic potential of foods derived from genetically engineered crop plants. Crit. Rev. Food Sci. Nutr. 36, S165-S186.
[38] FAO/WHO (2001). Evaluation of allergenicity of genetically modified foods. Report of a joint FAO/WHO expert consultation on allergenicity of foods derived from biotechnology. Rome, Italy: FAO/WHO (www.who.int/foodsafety/publications/biotech/en/ec_jan2001.pdf)
[39] FAO/WHO (2003). Codex principles and guidelines on foods derived from biotechnology. Rome, Italy: FAO/WHO Food Standards Programme (ftp.fao.org/codex/standard/en/CodexTextsBiotechFoods.pdf).
[40] Taylor, S.L., Hefle, S.L., Bindslev-Jensen, C., Bock, S.A., Burks, A.W., Christie, L., et al. (2002). Factors affecting the determination of threshold doses for allergenic foods: How much is too much? J. Allergy Clin. Immunol. 109, 24-30.
[41] Taylor, S.L. (2002). Protein allergenicity assessment of foods produced through agricultural biotechnology. Ann. Rev. Pharm. Tox. 42, 99-112.
[42] Bannon, G., Fu, T.J., Kimber, I. and Hinton, D.M. (2003). Protein digestibility and relevance to allergenicity. Environ. Health Persp. 111, 1122-1124.
[43] Aalberse, R.C. (2000). Structural biology of allergens. J. Allergy Clin. Immunol. 106, 228-238.
[44] Brusic, V., Petrovsky, N., Gendel, S.M., Millot, M., Gigonzac, O. and Stelman, S.J. (2003). Computational tools for the study of allergens. Allergy 58, 1083-1092.
[45] Brusic, V. and Petrovsky, N. (2003). Bioinformatics for characterisation of allergens, allergenicity and allergic crossreactivity. Trends Immunol. 24, 225-228.
[46] Wiltshire, S., O'Malley, S., Lambert, J., Kukanskis, K., Edgar, D., Kingsmore, S.F., et al. (2000). Detection of multiple allergen-specific IgEs on microarrays by immunoassay with rolling circle amplification. Clin. Chem. 46, 1990-1993.
[47] de Lalla, C., Sturniolo, T., Abbruzzese, L., Hammer, J., Sidoli, A., Sinigaglia, F., et al. (1999). Cutting edge: Identification of novel T cell epitopes in Lol p5a by computational prediction. J. Immunol. 163, 1725-1729.
[48] Furmonaviciene, R. and Shakib, F. (2001). The molecular basis of allergenicity: Comparative analysis of the three dimensional structures of diverse allergens reveals a common structural motif. J. Clin. Pathol.-Mol. Pathol. 54, 155-159.
[49] Bredehorst, R. and David, K. (2001). What establishes a protein as an allergen? J. Chrom. B 756, 33-40.
[50] Marti-Renom, M.A., Fiser, A., Madhusudhan, M.S., John, B., Stuart, A., Eswar, N., et al. (2003). Modeling protein structure from its sequence. In Current Protocols in Bioinformatics, Supplement 3, New York: Wiley, pp. 5.1.1.-5.1.32.
[51] Ivanciuc, O., Schein, C.H. and Braun, W. (2002). Data mining of sequences and 3D structures of allergenic proteins. Bioinformatics 18, 1358-1364.
[52] Pellequer, J.L., Westhof, E. and Van Regenmortel, M.H.V. (1991). Predicting location of continuous epitopes in proteins from their primary structures. Meth. Enzymol. 203, 176-201.
[53] Van Regenmortel, M.H.V. and Pellequer, J.L. (1994). Predicting antigenic determinants in proteins - looking for unidimensional solutions to a 3-dimensional problem. Pept. Res. 7, 224-228.
[54] Hileman, R.E., Silvanovich, A., Goodman, R.E., Rice, E.A., Holleschak, G., Astwood, J.D., et al. (2002). Bioinformatic methods for allergenicity assessment using a comprehensive allergen database. Int. Arch. Allergy Immunol. 128, 280-291.
[55] Hopp, T.P. and Woods, K.R. (1981). Prediction of protein antigenic determinants from amino-acid-sequences. Proc. Natl Acad. Sci. USA 78, 3824-3828.

[56] Kleter, G.A. and Peijnenburg, A.A.C.M. (2002). Screening of transgenic proteins expressed in transgenic food crops for the presence of short amino acid sequences identical to potential, IgE - binding linear epitopes of allergens. BMC Struct. Biol. 2, 8.
[57] Stadler, M.B. and Stadler, B.M. (2003). Allergenicity prediction by protein sequence. FASEB J. 17
[58] Zorzet, A., Gustafsson, M. and Hammerling, U. (2003). Prediction of food allergenicity: A bioinformatic learning systems approach. *In Silico* Biol. 2, 0048.
[59] Soeria-Atmadja, D., Zorzet, A., Gustafsson, M.G. and Hammerling, U. (2004). Statistical evaluation of local alignment features predicting allergenicity using supervised classification algorithms. Int. Arch.Allergy Immunol. 133, 101-112.
[60] Brusic, V., Millot, M., Petrovsky, N., Gendel, S.M., Gigonzac, O. and Stelman, S.J. (2003). Allergen databases. Allergy 58, 1093-1100.
[61] Ivanciuc, O.I., Schein, C.H., Venkatarajan, M. and Braun, W. (2003). Prediction of food allergenicity using the SDAP database and computational tools. Biophys. J. 84, 305A.
[62] Ivanciuc, O., Schein, C.H. and Braun, W. (2003). SDAP: Database and computational tools for allergenic proteins. Nucl. Acids Res. 31, 359-362.
[63] Gendel, S.M. (2002). Sequence analysis for assessing potential allergenicity. Ann. N. Y. Acad. Sci. 964, 87-98.
[64] Immervoll, T. and Wjst, M. (1999). Current status of the asthma and allergy database. Nucl. Acids Res. 27, 213-214.
[65] Wjst, M. and Immervoll, T. (1998). An internet linkage and mutation database for the complex phenotype asthma. Bioinformatics 14, 827-828.
[66] Fiers, M.W.E.J., Kleter, G.A., Nijland, H., Peijnenburg, A.A.C.M., Nap, J.P. and van Ham, R.C.H.J. (2004). Applications note: Allermatch, a webtool for the prediction of potential allergenicity according to current FAO/WHO Codex Alimentarius guidelines, submitted.
[67] Ronteltap, A., van Schaik, J., Wensing, M., Rynja, F.J., Knulst, A.C. and de Vries, J.H.M. (2004). Sensory testing of recipes masking peanut or hazelnut for double-blind placebo-controlled food challenges. Allergy 59, 457-460.
[68] Penninks, A., Knippels, L. and Hammerling, U. (2001). Determination of protein allergenicity: Studies in rats. Tox. Lett. 120, 171-180.
[69] Kimber, I., Betts, C.J. and Dearman, R.J. (2003). Assessment of the allergenic potential of proteins. Tox. Lett. 140, 297-302.
[70] Helm, R.M. (2002). Food allergy animal models - an overview. Ann. N. Y. Acad. Sci. 964, 139-150.
[71] Crevel, R. and Madsen, C. (2004). Assessing the allergenicity of novel and GM foods. In Plant Food Allergens. (Mills, E.N.C. and P. R. Shewry, P.R. Eds), Oxford, UK: Blackwell, 171-195
[72] Health Council of The Netherlands (2002). Herbicide-tolerant maize (NK603). The Hague: Health Council of The Netherlands (www.gr.nl/pdf.php?ID=621).
[73] Health Council of The Netherlands (2003). Insect-resistant and herbicide-tolerant maize (1507). The Hague: Health Council of The Netherlands (www.gr.nl/pdf.php?ID=908)
[74] Kuiper, H.A. and Kleter, G.A. (2003). The scientific basis for risk assessment and regulation of genetically modified foods. Trends Food Sci. Techn.14, 277-293.
[75] Kuiper, H.A., Kok, E.J. and Engel, K.H. (2003). Exploitation of molecular profiling techniques for gm food safety assessment. Curr. Opin. Biotechnol. 14, 238-243.
[76] Helm, R.M. (2003). Food biotechnology: Is this good or bad? Implications to allergic diseases. Ann. Allergy Asthma Immunol. 90, 90-98.

3. Ecological Genomics: Plants, Genes and Environments

Genomics for Biosafety in Plant Biotechnology
J.P.H. Nap, A. Atanassov and W.J. Stiekema (Eds.)
IOS Press, 2004

Biotechnology Meets Ecology: Insect-Resistant Transgenic Plants

Ruud A. DE MAAGD
Plant Research International,
Wageningen University and Research Centre,
P.O. Box 16, 6700 AA The Netherlands;
email: ruud.demaagd@wur.nl

Abstract. A major fraction of the genetically modified (GM) crops grown commercially today contain *Bacillus thuringiensis* (Bt) proteins for insect resistance. Because of their interactions with insects, such GM crops may have an impact on agro-ecosystems. A first issue is the potentially detrimental effect of Bt crops on non-target organisms with beneficial ecological roles. To assess such effects, novel methodologies or ecological genomics are being developed. Another important issue is development of resistance in the target insect populations. Several resistance management strategies are proposed, the most promising of which appears to be the pyramiding of different resistance genes. Modern genomics technologies as mRNA profiling will help increase understanding the molecular basis of resistance to Bt as well as the ecological impact of non-target effects of Bt. This can help design molecular or agronomic strategies to detect, delay, reduce or eliminate undesired or potentially adverse effects.

Introduction

The bacterium *Bacillus thuringiensis* (Bt) has been a valuable source of insecticidal proteins for use in agriculture. These proteins are used in conventional sprays or in genetically modified (GM) crops. The ecology of the bacterium is unclear. It can be isolated from soil, grain, insect cadavers as well as plant surfaces and may be an opportunist pathogen. Insecticidal Bt proteins are able to kill a large variety of host insects and even nematodes, but exhibit a high degree of specificity towards particular species. The molecular mode of action of Bt toxin is thought to be the puncturing cells in the insect's midgut. GM crops expressing Bt proteins were first commercialised in 1996 and over 16 companies have since been involved in Bt crops.

Insect resistance is now the second-most important trait in GM crops grown commercially world-wide. GM Bt crops are thought to offer novel opportunities for the management of insect pest populations without the use of agrochemicals. There are, however, quite some concerns about the desirability, ecological sustainability and biosafety of Bt crops. A major concern is evolution of resistance in target insect populations. Also, the ecological consequences of potential effects on non-target insects are topic of debate. We here give an overview of the current state of the biosafety assessment of Bt crops and identify research areas were modern genomics technologies may be or become important. The EU-funded Bt-BioNoTa project is given as example where modern genomics technologies are used in the ecological setting of Bt biosafety research.

1. *Bacillus thuringiensis:* A Natural Pesticide

Bacillus thuringiensis (here abbreviated as Bt) is a gram-positive bacterium which can be found in many habitats in Nature, such as in soil, on plants, in grain storage dust, and in dead or diseased insects. Its ability to infect and kill insects already drew attention from scientists in the early 20th century, and it has been utilised in agriculture for decades.

In unfavourable conditions Bt forms spores. Unique to Bt, and a few related species, is that at the time of sporulation it also produces crystals, which contain one or several proteins, outside the spores. The crystal proteins (abbreviated: Cry proteins), also called delta-endotoxins, are the major causes of the pathogenicity of Bt for insects. When an insect ingests spores and crystals, the crystals dissolve in the gut and the crystal proteins can kill the cells lining the gut. A sufficient dose of the crystals alone is often enough to kill the insect. However, besides the crystal proteins, Bt produces a large array of virulence factors that under natural conditions are thought to promote the infection: other protein- and non-protein toxins, and enzymes that weaken the defences of the insect [1].

An important issue in the study and use of Bt is specificity: each strain is active only against a small number of insect species, usually within the same taxonomic order. This specificity is mainly determined by the crystal proteins that the strain produces. A strain can contain up to five different crystal proteins. Traditionally Bt strains were only effective against larvae of moths and butterflies (*Lepidoptera*), and most crystals are in that group. Strains effective against larvae of flies and mosquito's (*Diptera*) are known since 1976, and strains against beetles and weevils (*Coleoptera*) since 1983 [2]. More recently, strains that are active against wasps and bees (*Hymenoptera*) or nematodes were identified. Given the specificity of Bt crystals, any discussion about Bt and its particular use should specify the Bt strain or -in case of GM crops- Bt gene used.

1.1 Crystal Proteins, Delta-endotoxins or Bt-toxins: Diversity and Mode of Action

Most of the differences in specificity of natural Bt strains come from their difference in *cry* gene content. There are to date over 100 different crystal protein genes known. A full, updated list with further links can be found on the Internet [3]. Each Cry protein is active against a small number of insect species, usually within one insect order, although there are known a few proteins that affect species in two or even three different orders.

Important for understanding the basis of specificity of Cry proteins is their mode of action. Each step in this may affect the activity of the toxin, and these may differ between the spore/crystal sprays and transgenic plants expressing a Bt-toxin. In a crystal, the toxins are present in an insoluble form, the protoxin, which is larger than the protein that actually kills the target: the activated toxin. Upon ingestion, protoxins have to dissolve in the gut fluid of the target insects. The ability to dissolve, or not, under specific gut conditions plays a role in the eventual toxicity of the protein. Next, proteases of the insect digest away parts of the protoxin to generate a very protease-resistant core, the activated toxin.

The ability of the protoxin to be properly activated by the enzymes in the gut of a particular insect also plays a role in its toxicity and specificity. The active toxin subsequently binds to specific receptors on the surface of the gut epithelial cells. It is this interaction that is considered to be the major factor in determining specificity at the species level. Without receptors, an insect will not be sensitive to a particular toxin. Different Cry toxins may recognise different receptors, partially overlapping receptors, or identical receptors in one insect. Overlap in receptor recognition is usually correlated with the level of similarity between the toxins. Finally the toxin inserts into the membrane, forming pores that will kill the epithelial cell. When this occurs in sufficient quantity, the insect dies.

1.2 Other Insecticidal Proteins

Currently, the Cry proteins are the only insecticidal proteins used in commercialised transgenic plants, with the exception of one case in China, where the application concerns a protease inhibitor. In addition, there are a number of other insecticidal proteins that may see commercialisation in the near or further future. Probably already in 2004, transgenic corn expressing a new class of Bt crystal proteins, conferring resistance to the Western corn rootworm, will be commercialised in the U.S. Western corn rootworm is an important root pest of corn in the United States and more recently, in south-central Europe [4]. These new Cry proteins are structurally unrelated to earlier isolated Cry proteins.

Another class of insecticidal proteins produced by Bt is not formed during sporulation, but during the preceding vegetative growth of the bacterium. These proteins do not form a crystal but are secreted into the medium. These vegetative insecticidal proteins (Vip) form two distinct toxins. Both act on the midgut epithelium of target insects, but less is known about their molecular mode of action. Vip3, which kills midgut epithelial cells by inducing apoptosis, renders transgenic corn resistant to fall armyworm and black cutworm [5]. Many other candidate genes for insect resistant transgenic plants exist, such as genes encoding protease inhibitors, lectins, amylase-inhibitors, cholesterol oxydase or enzymes producing toxic metabolites [6]. For some of these genes, introduction into transgenic plants was demonstrated to result in insect resistance, but their application may still be far from commercialisation.

2. Applications of Bt in Agriculture

2.1 Insecticidal Sprays

Bt can be applied as a spray on plants. Such a spray consists of some form of spore/crystal-mixture. Applications of Bt-sprays are found in three major areas:

- Forest pest control. Particularly in North America, aerial Bt-sprays are used extensively for control of forest pests (spruce budworm, gypsy moth).
- Mosquito control. Particularly in the Middle East and Africa, but also in Europe (Rhine valley, Germany), sprays are used for the control of mosquitoes such as the vectors for diseases.
- Organic agriculture. Bt sprays are considered a natural pesticide. Therefore, it is one of the few pesticides that can be used in organic agriculture. Such sprays are used particularly in the cultivation of horticultural crops.

Despite its attractiveness as a natural pesticide, Bt has never conquered a large share of the global pesticide market. Although it is the most widely used biological pesticide, it takes up only about 1% of the total insecticide market. Reasons for the limited use of BT-sprays in agriculture are:

- Low persistence. The crystal protein is rapidly inactivated by solar UV-radiation.
- Limited activity spectrum. Each Bt strain is active only against a few pest species, so one product is never sufficient for all pests encountered in the field.
- Many important pest species are insensitive to all known Bt strains. Bt sprays, as many chemical insecticides, are not very effective against insects that bore into the crop tissue.

Table 1. Commercial Bt-crop varieties approved for environmental releases anywhere in the world up to 2002. Note that some varieties have subsequently been withdrawn. An event name refers to the particular individual transformed plant from which all others originate by crossing and multiplication.

Crop	Gene	Event name	Variety name	Target pests	Company
Corn	*cry1Ab*	176[1]	Knockout[2] NatureGard	European corn borer (*Ostrinia nubilalis*), Southwestern corn borer (*Diatraea grandiosella*) and Mediterranean corn borer (*Sesamia nonagroides*)	Syngenta Seeds (Novartis) Dow AgroScience (Mycogen Seeds)
	cry1Ab	Bt11	YieldGard[2]	See above	Syngenta Seeds (Northrup King)
	cry1Ab	Mon810 Mon809	YieldGard[2]	See above	Monsanto Company Pioneer Hi-Bred
	cry1Ac	DBT418	BT-Xtra[3]	See above	DeKalb Genetics
	cry9Ca	CBH351	Starlink[4]	See above	Aventis CropScience
	cry1Fa	TC1507	Herculex	See above + fall armyworm (*Spodoptera frugiperda*) and black cutworm (*Agrotis ipsilon*)	Dow AgroScience Pioneer Hi-Bred (Dupont)
	cry3Bb	Mon863		Corn Rootworm (*Diabrotica* sp.)	Monsanto Company
Cotton	*cry1Ac*	Mon531/ 757/1076	BollGard[2] InGard	Tobacco budworm (*Heliotis virescens*),	Monsanto Company
			BXN	Cotton bollworm (*Helicoverpa zea, H. armigera or H. punctigera*) and Pink Bollworm (*Pectinophora gossypiella*)	Monsanto Company
	cry1Ac	31807/ 31808		see above	Calgene
	cry1Ac + cry2Ab	15985	BollGardII	see above	Monsanto Company
	cry1Ac[6]			see above	Chinese Academy of Agricultural Sciences
	cry1Ac + CpTi[7]			see above	Chinese Academy of Agricultural Sciences
Tomato	*cry1Ac*[5]	5345		see above	Monsanto Company
Potato	*cry3Aa*	Many[8]	NewLeaf	Colorado potato beetle (*Leptinotarsa decemlineata*)	Monsanto Company

[1]Registration will not be continued. Sold until 2003 planting season.
[2]Approved for release by the European Union.
[3]Registration of BT-Xtra was withdrawn in 2000.
[4]Starlink was registered only for animal feed use, and its registration was withdrawn in 2000.
[5]Approved but not yet on the market.
[6]Sometimes referred to as a hybrid toxin gene (domains I and II of Cry1Ab, domain III of Vry1Ac), the encoded protein differs very little from Cry1Ac itself.
[7]This variety contains besides cry1Ac a gene for Cowpea trypsin inhibitor, which is supposed to prevent the development of resistance to Cry1Ac in *H. armigera*.
[8]Many different events exist, some combining Cry3Aa with genes for resistance to Potato virus Y or Potato leafroll virus.

2.2 Transgenic Plants Expressing Bt-toxin Genes and their Commercial Uses

Some of the problems associated with Bt sprays could be solved by having present the toxin in the plant tissue itself. Toxin production can be continuous or inducible by chemicals or insect feeding. This would make repeated application of sprays unnecessary and provide a measure of protection for the toxin against inactivation by environmental factors. Moreover, plant-boring insects would also ingest it. The first transgenic plant expressing a Bt-toxin, conferring resistance to insects, was reported in 1987 [7]. Since then many plant species have been transformed with a variety of *cry* genes. An overview of the plant/gene-combinations which are currently being studied and thus an indication of which might be commercially released in the future can be found for the EU at the Internet site of the Joint Research Centre (notifications of deliberate field trials under EU Directive 2001/18/EC) [8] and for the USA at the Internet site of the EPA [9].

Most Bt-crops on the market are products of a small number of multinational companies or their subsidiaries. These crops and their commercial names as used in the USA, and some other countries, are listed in Table 1. A bit further in the future are the introduction of corn rootworm resistant corn varieties containing a novel binary toxin gene from *B. thuringiensis* (Dow AgroScience), and cotton expressing the Vip3A protein (Syngenta). The only other substantial effort in Bt-crop development is in China, where independently Bt-cotton was developed and large-scale field trials with GM rice, tobacco, potato, soybean and poplar take place. At the International Rice Research Institute (IRRI; Philippines) commercial rice varieties with a Bt-fusion gene were developed, that are close to introduction [10].

2.3 Differences between Bt-sprays and Bt-Plants

The main differences between Bt-sprays and Bt-plants are summarised in Table 2. To achieve expression of Bt-toxins in plants by insertion of a *cry* gene in the cell's nuclear genome a number of modifications to the gene have to be made, which are all consequences of the differences in regulation of gene expression between plants and bacteria. In contrast to sprays, transgenic plants usually express a single *cry* gene, although Monsanto has recently released GM cotton containing two different *cry* genes. Sprays are inherently non-persistent, so the activity of the toxin on a crop plant diminishes gradually after a spray, whereas most transgenic plants on the market to date express a Bt-toxin throughout the largest part if not their entire growth season. This makes Bt-crops more a pest prevention method than a measure after a pest arises, as pesticide applications usually are.

While the natural crystals contain one or several full size (protoxin) Cry proteins, the longer type of Cry1 proteins are expressed in plants as truncated (shorter) versions of the protoxin. In these truncated versions, the larger part of the C-terminal extension (which is usually clipped off by gut proteases) is lacking. Since this part is involved in crystallisation, and since toxin concentration in plant cells (typically no more than 2% of total soluble protein) is low compared to sporulating Bt cells, the toxin in plant cells is probably not present in a crystalline, but in a soluble form. Furthermore, the lack of most of the C-terminal extension bypasses part of the activation process by gut proteases that occurs with the natural protoxins. A statement often found in literature, that Bt-crops produce an activated toxin, is therefore wrong. At least some target gut protease action is required for the plant-derived toxin to become active [11]. Lack of crystal formation and the natural activation process, both of which are known to play some role in determining insect specificity (as discussed above), could therefore in theory lead to some loss of insect specificity of the toxin present in transgenic plants.

Table 2. Major differences between the Bt-toxins in Bt sprays and in Bt-plants currently in use.

	Bt-sprays	**Bt-plants**
Number of different toxins present	Multiple Cry proteins and other virulence factors	One Cry protein (in the future possibly two)
Non-Bt formulation components	Present	Absent
Physical property of Cry protein	Crystalline	Soluble
Degree of activation of the expressed protein	Yet to be activated (full protoxin)	Partially activated (Cry1A proteins) or not yet activated (Cry3's)
Presence of the toxin	Intermittent	Continuous

However, receptor binding, which is considered by most scientists the major limiting factor in host range, is not likely to have changed. In general, differences in the level of toxicity between protoxins and *in vitro* solubilised and activated toxins do exist, but in most cases without a major difference in insect specificity. Furthermore, Bt-sprays show a wider host range than the single protein in a transgenic plant, because they often contain multiple *cry* genes and other virulence factors (Table 2).

3. Potential Benefits and Perceived Risks of Bt-crops

Discussions on the acceptance of Bt-crops, or transgenic crops in general, will probably more and more involve a cost/benefit-analysis for a particular crop. The perceived benefits of Bt-crops are not the focus of this paper and will therefore be treated only briefly. In general, the main benefits of the use of Bt-crops as compared to a conventional cropping system are presented as:

- Increased crop yield as a result of reduced insect damage.
- Reduced application of chemical insecticides or replacement by more benign ones. Lower insect control costs and less damage to the environment and the health of workers; reduction of chemical residues on food and in run-off.
- A secondary, additional benefit (for maize) may be reduced levels of mycotoxins (fungal compounds toxic for animals and humans) as a result of less fungal infestations that normally often follow damage to the corn cob by herbivorous insects [12].

Critics of the introduction of Bt-crops usually focus on two risk issues that are specific for transgenic Bt-plants. Other risks that are more generally perceived for transgenic crops, such as transgene spread to wild relatives and the use of antibiotic-resistance markers could also apply for Bt-crops. The two more specific issues are:

- Effects, direct or indirect, of the toxin in the crop on non-target animals, particularly arthropods including beneficial insects.
- Development of resistance to Bt-toxin in the target insects. Insects are extremely adaptable and have developed resistance to many chemical insecticides. Bt-toxins are probably not different in this respect. It is feared that the large-scale introduction of Bt-crops will generate a strong selection pressure on the target insects, increasing the speed and likelihood with which resistance to the toxin will develop.

4. Non-Target Effects of Bt-crops

4.1 Problem Definition

Transgenic plants expressing a Bt-toxin are designed to kill or severely limit the population of one or several important pests. In the field, crops not only support these herbivorous pests, but also non-target herbivores (for which increased mortality may be undesirable), and arthropods or higher animals that feed on both target and non-target herbivores (predators and parasitoids). In ecology, this is called a tritrophic system. Up to five trophic levels may exist. The third trophic level may be considered a part of the plant's defence system (predators and parasites can limit a damaging pest) and as such can play an important role in pest management [13]. Therefore, effects on tritrophic interactions should be considered in order to asses the environmental safety and sustainability of any cropping system including, but not limited to, those involving Bt-crops. In fact, for much of this scientific evaluation, Bt-toxin mediated insect resistance can be considered to be like other types of host plant resistance, in that it may have synergistic, additive, or antagonistic effects on natural enemies. Many examples of all three types of effects exist for conventionally bred resistant cultivars [14].

Other herbivores (pest or non-pest) may also ingest the toxin and possibly accumulate it. Both target pests as well as non-target herbivores may pass on the toxin to their respective predators or parasites. That there will be an effect of a specific herbivore pest being killed, on the population of a parasitoid or predator that is exclusively dependent on this species can be easily imagined. This type of effect, or an effect of plant defence on predators or parasitoids through a target herbivore can occur with any successful control strategy for a particular pest, whether it is natural resistance of the plant, or by chemical or biological pest control measures [14]. Included in the second trophic level are soil dwelling organisms involved in degradation of dead plant material, such as detritus feeders (earthworms, springtails etc.) and soil microbes. Bt-toxin from transgenic plant material can persist for some time in the soil, depending on the soil conditions.

A group of organisms that is less obvious, because they are never or only intermittently present on the crop plant, are strictly speaking non-target herbivores. Insects that collect pollen (pollinators such as honeybees and bumblebees, which feed pollen to their larvae, while adults feed on nectar) and insects that may incidentally ingest pollen that has spread from the crop unto their host plant, such as the Monarch butterfly larvae on milkweed plants in a corn field. If the toxin is expressed in substantial amounts in the pollen (such as in Corn event 176 – see table 1), there may be routes of exposure to the toxins for these insects. Nectar usually contains only very small amounts of protein.

In discussions about negative effects of transgenic crops, the terms "risk" and "hazards" are sometimes not clearly defined, often leading to confusion. The term "hazard" identifies an effect that may occur (or a "worst-case scenario"), without quantifying or including the possibility or likelihood that this effect will actually occur. "Risk" is the product of "hazard" and "likelihood" (or: "exposure"), thus not only including the possible effect, but also the likelihood that it will actually happen: risk = hazard x likelihood.

Risk assessment may be considered an attempt to quantify a particular risk. A hazard may be identified, but if the likelihood of occurrence is zero, so is the risk associated with the hazard. If there is no hazard, the risk will be zero even if the likelihood is high. With this in mind it is easy to see that in the context of this literature review, many laboratory studies identify hazards (for example a non-target insect being sensitive to a toxin), while the likelihood and therefore risk is unknown. Field trials attempt to both identify hazards as well as quantify the risks of these and of earlier identified hazards.

4.2 Non-Target Effects Described in the Literature

Experiments described in the literature on non-target effects of Bt-toxin and Bt-plants can be classified roughly in two basic types: laboratory and field trial experiments. A special case are the "caged plant" experiments where plants, in the field, are covered by nets to prevent migration of insects to or from the plants during the experiment, in order to study particular insect or insect combinations in a semi-controlled environment. Laboratory experiments may have been performed with purified toxin in an artificial diet or as an additive to a natural diet, or with transgenic plants expressing the Bt-toxin.

Laboratory experiments are usually (when properly) designed to study simple bitrophic or tritrophic interactions under controlled conditions and do not usually mimic the complex interactions in a natural or agricultural setting. For example, they usually do not take into account migrations of insects in and out of fields, the presence of other organisms (insects), and the choice of food source that some herbivores, predators and parasitoids have, while others are bound to one plant or prey species. Although this by no means downplays the role of laboratory experiments as useful and necessary approaches for identifying possible hazards of Bt-toxins or the plants containing them, extrapolating results to the field situation requires extreme caution.

In 1998 a lot of media attention was drawn to a study by a Swiss group, on the effects of transgenic corn on an insect predator, green lacewing [15]. Since then, this study has often been inappropriately quoted, referred to and/or its contents misrepresented or –interpreted both by opponents as well as by proponents of transgenic plant release. Green lacewing larvae are important predators of pest insects, particularly of aphids, and insect eggs, and probably play an important role in the agro-ecosystem to suppress populations of agricultural pests [16]. Lacewing larvae are reared commercially for biological control, particularly for use in greenhouses. Moreover, lacewings are commonly tested for side effects of pesticides, in the process of registration for use of these pesticides [17]. In the Swiss study [15], lacewing larvae were fed with caterpillars of two species, European corn borer (ECB, *Ostrinia nubilalis*) and cotton leafworm (CLW, *Spodoptera littoralis*), which in turn both had fed on transgenic Cry1Ab-corn or as a control, on non-transformed corn (schematically shown in Figure 1).

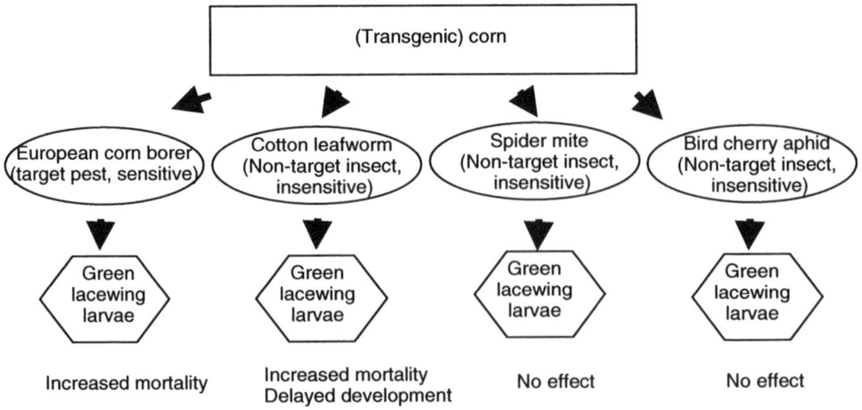

Figure 1. Schematic representation of the experimental data on tritrophic effects of Bt-corn on lacewing larvae through feeding on European corn borer or cotton leaf worm larvae [15], on spider mites [18] or on bird cherry aphids [19].

Whereas ECB is the target insect for the transgenic corn (ECB is very sensitive to Cry1Ab), CLW is not considered a target insect for this crop (although it has a lower degree of sensitivity to Cry1Ab). It was found that for both prey species, caterpillars that had fed on the Cry1Ab-corn caused significantly higher mortality of lacewing larvae during the larval life stage. Also, there was delayed development of the predator fed on ECB, but not when fed on CLW. While ECB-larvae are sensitive to the toxin in the corn, and therefore it might be considered not surprising that they constitute a non-optimal diet for their predators when they are dying from intoxication, the effect caused by CLW-larvae on the predator was much more unexpected. The nature of this effect is still unclear. Although CLW was shown to be slightly sensitive to Cry1Ab and showed growth inhibition in response to ingestion of toxin from artificial diet, it is unknown whether this by itself could cause the effects on lacewing larvae [20]. The authors suggested that the toxin be possibly transmitted through the prey to the lacewing, which in turn is sensitive to the toxin.

It was shown in a subsequent study that lacewing larvae are indeed sensitive to Cry1Ab at relatively high doses in an artificial diet, but it seems unlikely that the lacewing larvae really obtain such doses when eating caterpillars from transgenic plants [20,21]. Other studies found no such direct effect of Cry1Ab on lacewing larvae, but those studies may be flawed in that the method or duration of exposure to the toxin was insufficient. Pollen or insect eggs were coated with toxin, while lacewing larvae suck their prey and may thus not have ingested the toxin [22,23]. In an another, more recent study, the negative effect on lacewings when fed CLW from Bt-corn was confirmed [18]. In addition, another prey species, *Tetranychus urticae* (spider mite) feeding on Bt-corn had no negative effect on *C. carnea*, while containing more corn-derived Cry1Ab then CLW feeding on the same plants. This, and the fact that in contrast to CLW, T. urticae itself is not affected by Cry1Ab, makes it even more likely that poor prey (CLW) quality plays a more important role in the negative effect on *C. carnea* then direct toxicity of the protein as it is transferred from corn via prey to *C. carnea*.

Regardless of the mechanism, while lacewings will probably not encounter many ECB larvae in the transgenic fields (after all, those are killed by the Bt-corn), the negative effect of transgenic plants through a non-target insect (CLW) on the third trophic level would constitute a hazard of such plants. Whether transgenic plants harming lacewing larvae in the field constitutes an actual risk of those plants, depends on the level of exposure (= likelihood). In the field, lacewing larvae have a choice of prey species to feed on, and actually prefer aphids and insect eggs [24]. Lepidopteran larvae are considered poor prey for *C. carnea* leading to higher mortality and slower development rate [18,25]. So far, (limited) field studies have shown no negative effects of transgenic Cry1Ab-expresssing corn on populations of insect predators in general, or on lacewings in particular [23,26,27].

5. The Bt-BioNoTa Project

Bt-BioNoTa is a collaborative research project within the European 5th Framework. The three-year project consists of a consortium of 5 European research groups. In this three-year project, that started the beginning of October 2000, the non-target effects of four crop/toxin-combinations will be studied at different organisational levels, in laboratory, glasshouse, and field experiments. Specific topics of interest are:

- The mode of action of Bt-toxin from transgenic crops in an insect predator.

  Understanding the mode of action of Cry1Ab expressed in transgenic maize in the green lacewing, an important insect predator, using target binding and pore formation assays.

In order to be effective, the refuge strategy needs to meet several assumptions that ensure that resistant and susceptible adults do mate in a sufficient manner. Refugia may be parts of the crop field, set aside for use with non-transgenic varieties of the crop. In this case it is important that there are no insect control measures taken in that part, or if there are, the refuge part should be bigger in order to allow sufficient susceptible insect numbers to develop. Alternatively, one could rely on wild plant or other crop refugia if the pest can feed in sufficient numbers on these plants. Again it should be established that sufficient mixing of adults for mating is taking place.

The method currently in place in the United States for Bt-corn and Bt-cotton is a specific variation of the refuge strategy, namely the high dose/refuge-strategy. In this strategy the larger part of the field contains plants expressing the toxin at a high level, supposedly 25 or more times the amount needed to kill all sensitive larvae. Again this strategy is based on a number of assumptions. Important is the ability of the high dose to kill larvae that are heterozygous for the resistance allele, in other words that resistance is functionally recessive.

An alternative resistance management strategy, which is likely to find more applications in the future, is gene pyramiding. This may include crops with two different Cry proteins. Alternatively it may include a Cry protein with an altogether different type of resistance, natural or transgenic. Modelling studies suggest that gene pyramiding is much more effective in delaying the evolution of a resistant insect population, even with smaller refuge sizes [38,39]. In the US, Monsanto has developed BollGard II, a GM cotton expressing Cry1Ac as well as Cry2Ab (Table 1) [40-42]. The Chinese Academy of Agricultural Sciences has developed transgenic cotton varieties containing Cry1Ac and in addition a protease inhibitor derived form Cowpea (CpTi) [43,44], which is used in the field. For these varieties there is some experimental evidence that the presence of two genes delays resistance development [45,46]. Transgenic broccoli varieties with pyramided Cry1Ac and Cry1Ca genes were resistant to two *P. xylostella* lines selected for resistance to one of these two toxins [47]. Side-by-side comparison of pyramided and single-gene lines confirms that gene pyramiding is indeed delaying resistance development [48].

7. Future Outlook

Despite shortcomings in some studies, the large amount of available data have not shown any negative effect of Bt-crops on non-target insects in the field, nor were any of the negative effects suggested by laboratory experiments deemed likely to occur in the field. The only general exception to this conclusion is in the case of specialised parasitoids with a host range restricted to the insect that is a target pest of the Bt-crop. Not unexpectedly, the successful suppression of an insect pest species, also negatively affects a parasitoid that exclusively depends on it.

It is clear that lessons can be learned from the above-mentioned shortcomings to improve the risk assessment of transgenic crops. Non-target organisms considered to be at risk from a particular Bt-crop should be studied in the laboratory where possible. Such studies should include due consideration for the natural routes of exposure [49]. Although one may require testing more species than currently used, it would be unrealistic to require testing of every conceivable organism that could be present in a field. Therefore, it is important that key species are identified and studied. Furthermore, laboratory studies will never be able to predict with sufficient certainty what will actually happen in the field. The laboratory studies are useful as methods for identifying hazards, but should not be used not for measuring actual risk. For the latter, field experiments are indispensable.

Although the more obvious risks could be identified in pre-release field trials, the more certainty is wanted the bigger the experiments become. Post-release monitoring could be a solution for the measuring of smaller, long-term risks. Taking into consideration the available data and past experiences, we support the recommendations of several (groups of) scientists to adopt a three-tiered procedure for the environmental safety assessment of transgenic crops with insect resistance [50]. In practice many of the parts of this procedure are already in place, but should probably be a part of the regulation of introduction of such crops. In the European Union, this approach will be adopted as part of the new regulatory framework. The three tiers of the recommended procedure are:

- Identification of species that could directly or indirectly be affected by the transgenic crop, because their habitat overlaps with it (which includes overlap with the area where pollen might drift). Laboratory experiments using realistic, natural exposure routes should establish whether the toxin constitutes a possible direct hazard for the organism.
- Pre-release field experiments to follow the previously identified species and identify and measure short-term effects.
- Post-release monitoring of transgenic crop fields to identify and measure long-term effects on non-target organism populations.

Although one could tentatively conclude that the risks of non-target effects of the current Bt-crops are small indeed, it would be prudent to expect that some effects will show up after extended, long-term monitoring. Any successful crop-protection method will probably have some non-target effects, and it would be fair to compare such effects of Bt-crops with those of other cropping systems. Chemical insecticide treatment, with well known wide-scale non-target effects is the most obvious alternative that is usually brought up for comparison, but other systems like biological control should not be neglected. The severity of non-target effects would also have to be compared against the background of "natural" fluctuations occurring over the years as for example as a result from choosing another crop. Given these variables it seems unrealistic to say that any change in non-target population levels as a result of Bt-crop use is unacceptable.

From existing literature and experience with chemical insecticides it seems clear that there is a hazard of insect populations developing resistance to toxins in Bt-crops. This endangers the durability of Bt-crops as well as of Bt-sprays when used against the same insects, although one might argue that if Bt-sprays were used more frequently then they are now instead of or simultaneous with Bt-crops, this would also lead to resistance development. In order to minimise the risk of resistance development, commercialisation of Bt-crops should be accompanied by regulations for insect resistance management strategies. Research on the ecology and genetics of the pest insect (behaviour, resistance allele frequencies) and of the pest/crop-interaction should provide the necessary data for the strategy and the strategy should be adjusted as new data appear. The insect resistance management strategy should include monitoring for resistance and a contingency plan for cases where resistance does appear to increase.

Finally, knowing and even quantifying the risks of environmental effects due to the release of genetically modified crops is only one side of the balance in the decision process leading to approval or rejection of this release. The risks will have to be weighed against the potential economic and environmental benefits of that introduction.

Acknowledgements. This work was financed by the Dutch Ministry of Agriculture, Nature Management and Fisheries (DWK program 347) and the European Union (project Bt-BioNoTa, QLK3-CT-2000-00547).

References

[1] de Maagd, R.A., Bravo, A., Berry, C., Crickmore, N. and Schnepf, H.E. (2003). Structure, diversity and evolution of protein toxins from spore-forming entomopathogenic bacteria. Ann. Rev. Genet. 37, 409-433.

[2] Beegle, C.C. and Yamamoto, T. (1992). History of *Bacillus thuringiensis* berliner research and development. Can. Entomol. 124, 587-616.

[3] Crickmore, N. *Bacillus thuringiensis* Toxin Nomenclature. (http://www.biols.susx.ac.uk/Home/Neil_Crickmore/Bt/)

[4] Moellenbeck, D.J., Peters, M.L., Bing, J.W., Rouse, J.R., Higgins, L.S., Sims, L. et al. (2001). Insecticidal proteins from *Bacillus thuringiensis* protect corn from corn rootworms. Nat. Biotechnol. 19, 668-672.

[5] Estruch, J., Gw, W., Desai, N., Koziel, M. and Nye, G. (2000). Class of proteins for the control of plant pests. US Patent 6,137,033.

[6] Carozzi, N. and Koziel, M. (1997). Advance in insect control: the role of transgenic plants. London, UK: Taylor and Francis.

[7] Vaeck, M., Reynaerts, A., Höfte, H., Jansens, S., Beuckeleen, M.D., Dean, C. et al. (1987). Transgenic plants protected from insect attack. Nature 328, 33-37.

[8] Anonymous. Deliberate field trials notified under Directive 2001/18/EC. (http://gmoinfo.jrc.it/)

[9] Anonymous. Biopesticide New Active Ingredients Approved 1998-2000. (http://www.epa.gov/oppbppd1/biopesticides/regtools/index.htm)

[10] Tu, J.M., Zhang, G.A., Datta, K., Xu, C.G., He, Y.Q., Zhang, Q.F. et al. (2000). Field performance of transgenic elite commercial hybrid rice expressing *Bacillus thuringiensis* delta-endotoxin. Nat. Biotechnol. 18, 1101-1104.

[11] Bravo, A., Sanchez, J., Kouskoura, T. and Crickmore, N. (2002). N-terminal activation is an essential early step in the mechanism of action of the *Bacillus thuringiensis* Cry1Ac insecticidal toxin. J. Biol. Chem. 277, 23985-23987.

[12] Dowd, P.F. (2000). Indirect reduction of ear molds and associated mycotoxins in *Bacillus thuringiensis* corn under controlled and open field conditions:utility and limitations. J. Econ. Entomol. 93, 1669-1679.

[13] Poppy, G. (1997). Tritrophic interactions: improving ecological understanding and biological control? Endeavour 21, 61-65.

[14] Panda, N. and Khush, G. (1995). Host plant resistance to insects. Oxon, UK: CAB International.

[15] Hilbeck, A., Baumgartner, M., Fried, P.M. and Bigler, F. (1998). Effects of transgenic *Bacillus thuringiensis* corn-fed prey on mortality and development time of immature *Chrysoperla carnea* (Neuroptera: Chrysopidae). Environ. Entomol. 27, 480-487.

[16] New, T. (1975). The biology of *Chrysopidae* and *Hemerobiidae* (Neuroptera), with reference to their usage as biocontrol agents: a review. Trans. Royal Entomol. Soc. London 127, 115-140.

[17] Croft, B. (1990). Arthropod biological control agents and pesticides. New York, USA: John Wiley.

[18] Dutton, A., Klein, H., Romeis, J. and Bigler, F. (2002). Uptake of Bt-toxin by herbivores feeding on transgenic maize and consequences for the predator *Chrysoperla carnea*. Ecol. Entomol. 27, 441-447.

[19] Lozzia, G.C., Furlanis, C., Manachini, B. and Rigamonti, I.E. (1998). Effects of Bt corn on *Rhopalosiphum padi* L. (Rhynchota Aphididae) and on its predator *Chrysoperla carnea* Stephen (Neuroptera Chrysopidae). Boll. Zool. Agr. Bachic. Ser. II 30, 153-164.

[20] Hilbeck, A., Moar, W.J., Pusztai-Carey, M., Filippini, A. and Bigler, F. (1998). Toxicity of *Bacillus thuringiensis* Cry1Ab toxin to the predator *Chrysoperla carnea* (Neuroptera: Chrysopidae). Environ. Entomol. 27, 1255-1263.

[21] Head, G., Brown, C.R., Groth, M.E. and Duan, J.J. (2001). Cry1Ab protein levels in phytophagous insects feeding on transgenic corn: implications for secondary exposure risk assessment. Entomol. Exp. Appl . 99, 37-45.

[22] Sims, S.R. (1995). *Bacillus thuringiensis* var. kurstaki (CryIA(c)) protein expressed in transgenic cotton: Effects on beneficial and other non-target insects. Southwest. Entomol. 20, 493-500.

[23] Pilcher, C.D., Obrycki, J.J., Rice, M.E. and Lewis, L.C. (1997). Preimaginal development, survival, and field abundance of insect predators on transgenic *Bacillus thuringiensis* corn. Environ. Entomol. 26, 446-454.

[24] Obrycki, J.J., Hamid, M.N., Sajap, A.S. and Lewis, L.C. (1989). Suitability of corn insect pests for development and survival of *Chrysoperla carnea* and *Chrysopa oculata* (Neuroptera: Chrysopidae). Environ. Entomol. 18, 1126-1130.

[25] Klingen, I., Johansen, N.S. and Hofsvang, T. (1996). The predation of *Chrysoperla carnea* (Neurop. Chrysopidae) on eggs and larvae of *Mamestra brassicae* (Lep. Noctuidae). J. Applied Entomol. 120, 363-367.

[26] Orr, D.B. and Landis, D.A. (1997). Oviposition of European corn borer (Lepidoptera: Pyralidae) and impact of natural enemy populations in transgenic versus isogenic corn. J. Econ. Entomol. 90, 905-909.

[27] Manachini, B., Agosti, M. and Rigamonti, I. (1999). In: XI Symposium on Pesticide Chemistry, pp. 873-882 (Del Re, A.A.W., Brown, C.D., Capri, E., Evans, S.P. and Trevisan, M., Eds), Cremona, Italy.

[28] Mellon, M. and Rissler, J. (1998). Now or Never: serious new plans to save a natural pest control. Cambridge, MA: Union of Concerned Scientists.

[29] McGaughey, W.H. (1985). Insect resistance to the biological insecticide *Bacillus thuringiensis*. Science 229, 193-195.

[30] Ferré, J. and Van Rie, J. (2002). Biochemistry and genetics of insect resistance to *Bacillus thuringiensis*. Ann. Rev. Entomol. 47, 501-533.

[31] Tang, J.D., Shelton, A.M., Vanrie, J., Deroeck, S., Moar, W.J., Roush, R.T. et al. (1996). Toxicity of *Bacillus thuringiensis* spore and crystal protein to resistant Diamondback moth (*Plutella xylostella*). Appl. Environ. Microbiol. 62, 564-569.

[32] Ferré, J., Real, M.D., van Rie, J., Jansens, S. and Peferoen, M. (1991). Resistance to the *Bacillus thuringiensis* bioinsecticide in a field population of *Plutella xylostella* is due to a change in a midgut membrane receptor. Proc. Natl Acad. Sci. U.S.A. 88, 5119-5123.

[33] Tabashnik, B.E., Cushing, N.L., Finson, N. and Johnson, M.W. (1990). Field development of resistance to *Bacillus thuringiensis* in diamondback moth. J.Econ.Entomol. 83, 1671-1676.

[34] Moar, W.J., Pusztai-Carey, M., Faassen, H.V., Bosch, D., Frutos, R., Rang, C. et al. (1995). Development of *Bacillus thuringiensis* Cry1C resistance by *Spodoptera exigua* (Hubner) (Lepidoptera, Noctuidae). Appl. Environ. Microbiol. 61, 2086-2092.

[35] Liu, Y.B., Tabashnik, B.E., Masson, L., Escriche, B. and Ferré, J. (2000). Binding and toxicity of *Bacillus thuringiensis* protein Cry1C to susceptible and resistant diamondback moth. J. Econ. Entomol. 93, 1-6.

[36] Zhao, J.Z., Collins, H.L., Tang, J.D., Cao, J., Earle, E.D., Roush, R.T. et al. (2000). Development and characterization of diamondback moth resistance to transgenic broccoli expressing high levels of Cry1C. Appl. Environ. Microbiol. 66, 3784-3789.

[37] Gould, F., Kennedy, G., Johnson, T., Riggin Bucci, T. and Arpaia, S. (1998). Impact of natural enemies on the evolution of pest resistance to engineered crops. IOBC Newsletter 1, 16-17.

[38] Roush, R. (1997). Managing resistance to transgenic crops. In: Advances in Insect Control: the role of transgenic plants (Carozzi, N. and Koziel, M., Eds). London: Taylor and Francis, pp. 271-294.

[39] Roush, R.T. (1997). Bt-transgenic crops: just another pretty insecticide or a chance for a new start in resistance management? Pest. Science 51, 328-334.

[40] Adamczyk, J.J., Adams, L.C. and Hardee, D.D. (2001). Field efficacy and seasonal expression profiles for terminal leaves of single and double *Bacillus thuringiensis* toxin cotton genotypes. J. Econ. Entomol. 94, 1589-1593.

[41] Edge, J., Benedict, J., Carroll, J. and Reding, H. (2001). Bollgard Cotton: an assessment of global economic, environmental, and social benefits. J. Cotton Sci. 5, 121-136.

[42] Stewart, S.D., Adamczyk, J.J., Knighten, K.S. and Davis, F.M. (2001). Impact of Bt cottons expressing one or two insecticidal proteins of *Bacillus thuringiensis* Berliner on growth and survival of noctuid (Lepidoptera) larvae. J. Econ. Entomol. 94, 752-760.

[43] Guo, S.D., Cui, H.Z., Xia, L.Q., Wu, D.L., Ni, W.C., Zhang, Z.L. et al. (1999). Development of bivalent insect-resistant transgenic cotton plants. Sci. Agricul. Sin. 32, 1-7.

[44] Li, F., Cui, J., Liu, C., Wu, Z., Li, F., Zhou, Y. et al. (2000). The study of insect-resistant transgenic cotton harbouring double-gene and its insect-resistance. Sci. Agricul. Sin. 33, 46-52.

[45] Zhao, J.-Z., Fan, X., Shi, X., Zhao, R. and Fan, Y. (1997). Gene pyramiding: an effective strategy of resistance management for *Helicoverpa armigera* and *Bacillus thuringiensis*. Res. Pest Manag. 9, 19.

[46] Zhao, J.Z., Fan, Y.L., Fan, X.L., Shi, X.P. and Lu, M.G. (1999). Evaluation of transgenic tobacco expressing two insecticidal genes to delay resistance development of *Helicoverpa armigera*. Chinese Science Bull. 44, 1871-1874.

[47] Cao, J., Zhao, J.Z., Tang, J.D., Shelton, A.M. and Earle, E.D. (2002). Broccoli plants with pyramided cry1Ac and cry1C Bt genes control diamondback moths resistant to Cry1A and Cry1C proteins. Theor. Applied Genet. 105, 258-264.

[48] Zhao, J.Z., Cao, J., Li, Y.X., Collins, H.L., Roush, R.T., Earle, E.D. et al. (2003). Transgenic plants expressing two *Bacillus thuringiensis* toxins delay insect resistance evolution. Nat. Biotechnol. 21, 1493-1497.

[49] Jepson, P.C., Croft, B.C. and Pratt, G.E. (1994). Test systems to determine the ecological risks posed by toxin release from *Bacillus thuringiensis* genes in crop plants. Mol. Ecol. 3, 81-89.

[50] de Maagd, R.A. (2002). Environmental impact of insect-resistant crop plants expressing a Bt-toxin. II. Non-target effects (updated) and Bt-resistance in target insects. Note 218, Wageningen, The Netherlands: Plant Research International.

Genomics for Biosafety in Plant Biotechnology
J.P.H. Nap, A. Atanassov and W.J. Stiekema (Eds.)
IOS Press, 2004

Influence of a Contaminated Environment on the Stability of Plant Genomes

Lyubov ZELENA and Boris SOROCHINSKY*
Institute of Cell Biology and Genetic Engineering,
National Academy of Sciences of Ukraine, Zabolotnogo 148, Kiev 03143, Ukraine;
**email: bvs@public.icyb.kiev.ua*

Abstract. Genetic distances between control and irradiated plant populations were analyzed with RAPD-PCR. It was shown that the level of genetic divergence of irradiated plants depends on the level of radioactive contamination. This phenomenon must be taken into account when assessing environmental genotoxicity.

Introduction

There are considerable international concerns about the environmental risk of transgenic crops. There is concern that such a crop itself will alter the 'natural balance' and there is concern that the transgene will escape into wild relatives through hybridization and increase for example invasiveness. Important questions concerning the environmental issues of GM plants are currently missing. They concern among others the following: (1) what is the evolutionary dynamics of transgenic crops? (2) can a contaminated environment increase the environmental risk of transgenic crop?

Different chemical pollutants (heavy metals and organic compounds), background irradiation and probably global climate change may be environmental factors that are also important for the environmental risk assessment of GM plants. The aim of this study was to analyze genome divergence in non-transgenic irradiated plant populations. We have focussed on Scotch pine populations in the Chernobyl exclusion zone in the Ukraine.

The accident on 26 April 1986 at the fourth unit of the Chernobyl Nuclear Power Plant (ChNPP) resulted in the largest accidental release of radioactive materials reported. The accident at Chernobyl was a great tragedy, but its after-effects provide an unprecedented opportunity to gather information on the genetic effects of nuclear waste. The fallout from Chernobyl is quite distinct from that of other nuclear accidents or the bombing of Hiroshima and Nagasaki, both in the extent and diversity of pollutants. Prediction from one to the other is, therefore, not likely to be very informative.

The meltdown of the reactor released more than 2000 petaBq (PBq; 2000×10^{15} Bq) total radioactivity. Many regions within the Europe and former Soviet Union were heavily contaminated by radioactive fallout. Even now the maximal levels of radioactivity inside the Chernobyl zone (excluding unit 4 of the ChNPP) is up to 10,2 PBq for ^{137}Cs and 6,1 PBq for ^{90}Sr and the level of background gamma-irradiation in many parts of contaminated zone may correspond to values still as high as 1 mR/h - 20 mR/h. Plant communities may be good experimental models for genetic monitoring of low level irradiation because of the possibilities to calculate the level of absorbed doses and to describe the "genetic history" of non-mobile irradiated populations. There are only a few publications where the consequences of the Chernobyl accident for plant communities were analyzed in terms of molecular genetic studies [1,2]. Both the quantitative and the qualitative results from these studies were unexpected. In these analyses, the observed increase in mutation rates is not fully consistent with the expected consequences of radiation damage, which mostly produces length changes in DNA that result from double-stranded breaks. Increased mutation rates due to radiation exposure must increase the amount of genetic diversity within populations.

1. Experimental Approach

1.1 Plant and DNA Material

Morphologically normal and dwarf needles were collected from ten 7-12 year old *Pinus sylvestris* trees growing near the ChNPP. Needles (1- and 2-year) were harvested from trees growing in several regions of the ChNPP zone having different levels of radioactive contamination. Lyutezh site (control region), the Zapole site (D1, region with low level contamination), the Autobase site (D2, site with middle level contamination) and the Chistogalivka site (D3, most contaminated site) [3].

Genomic DNA was isolated from needle tissues by a modified CTAB procedure [4]. The DNA sample was treated with 1 μl of RNase A (10 mg/ml, Sigma) to destroy RNA at 37^0C for 1 hr. The concentration of the genomic DNA was determined by spectrophotometry at 260 nm (DU-50; Beckman). Samples were diluted to 15 ng/ μl with sterile deionised water. DNA and stored in a freezer (-20^0C) until use. Needles were bulked for DNA isolation from 7-10 individual Scots pine trees of each of the four sites (1- and 2-year needles were separated). Genomic DNA from normal and dwarf needles was extracted for each tree separately.

1.2 RAPD Analysis

Different PCR primers were tested for production of polymorphic RAPD markers with the Scots pine DNAs. For each sample, a RAPD band profile was recorded. Presence of a band was scored as plus (+) while absence of the band was scored as minus. The PopGen32 program was used for the statistical analysis of data. The optimum PCR reaction mix, developed with Scots pine template DNA, consisted of 15 ng template DNA, 2,5μl 10 x reaction buffer (Promega), 2 mM $MgCl_2$ (Promega), 0,3 μM primer (single, ten base oligonucleotide), 200 μM each dNTP, 1,5 U Taq DNA polymerase (Sigma) in a final volume of 25 μl, overlaid with mineral oil. The amplification profile was 2 min at 94 ^{0}C for starting and then taken through 40 cycles of 1 min at 94^0 C, 1 min at 35^0 C and 2 min at 72^0 C. A final extension of 7 min at 72^0C followed before incubating at 4^0C until the products were loaded. Reactions were amplified in thermocycler "Tercik" (Russia).

A total of 18 primers including 17 provided from the publication of Tulsieram et al. [5] and OPA6 from Operon Technology were screened. RAPD products were subjected to electrophoresis on 1,7% (w/v) agarose gels in Tris-borate EDTA buffer at 3 V/cm, stained with ethidium bromide, visualized under UV light and photographed. Molecular weights for the RAPD fragments were determined by comparison with the molecular weight of a reference marker produced by a standard 1 kb-ladder from Promega. RAPD amplifications were repeated at least three times and only the consistent PCR products were scored.

obtained in our study, age-related changes may occur not only in morphology and physiology of conifers needles but in genome as well.

Differences in morphology, histology, and number of mitotic abnormalities and chromosomal aberrations between control and dwarf needles were observed in previous studies [9, 10]. Four selected primers produced 23 fragments with a size range from 300 to 1500 bp. From 23 PCR products, 18 were polymorphic between control and dwarf pine needles. The level of polymorphic loci within each group analyzed was high and amounted 78.3 and 73.9 % for control and dwarf needles respectively. The genetic distance index calculated according to Nei was 0.0199.

3. Conclusions and Future Outlook

The results obtained support the hypothesis that a radioactively contaminated environment may affect the stability of a plant genome. There is a strong correlation between the level of radioactive contamination and the genetic divergence observed among the plant populations. Such genomic differentiation may form the basis of an increased micro-evolution. In effect, millennia of change are being compressed into a few years. Of course, whether potentially dramatic increases in mutation rate can be sustained by the populations concerned is an open question that should be investigated by continued analyses and monitoring.

References

[1] Kovalchuk, O., Kovalchuk, I., Arkhipov, A., Hohn, B. and Dubrova, Y. (2003). Extremely complex pattern of microsatellite mutation in the germline of wheat exposed to the post-Chernobyl radioactive contamination. Mutat. Res. 525, 93-101.

[2] Kovalchuk, O., Dubrova, Y., Arkhipov, A., Hohn, B. and Kovalchuk, I. (2000). Wheat mutation rate after Chernobyl. Nature 407, 583-584.

[3] Ukrainian Institute of Agricultural Radiology (2001). Contamination of the ChNPP 30-km zone, CD 1.

[4] Doyle, J.J. and Doyle, J.L. (1990). Isolation of plant DNA from fresh tissue. Focus 12, 13-15.

[5] Tulsieram, L.K. Glaubitz, J.C., Kiss, G. and Carlson, J.E. (1992). Single-tree genetic linkage mapping in conifers using haploid DNA from megagametophytes. Bio/Technol. 10, 686-690.

[6] Brickham, J.W., Sandhlu, S., Hebert, P.D.N., Chikhi, L. and Athwal, R. (2000). Effects of chemical contaminants on genetic diversity in natural populations: implications for biomonitoring and ecotoxicology. Mutat. Res. 463, 33-51.

[7] Day, M.E., Greenwood, M.S. and White, A.S. (2001). Age-related changes in foliar morphology and physiology in red spruce and their influence on declining photosynthetic rates and productivity with tree age. Tree Physiol. 21, 1195-1204.

[8] Sellin, A. (2001). Morphological and stomatal responses of Norway spruce foliage to irradiance within a canopy depending on shoot age. Env. Exp. Bot. 45, 115-131.

[9] Kozubov, G.M. and Taskaev, A.I. (1994). Radiobiological and Radioecological Investigations of Woody Trees. St.-Pb.:Nauka Publ, 256 p.

[10] Zelena, L.B., Sorochinsky, B.V. and Grodzinsky, D.M. (2002). Characteristics of karyotype of pine tissues formed morphological abnormalities in Chernobyl zone. Rep. Natl Acad. Sci. Ukraine 1, 171-174.

Genomics for Biosafety in Plant Biotechnology
J.P.H. Nap, A. Atanassov and W.J. Stiekema (Eds.)
IOS Press, 2004

Ecological Risk Assessment of Genetically Modified Crops

Constantin RADU[1*], Michael HORAK[2] and Thomas NICKSON[2]
[1]*Monsanto CR SRO, Krakovska 9, 110 00 Prague 1, Czech Republic;*
[2]*Monsanto, St. Louis, Missouri, United States of America;*
**email: constantin.radu@monsanto.com*

Abstract. Safe, effective and beneficial development of plant biotechnology can only occur if there is public understanding that adequate regulations are in place and no unacceptable harm will occur. Assessing the potential for risk in case of genetically modified (GM) plants has to be done in the context of existing agricultural activities, which may also have the potential to cause adverse environmental effects. The environmental risk assessment model that has been developed and used for the last decade is scientifically sound and satisfies the needs of societies. A valuable tool will be the further development of genomics' instruments that will help determine functions and understanding of interactions between gene products. Society needs to realize that acceptance of the environmental impact of any human activity must be acknowledged. If society accepts the impact from current agricultural practice, it should address the question how biotechnology may affect that impact.

Introduction

In common language and understanding, risk is a chance that something could happen. The "something" could be good, but is usually interpreted as bad or harmful. Scientifically, risk is the probability that something harmful could occur as a result of some activity or continued, unmanaged condition. Science-based ecological risk assessment is used to understand and estimate the potentially harmful outcome to the environment stemming from a choice of activities. An important aspect is that such a choice is not risk free. As such, action and inaction are associated with environmental risk. This is particularly evident in ecological risk assessment of agricultural production systems and the choices made in the course of producing sufficient food for the people of the earth.

Today, modern agro-biotechnology is at the heart of an intensive public and political debate that is risk-centered and often emotional. Similar controversies were virtually unknown to traditional breeders 20 years ago. The present controversy encompasses disparate concerns. For example, the fear that gene flow from GM crops will result in irreparable ecological damage. Others worry that the economic and ecological benefits of GM crops are being unnecessarily delayed due to excessive regulatory burdens based on theoretical speculations with low probability of occurrence. Perhaps the debate surrounding the risks associated with GM crops can be summarized as disparate opinions of the importance of precaution versus the cost of disproportionate precaution. The scientific community is heavily immersed in this passionate and global controversy. This affects the important discussion of environmental risk and how environmental risk is assessed for current products and those that will be developed in the post-genomics era.

1. Agricultural Perspectives Today

Agriculture has been an essential component of the daily life of man for thousands of years. Over the past five decades, agricultural practices and technologies have been at the center of controversy starting with the well-known and celebrated publication of Rachel Carson's *Silent Spring*. There is an increased awareness that the world is facing increasing pressure from habitat destruction, population growth, pollution, global warming, and over harvesting. Simply stated, there is not enough land available for food production without incurring significant damage to the mega-diverse tropical rain forests, which are already poor agricultural lands. New methods to produce sufficient food to feed over six billion people are necessary to avoid further environmental degradation and social upheaval [1].

Agriculture as we practice it in the developed and developing world alike is degrading the environment. For decades, tillage of soil was used as effective means to control weeds. However, such cultivation can lead to soil degradation, and has the potential to cause erosion, reduce soil quality and harm soil organisms [2]. Pollution (phosphates, nitrates, air quality) and habitat alterations through land reclamation (diking fields) are other examples of negative effects of current-day agriculture [3]. However, associated with current and past agriculture practices, several neutral-to-positive changes are taking place within agro-ecosystems that are efficiently producing food, and oxygen, while also establishing improved wildlife habitats. Tools being developed through modern biotechnology are showing great potential to contribute to more sustainable crop production systems through their environmental and economic benefits. In the future, GM crops developed from advances in genomics will also be evaluated for their potential to contribute to sustainable agriculture. Land management by human beings is by far the single most destructive activity to biodiversity. Activities such as urbanization and agriculture alter vast tracts of land to meet human needs. Crop and animal agriculture provide the food needed. Today's agriculture provides sufficient food for the population of the earth. Current practices become less efficient over time due to degradation of soil, susceptibility of crops and cropping systems to new pests, and other factors that reduce yield. Without improving yield in ways that mitigate the agricultural 'footprint', it is predicted that over the next 50 years, a third of all remaining natural tropical and temperate ecosystems will be lost to agriculture.

Historic developments and current land availability indicate that the natural areas in South America and sub-Saharan Africa are the most vulnerable. More than half of the earth's natural areas is found in just seven countries. Most of these countries contain rain forest that should be considered extremely diverse. Also countries with adjacent subtropical areas have rich and delicate ecosystems [4,5]. Much of the land conversion for agriculture will concern forests and rain forests, for example by controlled incineration. This will affect species that depend on such ecosystems with respect to habitat or food resources. Such forest loss and habitat destruction may have short-term socio-economic benefits, especially in developing countries, but will present long-term effects on biodiversity that are undesirable. The risk of extinction in birds estimated by BirdLife International using the World Conservation Union's Red List of threatened avian species indicates that as much as 350 species (which is 3.5 percent of all bird species of the world) could disappear in the next four to five decades [5,6]. Environmental degradation must be slowed down and eventually reversed for the future sustainability of mankind on earth. In addition to population growth and poverty, agricultural management practices are the main cause of this degradation. It appears that more tools and more knowledge are needed to address the problems associated with agriculture. Plant biotechnology is such a tool that merits careful consideration as to how it can increase agricultural productivity and efficiency on fixed land assets with less impact on the environment.

2. The Pipeline of Agricultural Biotechnology

Modern agro-biotechnology, commonly referred to as genetic modification (GM) or genetic engineering (GE), allows researchers to remove genes and genetic elements between species without the need for sexual compatibility. When applied to agriculture, this technology has complemented traditional breeding in the development of improved varieties of plants containing new and improved traits such as pest/disease/stress resistance, tolerance to environmentally superior herbicides, improved nutritional quality or the ability to produce value added materials in plants (Figure 1).

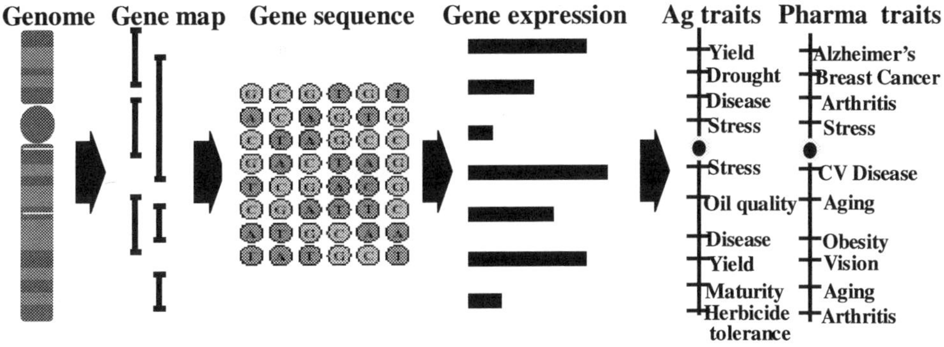

Figure 1. New tools for development indicate a move from chemistry to biology

The first hurdle confronting GM crops has been to demonstrate that they fit into current agricultural systems. Growers had to see how they could use such crops and how they would benefit. If growers would not adopt GM crops, GM crops would not have an economical future and would not be developed. Growers who have realized economic and quality-of-life benefits have driven the growth and acceptance of the first GM products. In fact, agricultural biotechnology has been the single most rapidly adopted technology in the history of agriculture [6]. Since their first commercialization in 1996, GM crops increased from 1.7 to 67.7 million hectares in 18 different countries by 2003 (Figure 2).

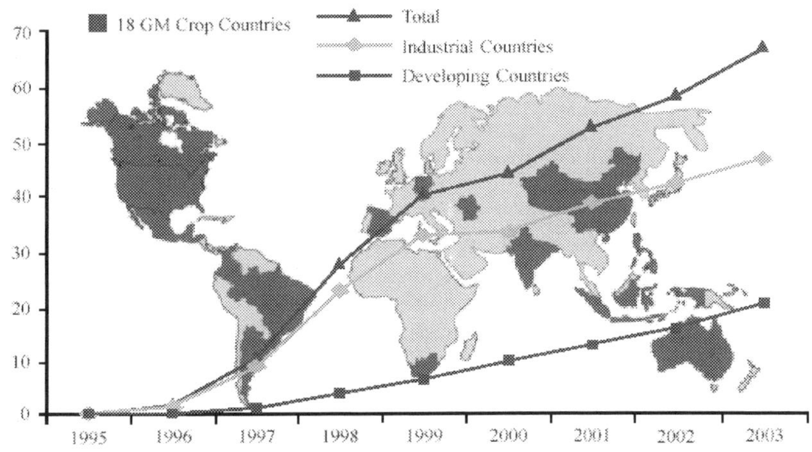

Figure 2. The global expansion of GM crops (from [7])

Today, biotechnology is an invaluable new tool in the development of crop varieties adapted to local needs, both in the developed and developing world, providing on case-by-case benefits for farmers, consumers and the environment [7,8]. The public at large, however, consists of consumers who view risk, agriculture and GM crops from a different perspective than growers. Producer and consumer perspectives together determine acceptability of GM crops. In this context, various biotechnology products are in the developmental pipeline of agro-biotechnology [5,8]. Three types or 'generations' of GM products can be distinguished.

2.1 First Generation GM Crops: Agronomic Traits

Agronomic traits, also known as input traits, are designed to make crop production more efficient or better adapted to specific environments. The major agronomic traits found in the marketplace today concern corn, cotton and potato with protection against insect pests, corn, cotton, soybean, canola with tolerance to herbicides, or crops with a combination of both ('stacked') [7]. These biotechnology crops have been rapidly adopted by farmers across the globe, in both the developing and developed world. They provide documented reduced pesticide use, increased yields and an overall highly significant economic impact on agriculture [7-13]. A range of GM crops (papaya, cassava, squash and potato) are resistant against viral diseases. Although most of these crops have not been very successful at the market level, their development is important for developing countries [14]. Improvements to the existing agronomic traits are being commercialized. Products with multiple Bt traits are more durable because they will delay the development of resistance in target insects and broaden protection to insect pests [15,16]. In the future, GM crops with increased tolerance to abiotic stresses such as drought or salt will be evaluated. Many of the genes responsible for these more complex traits will come from the use of modern genomics approaches. Abiotic-stress tolerant GM products have an enormous potential to increase agricultural efficiency while addressing important environmental concerns [17].

2.2 Second Generation GM Crops: Quality Traits

Quality traits, also known as output traits, enhance the nutritional quality and value of food, feed and fiber [18,19]. Calgene introduced the first GM product of this generation in 1994 in the US: the FlavrSavr tomato. Antisense technology resulted in a tomato with delayed ripening and improved flavor. However, for market-based reasons relating to quality, not to safety, this product is no longer commercially available. In the UK, Zeneca developed a paste from GM tomatoes that had lower water content. This product was sold by Sainsbury and Safeway supermarkets in the late nineties. Because the processors used less energy to produce the tomato paste, the product could be sold at a lower price. While the product was initially very successful in the UK, it was recalled from the shelves upon pressure from opponents of GM food. Different categories of future GM targets can be distinguished [5].

Reduced or eliminated allergens and antinutrients. In many common foods or food ingredients (such as in peanut, soybean, or rice), proteins inducing allergic reactions and other antinutritional factors (such as trypsin inhibitors) can be abundant. GM tools such as antisense RNA and gene 'knock-out' technologies enable to reduce or remove undesired, allergenic or toxic components from food [20,21]. Direct safety and health benefits show increased interest and acceptance from consumers and could prove crucial for the future development of the technology.

***In planta* produced health-promoting substances.** Plants with improved content in fatty acids profile are being developed to either supplement human dietary needs or to facilitate processing characteristics. Plants with the capacity to produce omega-3 fatty acids are among the more illustrative examples of the nutritional enhancement of food. Omega-3 fatty acids have strong cardio-protective effects. Their main food source is fish oil. With changes in the dietary habits, omega-3 fatty acid intake is becoming problematic in a number of population groups [22].

Vitamins and micronutrients. A balanced nutrition should, ideally, provide for adequate levels of micronutrients (minerals, vitamins). In many world areas, GM crops can contribute to achieve this goal. Golden Rice with increased levels of beta-carotene is currently being tested as an example of how vitamin A deficiency can be targeted [23,24]. Another example is canola with increased levels of beta-carotene in its oil [25]. Current supplements for essential amino acids could potentially be replaced by the development of corn and soybeans with increased levels of lysine and methionine [5].

Nutritionally enhanced crops for livestock production. Animal husbandry may greatly benefit from crops with characteristics that improve the efficiency, quality and/or quantity of animal protein production [26]. High oleic acid soybeans can contain more than 80 per cent oleic acid in their oil compared with 24 per cent for traditional soybean oil. Feeding high oleic full-fat soybean to cows and chicken provides for both an efficient energy source, and can result in lower saturated fat levels in milk and poultry [26]. Other enhancements concern the amino acid profile, the digestibility of feed (starches, oils, nitrogen), improved nitrogen digestibility, lowered phytase content or higher phytic acid amounts. These developments can also contribute to the reduction of nitrogen and phosphorus pollution of soil and water sources near animal facilities.

2.3 Third Generation GM Crops: Plant-produced Pharmaceuticals and Compounds

Future GM crops will replace current production systems based on fermentation or chemistry. Based on the efficiency of biomass production in plants, GM crops may be used to produce proteins and industrial compounds, such as enzymes that are used in detergent formulations, industrial processes or in food manufacturing. Plant-based biopolymer production for degradable and recyclable packaging, production of therapeutic proteins or the production of edible vaccines [27,28] is all under investigation. The use of plants as factories could realize considerable benefits. An approach to be carefully assessed is the ability of GM crops to establish non-polluting renewable production systems.

3. Ecological Risk Assessment: Ensuring the Safety of Current and Future Products

Risk assessment is built into the overall process of developing GM crops. This is described as the 'cradle-to-grave' approach, in which a product in continuously tested (Figure 3). Comprehensive testing of GM products begins early in product development and continues throughout the life cycle of the product. Testing starts in the laboratory and continues from the greenhouse to small-scale confined field trials. Of hundreds of plants (so-called 'transformation events'), most are eliminated in this early screening and selection process that is conducted by experts in plant breeding, genetics, biochemistry and other relevant disciplines. Numerous morphologic, agronomic, and other growth parameters are investigated and have to be approved prior to the release of a GM plant into the environment. At the end of this extensive and elaborate testing, typically only a few candidates out of the initial hundreds are selected to undergo ecological risk assessment and regulatory evaluation.

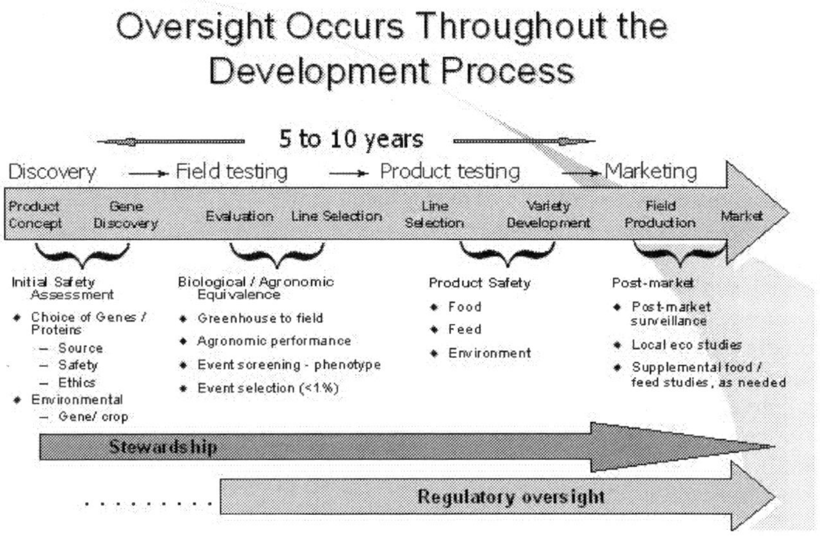

Figure 3. Life history of testing conducted on GM crops.

The formal risk assessment used to make regulatory decisions is indicated as 'Regulatory Oversight' in Figure 3. Risk assessment can be defined as the procedure used to assess the safety of a transgenic 'event'. Risk assessment involves the technology providers, those producing GM crop, and the regulators, those with the legal responsibility to assess and regulate GM products. The process is aimed to examine the possible harm that could arise from the use of the product and the likelihood of that happening. There four general principles to investigate that a GM crop is safe to the environment. Ecological risk assessment is, or should be: (1) science-based, (2) conducted on a case-by-case basis, (3) comparative and (4) independently reviewed. The best available science should be used to evaluate on a case-by-case basis the risks associated with the GM crop compared to their non-GM counterparts in such a way that the ecological risk assessment data are reviewed by independent parties with regulatory authority prior to decision making. Traditionally bred crops are considered the appropriate reference for comparison.

An assessment plan is developed that takes into consideration the nature of the crop, the trait, the receiving environment, the product and its proposed uses. The first step is a detailed characterization of the GM 'event' to identify those characteristics that are unfamiliar, novel and/or unexpected. Characterization involves detailed molecular analysis, expression analysis, compositional analysis as well as agronomic and phenotypic analysis. In defining the risk potentially posed by a transgenic crop, it is important to start from the crop species itself. Conventionally bred crops are thus the basis from which the comparison is made. The hypothesis to be tested is that the GM crop is not different except for the characters imparted by the introduced genes. Based on the results of the characterization, those aspects that are different (not familiar) compared to the conventional crop are subject of the risk assessment.

Based on the characteristics of the crop and the results of the risk assessment, risk management strategies could be developed. Post-market monitoring and other stewardship actions can be implemented. Stewardship activities are derived from the characteristics of the GM crop, its use cropping systems and other important factors that could affect the value and durability of the product. Industry works with growers to assess how well products are working, to investigate complaints and to provide vigilance to issues and concerns. Field tests are combined with independent research that yields new information to address upcoming issues and concerns. Data on product performance, customer satisfaction and benefits information are collected throughout the lifecycle of a GM product.

4. Decision Making in Ecological Risk Assessment

The risk assessment focuses only on those characteristics that are meaningfully different in the GM crop compared to the non-GM crop. For the products approved to date, the only novel characteristic of the GM crop being investigated is the introduced trait. Such an introduced trait is therefore evaluated for its potential to cause harm. Traits like herbicide tolerance [29] have no toxic or selective advantage (weediness) potential. No further testing is conducted on these environmentally benign traits. A trait such as insect resistance conferred by the *Bacillus thuringiensis* insecticidal protein [30,31] may have toxic effects on non-target organisms. Such trait is evaluated in a sequential manner, which is outlined in Figure 4. The outcome of the science-based ecological risk assessments of the GM crops currently being marketed has been a finding of no evidence of meaningful risk. That is, the environmental risks present from the GM crop are no greater than those posed by their non-GM counterparts.

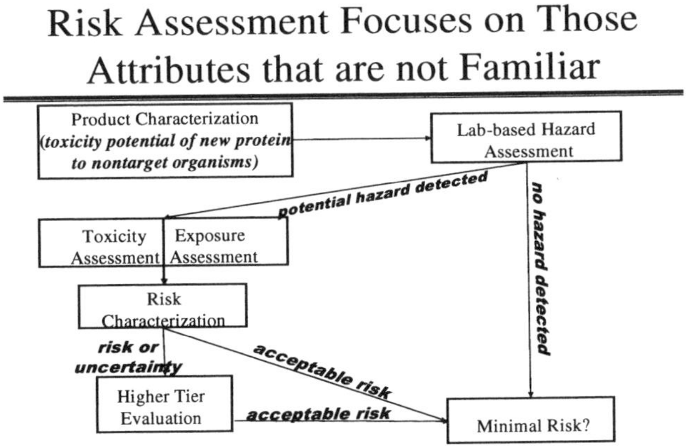

Figure 4. Sequential assessment model for the evaluation of GM crops.

Ultimately risk assessment conclusions are integrated with other important information in the regulatory decision. It is important to recognize that decisions must be made within a landscape of diverse values with imperfect knowledge and imperfect ability to predict the future. Decision makers must distinguish meaningful uncertainty associated with GM crops and the risks associated with delaying or making a 'no' decision. In the end a decision must be made that acknowledges that biotechnology cannot be made safer than biology. Moreover, there is also a risk in NOT deciding. The comparative assessment model (Figure 5) is seen by many regulatory systems as a good, if not the best, way to weigh the balance between existing technology and new technology.

Scientific advances continuously generate more useful tools and methods to be applied to the environmental problems facing people. A valuable tool will be the further development of genomics that will help determine functions and understanding of interactions between gene products. The genomics profiling methods will contribute to the scientific rigor safety assessment of GM crops, and will also point to new directions for the development of GM crops. GM crops are a tool that could be used to better address important environmental issues. When the ecological risk assessment shows no meaningful risk, a GM crop should be assessed for its potential to reduce inputs, its specificity and overall compliance with integrated management and agronomic practices. Based on these characteristics, decisions can be made that integrate science data and personal values.

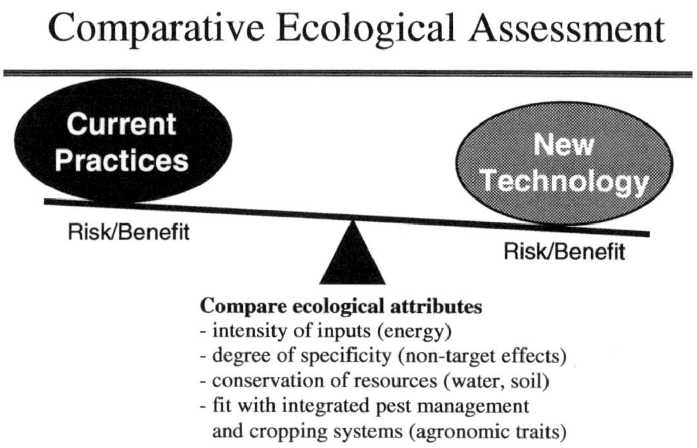

Figure 5. The decision making process has to balance risks between existing technology and new technology

New GM crops, including those developed through genomics technologies, will not require a new risk assessment model. The current approach is scientifically sound and addresses societal needs. Focus in the future will be to develop better experimental protocols to address issues like weediness, invasiveness, and nontarget impacts. Furthermore, more information is needed on how current crop production systems are affecting the environment in order to develop truly sustainable agriculture. Genomics technologies will here have a role as well.

5. Conclusions and Future Outlook

In conclusion, the current environmental risk assessment model is scientifically sound and appropriate for society. Improvements can be made in the future as the knowledge of ecological processes and agricultural baseline information increases. Exciting developments derived from the study of genomics are on the horizon of GM technology. These developments will impact safety assessments and could bring enhanced benefits to human nutrition and environmental sustainability. However, as with any new technology, it is necessary to proceed prudently and cautiously. Acknowledging that our current agricultural practices are degrading the environment, and threatening biodiversity especially in the developing world, environmental degradation must be slowed and eventually reversed for the future sustainability of humans on the earth. Related to population growth and poverty, agricultural management practices are a root cause of this undesired degradation. More tools and more knowledge need to be applied to the problems inherently associated with agriculture. If society accepts the impact of current agricultural practice, it should address the question how biotechnology could affect that impact. Plant biotechnology is therefore one tool that merits careful consideration as to how it can increase agricultural productivity and efficiency on fixed land assets with less environmental impact.

References

[1] UNEP (2002). Global Environmental Outlook (GEO) 3: Past, Present and Future Perspectives. Synthesis (www.grida.no/geo/geo3/english/overview/001.htm) .
[2] FAOSTAT (2003). Agricultural Data. Food and Agriculture Organization (FAO) (apps.fao.org/page/collections?subset¼agriculture).
[3] Pew Center (2003). Global warming in Depth: Environmental Impacts. (www.pewclimate.org/global-warming-in-depth/environmental_impacts/)
[4] Bruinsma, J. (Ed.) (2003). World Agriculture: Towards 2015/2030 - An FAO Perspective. London: Earthscan (www.fao.org/english/newsroom/news/2002/7833-en.html).
[5] Cockburn, A. (2004). Commercial plant breeding: what is in the biotech pipeline? J. Comm. Biotechnol. 10, 209-223.
[6] BirdLife International (2002). Globally Threatened Birds. Indicating Priorities for Action. (www.birdlife.net/action/science/species/globally_tbu/gtb_booklet.pdf).
[7] James, C. (2003). Global status of commercialized transgenic crops: 2003. ISAAA Briefs 30-2003. (www.isaaa.org).
[8] Harlander, S.H. (2002). The evolution of modern agriculture and its future with biotechnology. J. Amer. Coll. Nutr. 21, 161-165.
[9] Betz, F., Hammond, B. and Fuchs, R. (2000). Safety and advantages of *Bacillus thuringiensis*-protected plants to control insect pests. Reg. Toxicol. Pharmacol. 32, 156-173.
[10] Gianessi, L.P., Silvers, C.S., Sankula, S. and Carpenter, J.E. (2002). Plant Biotechnology: current and potential impact for improving pest management in US agriculture. An analysis of 40 case studies. Washington, DC: National Center for Food & Agriculture Policy (NCFAP). (www.ncfap.org/40CaseStudies.htm).
[11] Gianessi, L.P. and Carpenter, J.E. (1999). Agricultural Biotechnology: insect control benefits. Washington, DC: National Center for Food & Agricultural Policy (NCFAP). (www.ncfap.org/reports/biotech/insectcontrolbenefits.pdf)
[12] Heimlich, R.E., Fernandez-Cornejo, J., McBride, W., Klotz-Ingram, C., Jans, S. and Brooks, N. (2000). Genetically engineered crops: has adoption reduced pesticide use? Agric. Outlook 273, 13-17. (usda.mannlib.cornell.edu/reports/erssor/economics/ao-bb/2000/ao273.pdf)
[13] Phipps, R.H. and Park, J.R. (2002). Environmental benefits of genetically modified crops: global and European perspectives on their ability to reduce pesticide use. J. Anim. Feed Sci. 11, 1-18.
[14] Nuffield Council on Bioethics (2003). The use of genetically modified crops in developing countries - a follow-up Discussion Paper. London: Nuffield Council on Bioethics. (www.nuffieldbioethics.org/filelibrary/pdf/gm_crops_paper_final.pdf)
[15] Greenplate, J.T., Penn, S.R., Shappley, Z., Oppenhuizen, M., Mann, J., Reich, B. et al. (2000). Bollgard II efficacy: quantification of total Lepdopteran activity in a 2-gene product. Proc. Beltwide Cotton Conf. 2, 1039-1040.
[16] Marchosky, R., Ellsworth, P., Moser, H. and Henneberry, T. (2001). Bollgard and Bollgard II efficacy in near isogenic lines of 'DP50' upland cotton in Arizona. In Cotton: A College of Agriculture Report for 2001. (Silvertooth, J. Ed.), Tucson, USA: CALS. (ag.arizona.edu/pubs/crops/az1224/az12247c.pdf)
[17] Cheikh, N., Miller, P.W. and Kishore, G. (2000). Role of biotechnology in crop productivity in a changing environment. In Climate Change and Global Crop Productivity. (Reddy, K.R. and Hodges, H.F., Eds). New York: Oxford University Press, 488 pp.
[18] Mackey, M. (2002). The application of biotechnology to nutrition: an overview. J. Amer. Coll. Nutr. 21, 157S-160S.
[19] Falk, M.C., Chassy, B.M., Harlander, S.K., Hoban IV, T.J., McGloughlin, M.N. and Akhlaghi, A.R. (2002). Issues and opinions - food biotechnology - benefits and concerns. J. Nutr. 132, 1384-1390.
[20] Kleiner, K. (2002). Engineering safer soya. New Scientist 175, 7.
[21] Rabjohn, P., West, C. M., Connaughton, C., Sampson, H.A., Helm, R.M., Burks, A.W. et al. (2002). Modification of peanut allergen Ara h3: effects on IgE binding and T cell stimulation. Int. Arch. Allergy Immunol. 128, 15-23.

- phytoextraction, employing metal-accumulating plants to extract metals from contaminated soils [9];
- phytostabilization, using metal-tolerant plants, (known as "excluders") to reduce or exclude the environmental risk from heavy metals [1];
- phytodegradation, applying plants to degrade nonvolatile hydrocarbons, thus removing them from the environment [10].

The improvements in plants offered by genetic engineering will provide new possibilities for the phytoremediation of contaminated sites. For example, it has been detected that the transfer of a gene from a bacterium to a plant might improve the plant's capability to remediate polluted soils [11]. The strategy of creating transgenic plants based on the use of genes from prokaryotic organisms has been applied for many years now. The key element of this strategy is to select a microorganism, which could improve plant's capability to remediate polluted soils, to clone the bacterial genes involved in the process of bioremediation and introduce these into suitable plant hosts.

It has been discovered that a certain micro-organism, known as *Pseudomonas aeruginosa*, possesses unusual features. *Pseudomonas aeruginosa* subsists in petroleum-contaminated soils and participates in the biodegradation of aliphatic, aromatic and polyaromatic oil hydrocarbons [12]. It has been also verified that the manifestation of this property was connected with the presence of rhamnolipids [13] produced by *Pseudomonas aeruginosa*. Rhamnolipids, in which one or two molecules of rhamnose are linked to one or two molecules of beta-hydroxydecanoic acid, belong to the group of rhamnose-containing glycolipid biosurfactants [14] first described in 1949 [15]. Now, rhamnolipids are among the best-studied glycolipids. In 1994, Ochsner et al. cloned two genes from *Pseudomonas aeruginosa*, *rhlA* and *rhlB*, which encoded rhamnosyltransferase and participated in the biosynthesis of rhamnolipids [16]. Then rhamnolipids were ascertained to not only degrade oil but also participate in copper, cadmium and zinc removal from contaminated soils [17,18]. These facts have suggested an idea to create transgenic plants with *rhlA* and *rhlB* genes. We hypothesized that these transgenic plants because of the production of rhamnolipids would degrade petroleum products and demonstrate tolerance to heavy metals. In our view, the plants with such necessary properties could become an effective instrument for the phytoremediation of polluted sites.

1. Creation of Transgenic Plants for Remediation Purposes

The creation of oil-degrading and metal-resistant plants is on the current agenda of environmental biotechnology [1]. Although the results are still modest, this task has consumed much effort. Our research is directed towards producing GM plants that could efficiently demonstrate tolerance to heavy metals and convert oil hydrocarbons into non-toxic compounds. The key element of our studies is the strategy of using microbial genes for the creation of transgenic plants that can be applied for phytoremediation purposes. Our basic tools were *rhlA* and *rhlB* genes from *Pseudomonas aeruginosa PAO1*, which participate in rhamnolipid biosynthesis and, in a wider context, contribute to metal tolerance and oil degradation [16]. So far, no attempt has been made to engineer rhamnolipid expression in plants. Therefore, using the cloned *rhlA* and *rhlB* genes from *Pseudomonas aeruginosa*, we created three different constructs for constitutive expression in plant tissues, based on the plasmid pGreen0229 (Figure 1). We have transformed plants with all three types of constructs: with only the *rhlA* gene (pAS50) with only the *rhlB* gene pAS51) and a construct with both *rhlA* and *rhlB* genes (pAS52).

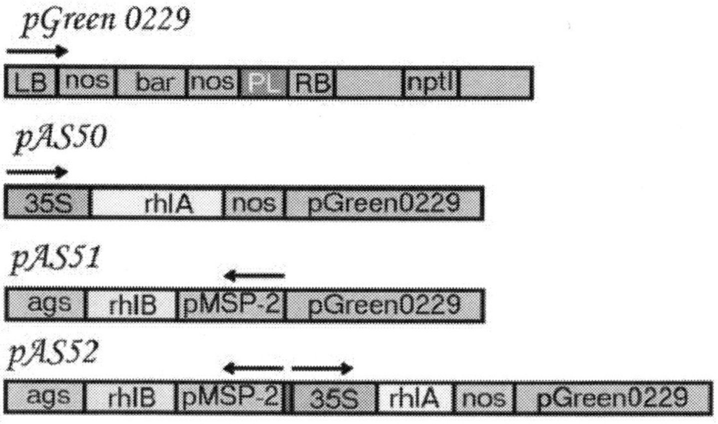

Figure 1. Schematic diagram of constructs used for transformation. All constructs were made on the basis of pGreen0229. The *rhlA* gene transcription is regulated by the 35S promoter and the nopaline synthase terminator. The *rhlB* gene is regulated by the monapine synthase super promoter and the agropine synthase terminator.

We have generated 30 independent lines of transgenic *Nicotiana tabacum*, (10 lines per construct), and 30 independent lines of transgenic *Arabidopsis thaliana*, (10 lines per construct), in order to study the effect of rhamnolipid overexpression *in planta* on metal resistance and oil degradation. As control for our experiments, we created plants with the empty vector.

The presence of transgenes in plants was confirmed by PCR and genomic Southern blot hybridisation (not shown). The expression level of the introduced rhlA and rhlB genes was determined by RT-PCR (Figure 2) and Northern blot hybridisation. For further experiments we have chosen the lines with the highest level of transgene expression.

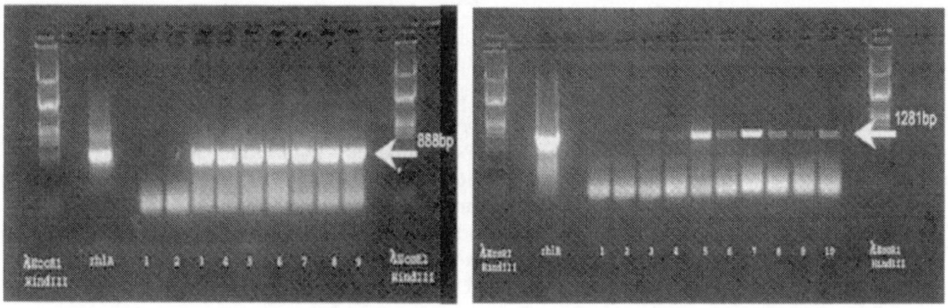

A B

Figure 2. Expression of *rhlA* and *rhlB* genes in *Arabidopsis thaliana* transgenic plants.
A. Photograph of gel electrophoresis of RT-PCR reactions using primers for the *rhlA* gene. 1, wild type control, 2, pGreen (empty vector), 3-6, transgenic plants with *rhlA* gene, 7-9, transgenic plants with the *rhlA+rhlB* genes.
B. Photograph of gel electrophoresis of RT-PCR reactions using primers for the *rhlB* gene. 1, wild type control, 2 – pGreen (empty vector), 3-7, transgenic plants with the *rhlB* gene, 8-10, transgenic plants with the *rhlA+rhlB* genes.

2. Analysis of the Production of Rhamnolipids

Rhamnolipids produced by *Pseudomonas aeruginosa* have been proven to be extracellular compounds [19]. Consequently, we performed TLC testing to examine both control and transgenic plants for their ability to secrete rhamnolipids into the root zone. As a control, we used the rhamnolipids secreted by *Pseudomonas aeruginosa PAO1* strain. The rhamnolipids were isolated from the plant medium with the help of the same method that was applied when rhamnolipids were isolated from the microbial suspension. TLC results demonstrated that transgenic plants did secrete glycolipids into the root zone (Figure 3). In contrast, the control plants did not secrete such compound (Figure 3). The rhamnolipids were quantified using the anthrone reagent at $\lambda = 490$ nm.

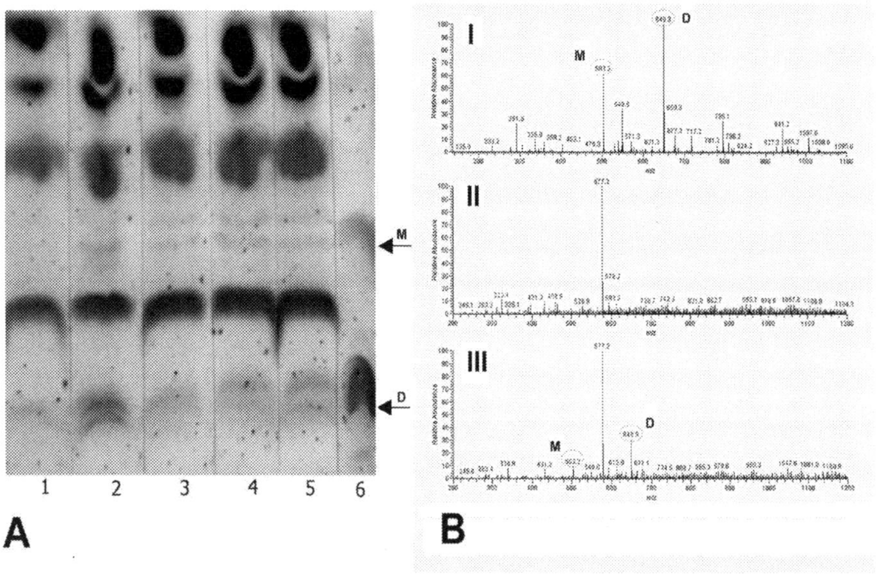

Figure 3. Rhamnolipid production by plants.
A. Thin Layer Chromatography of lipids from *Arabidopsis thaliana* plants. Lipid fraction from 1, wild type control, 2, control plant mixed with rhamnolipids from Pseudomonas aeruginosa, 3, transgenic plant with *rhlA* gene, 4, transgenic plant with *rhlB* gene, 5, transgenic plant with *rhlA+rhlB* genes, 6, supernatant from *Pseudomonas aeruginosa* grown on PPGAS medium. M, monorhamnolipids; D, dirhamnolipids.
B. Mass spectrometry of rhamnolipids. (I) rhamnolipid mix JBR-425 (Jeneil Biosurfactant Co.), (II) lipid extract from wild type control, (III) lipid extract from transgenic plants with *rhlA* gene. M, monorhamnolipids; D, dirhamnolipids.

Transgenic plants with the *rhlA* gene secreted mainly monorhamnolipids and a small quantity of dirhamnolipids (the concentration of both rhamnolipids was 10.3 $\mu g\ ml^{-1}$). TLC revealed that the transgenic plants with *rhlB* gene secreted mainly dirhamnolipids and a smaller amount of monorhamnolipids. The content of both rhamnolipids in the root zone was 9.1 $\mu g.ml^{-1}$. The transgenic plants carrying both the *rhlA+rhlB* genes also secreted monorhamnolipids and dirhamnolipids. Their concentration was about 14.6 $\mu g\ ml^{-1}$. This indicated that the presence of both *rlhA* and *rlhB* genes is required in order to be able to produce and secrete larger quantities of rhamnolipids.

We also verified if intracellular rhamnolipids were present in the plant tissues. Surprisingly, tests demonstrated the presence of large quantities of monorhamnolipids and dirhamnolipids in the tissues of the transgenic plants. According to the reference substances used, these intracellular substances are identical to rhamnolipids extracted from *Pseudomonas aeruginosa PAO1* (Figure 3A). In order to confirm that the substances produced by plants were rhamnolipids and were not the substances with the similar Rf, the mass spectrometry on mass spectrometer LCQ Deca XL^{plus} (Thermo Finnigan) was performed. The commercially available rhamnolipid mix JBR-425 (Jeneil Biosurfactant Co.) was used as a control. Mass spectrometry showed two major peaks with masses of 503.3 and 649.3 corresponding to mono- and di-rhamnolipids respectively (Figure 3BI). The peaks of the same masses were identified in transgenic plants (Figure 3BIII), but not in wild type control (Figure 3BII). It was confirmed that rhamnolipid production was determined by the inserted genes. It is not unexpected that these plants are able to synthesise the rhamnolipids, as both precursors (TDP-L-rhamnose and β-hydroxy-decanoyl-β-hydroxydecanoate) are present in plant tissue [20].

3. Biodegradation of Oil Hydrocarbons by Transgenic Plants

At present, oil hydrocarbons are among the most virulent organo-xenobiotics. These pollutants can translocate from roots to the above-ground parts of plants [3]. The toxic effect of oil hydrocarbons on plants ranges from the decrease in transpiration to plant mortality [21]. Biodegradation can be defined as a basic natural process that helps remove non-volatile hydrocarbons from the environment [10]. Our choice of rhamnolipids for biodegradation was governed by their ability to augment hydrocarbon bioavailability. It is a well-known fact that various biosurfactants can foster this process [16]. An increase in the bioavailability of hydrocarbons intensifies the degree of bioremediation of oil-contaminated soils.

We hypothesised that the use of plants with rhamnolipid expression could increase oil hydrocarbon biodegradation. Thus, we studied the rates of oil hydrocarbon degradation by measuring oil content in soil (more exactly, in the rhizosphere). The contamination level we applied was 2.4 % of crude oil per 1kg of wet soil. Period of plant cultivation lasted 45 days. We compared the quantities of oil hydrocarbons in soil at the beginning and at the end of the 45-day period both for non-GM and transgenic plants with three types of constructs. It turned out that in case of transgenic plants the amount of oil hydrocarbons in soil after this period dropped considerably. The percentage of oil degradation can be estimated with the degradation coefficient (C_d) in the equation: $C_d = (1 - A_r/A_i) \times 100\%$, where A_r is the remaining amount of oil hydrocarbons in soil, A_i is the initial amount of oil hydrocarbons in soil. In polluted soil (2.4% of oil) with non-GM control plants, the oil hydrocarbon concentration decrease was insignificant. This decrease is connected with the natural hydrocarbon degradation that occurs due to the influence of bacteria, sunlight, oxygen, and other factors. Control plants with empty vector indicated a slight rate of hydrocarbon biodegradation in the soil. However, in contrast to the control plants, transgenic plants containing rhamnolipid genes demonstrated rather high speed of soil reclamation from hydrocarbon pollution.

We obtained promising results in the plants containing the *rhlA* or the *rhlB* gene (Figure 4), suggesting that not the combination of the genes may be required. The degradation coefficient (C_d) for *rhlA*, rhlB and for *rhlA+rhlB*-containing plants was about 25-34% higher that for control plants (Fig. 4B). Such transgenic plants might be able to help cleaning up sites contaminated with different amounts of oil hydrocarbons.

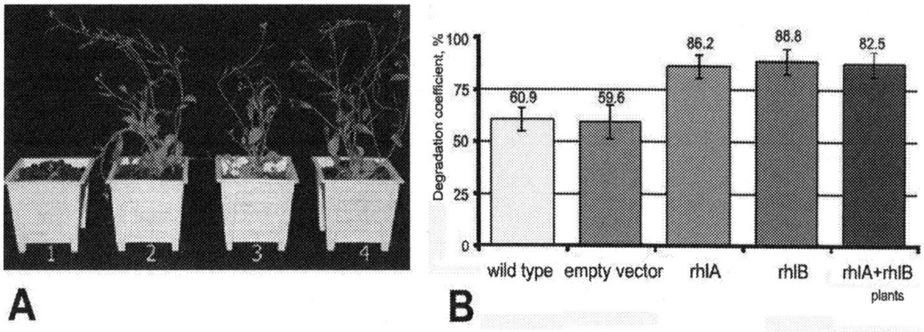

Figure 4. Analysis of *Arabidopsis* and tobacco plants for their ability to degrade crude oil. A. *Arabidopsis* plants in soil contaminated with 2.4 % crude oil.1, empty vector control; 2, transgenic plant with *rhlA* gene; 3, transgenic plant with *rhlB* gene; 4. transgenic plant with *rhlA+rhlB* genes. B. Oil hydrocarbon accumulation in *Arabidopsis* plants grown in soil with 2.4% crude oil.

4. Molecular Mechanism of Oil Degradation

Plants are known for a variety of means that can enhance the remediation of soils that contain organic pollutants [2]. Several studies [22,23] showed that the degree of phytoremediation appeared to be related to the plant species and most likely reflected differences in root architecture, plant metabolism, quantity and quality of root exudates, root radius, and soil type. Phytoremediation may largely result from the enhanced degradation of organic compounds in the rhizosphere caused by enhanced microbial activity [24] and by changes in the ratios of micro-organisms in the root zone as compared to the surrounding soil [2]. Experiments with ryegrass (*Lolium perenne* L.) showed that plant's growth in the contaminated soil led to the increase in microbial quantity and activity which in turn stimulated the biodegradation of hydrocarbons in the rhizosphere [24]. After 22 weeks of experiments the ryegrass increased the degradation level by 11 %. The results indicate that plant roots stimulate the degradation. *Pinus sylvestris* L. roots and mycorhizosphere development positively influence hydrocarbon utilization and results in a concomitant reduction in mineral oil levels in petroleum hydrocarbon contaminated soil [25]. In addition, plants themselves may mineralise organic pollutants [21].

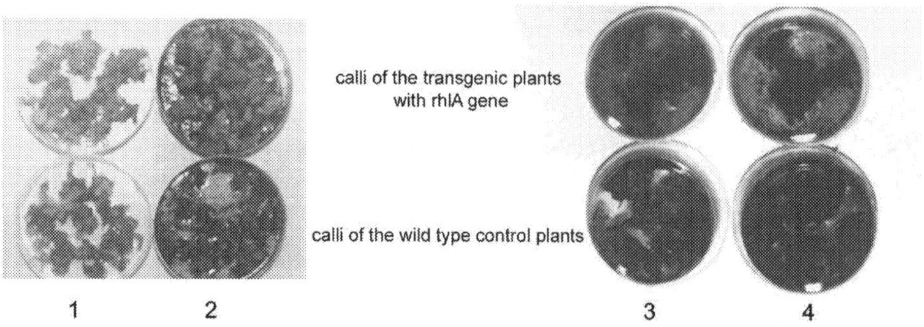

Figure 5. Tobacco callus development in the presence of crude oil
I, control MS medium. 2, MS morpho medium contaminated with 4% crude oil. 3, MS morpho medium contaminated with 3% of crude oil. 4, MS-morpho medium contaminated with 4% crude oil.

Our results indicated a slight rate of hydrocarbon biodegradation in the soil under both types of control plants. However, the rate of crude oil degradation in soil under all types of transgenic plants with rhamnolipid genes was much higher (up to 34 %). The artificial nature of soil mixtures (peat, sand, garden soil) and uniformity of soils types applied for both control and transgenic plants reject the hypothesis that links the degradation process with the presence of hydrocarbon-degrading bacteria in soil. The experiments performed with plant tissue culture growth on MS-medium supplemented with crude oil confirm that transgenic plants can tolerate oil hydrocarbons by themselves, without any help from bacteria (Figure 5).

We consider that the rhamnolipids excreted by our plants enhance the ability of plants to increase oil hydrocarbon degradation by rhizosphere and presumably by plants themselves. It is known that the more soluble organic compounds are the greater is the probability that plants will reduce the concentration of these substances [26]. The produced rhamnolipids are quite beneficial to bioremediation because they dissolve and immobilise hydrocarbons [27-29].

Zhang and Miller [19] found that not only did rhamnolipids increase apparent hydrocarbon solubility but also modify the cell surface, thus leading to increased hydrophobicity. Biosurfactants enhanced the biodegradation of slightly soluble organic compounds with the help of two mechanisms: firstly, biosurfactants could dissolve hydrophobic compounds within micelle structures, effectively increasing the apparent aqueous solubility of the organic compounds and its bioavailability for uptake by a cell [18,30]; secondly, biosurfactants could cause the cell surface to become more hydrophobic, thus increasing the direct physical contact between the cell and slightly soluble substrate [31]. Both mechanisms result in the transfer of the contaminants from the soil to the enzymes which catalyse the first conversion step [32]. Once the contaminant enters the cell, it is converted.

5. Transgenic Plants Facing Metal Contamination

An increase in heavy metal concentration in soil is proven to have visible effects on plants: it shortens shoots, yellows leaves and, eventually, kills the plant [33]. The toxic effects depend on the type of heavy metal, its bioavailability as well as on the plant. After preliminary tests with heavy metals we decided on experiments with copper. The basic plant culture for our tests was *Arabidopsis thaliana*. Experiments were performed with a flame atomic absorption spectrophotometer. Atomic absorption analysis can detect as much as 1020 mg copper in 1 kg of wet soil. Copper accumulation in plants was evaluated with the metal accumulation coefficient (C_{MA}) which is calculated according to $C_{MA} = [Cs]/ [C_r]$, where C_s is the copper concentration in shoot, C_r is the relevant copper concentration in the rhizosphere. The following experimental groups were investigated:

(a) uncontaminated control soil;
(b) contaminated soil;
(c) transgenic plants with *rhlA* gene;
(d) transgenic plants with *rhlB* gene;
(e) transgenic plants with *rhlA* and *rhlB* genes;
(f) control transgenic plants with empty vector;
(g) wild-type control plants.

Plants were tested both in uncontaminated and contaminated soils, readings taken from the measurements of both their root zones and shoots. We applied a concentration of 1067 mg Cu^{2+} per 1 kg of wet soil, which is considered toxic [33]. Our aim was to study the levels of

Cu concentration by measuring its content in rhizosphere and in shoots respectively. We took measurements at the beginning and in the end of the 45-day period of the plant growth. Results demonstrated that injection of 1067 mg Cu^{2+} into 1 kg of wet soil killed both non-GM, wild-type *Arabidopsis* plants and *Arabidopsis* plants transformed with the with the empty vector.

For transgenic *Arabidopsis* with the three types of constructs (*rhlA*, *rhlB*, *rhlA+rhlB*) the results were different. Transgenic *Arabidopsis* plants with *rhlA* gene and transgenic *Arabidopsis* plants with both genes could grow on the contaminated soil without exhibiting any toxic symptoms. Not only did they give seeds but also demonstrated great biomass production in the soil contaminated with copper. As for plants with *rhlB* gene, their growth in contaminated soil was suppressed. However, we observed no chlorosis. Atomic absorption analysis helped determine the levels of copper accumulation by measuring Cu content in all plants grown both in control soil and in contaminated soil (Figure 6A).

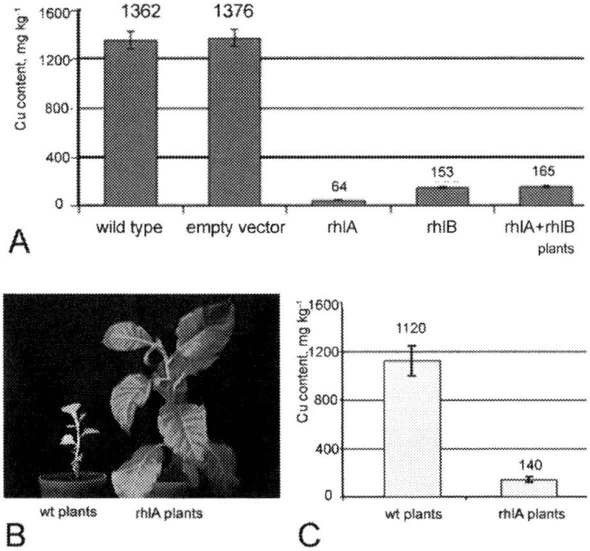

Figure 6. Analysis of Arabidopsis and tobacco for copper tolerance.
A. Copper accumulation in Arabidopsis plants grown in contaminated soil (1020 mg of Cu per 1 kg of wet soil). For comparison, plants grown in uncontaminated soil demonstrated the following levels of Cu accumulation: control plants 9.47 mg kg^{-1}; plants with empty vector 8.63 mg kg^{-1}; plants with *rhlA* 18.07 mg kg^{-1}; plants with *rhlB* 8.44 mg kg^{-1}; plants with *rhlA* and *rhlB* 8.73 mg kg^{-1}.
B. Tobacco plants grown in soil with the 912 mg kg^{-1} wet soil Cu amount. On the left, a non-GM control plant, on the right a transgenic plant with rhlA gene.
C. Copper accumulation demonstrated by tobacco plants grown in contaminated soil. For comparison, plants grown in uncontaminated soil (3 mg kg^{-1} wet soil) showed copper accumulation: control plants 42 mg kg^{-1}; plants with *rhlA* 39 mg kg^{-1}.

In uncontaminated control soil, the Cu accumulation coefficient (C_{MA}) was 3.5 in both transgenic and non-GM plants. In contaminated soil, C_{MA} was 1.3 for the control plants with empty vector and non-GM control plants. The plants with *rhlB* gene had C_{MA} equal to 0.15. Plants containing constructs with both *rhlA* and *rhlB* genes demonstrated C_{MA} at 0.16. In plants with *rhlA* gene, C_{MA} proved the best, less than 0.05. Evidently, in transgenic plants with the three types of constructs, the low figure of C_{MA} implies very small copper content in shoots in comparison with Cu content in the rhizosphere. Backup tests were made on *Nicotiana tabacum*. We verified control tobacco and *rhlA*-containing tobacco plants using the copper content, which was lower than in our tests on *Arabidopsis*. We applied the copper concentration of 912 mg kg^{-1} wet soil. Control tobacco plants survived in this copper concentration, but demonstrated slow growth, inability to form lateral roots, chlorosis of leaves and infertility. Reversibly, transgenic tobacco plants with *rhlA* gene grew without any toxic symptoms and gave normal seeds (Fig. 6B). Atomic absorption analysis showed that these GM tobacco plants with *rhlA* gene were able to exclude copper from their aboveground tissues (C_{MA} = 0.07) in contrast to the ordinary non-GM tobacco plants (C_{MA} =1.11) (Fig. 6C).

It was supposed that the plant tolerance to copper connected with the secretion of rhamnolipids. However, it was shown that mono-rhamnolipid can form complexes not only with copper, but also with a wide range of metals, including Al^{3+}, Pb^{2+}, Cd^{2+}, Zn^{2+}, Hg^{2+}, Co^{2+}, Ni^{2+} [34, 35]. So, it was decided to test transgenic plants on further seven metals (six heavy metals and one light metal). We applied the concentration levels equivalent to, or even exceeding, those present in industrially contaminated sites. All the transgenic plants demonstrated high tolerance to 6.6 g kg^{-1} Al^{3+}, 1.7 g kg^{-1} Bi^{3+}, 1.2 g kg^{-1} Zn^{2+}, 0.8 g kg^{-1} Pb^{2+}, 0.8 g kg^{-1} Cs^{+}, 0.5 g kg^{-1} Cr^{3+} and 0.5 g kg^{-1} Ni^{2+}. Generally, transgenic plants with the three types of constructs could survive in soil with a 300-fold higher Cu concentration than control.

The most interesting and promising results were seen in plants with *rhlA* gene. These metal-tolerant *Arabidopsis* and tobacco plants can apparently grow in very contaminated soil by avoiding any metal transport from roots to shoots. Thus, they behave as true excluders. We may therefore suppose that introduction of the *rhlA* gene in the ordinary plants allow them to exclude toxic heavy metals from their tissues and grow in extremely polluted soils. In future, plants with this gene may be used for phytostabilization on sites contaminated with heavy metals.

6. Mechanism of Resistance to Heavy Metals

Our study demonstrates that rhamnolipid-containing plants, especially plants with *rhlA* gene and plants with both genes, do not accumulate any considerable amount of heavy metals in their tissues. It is known that metal-tolerant plants employ a number of effective mechanisms for tolerating high concentrations of metals [36]. Thus, excluders possess specialised mechanisms, which restrict or prevent toxic metal uptake from getting into root cells. From these mechanisms we single out the one that, we suppose, is responsible for the metal-tolerant nature of our plants. This mechanism involves extracellular binding by root exudates.

Root cells may synthesize and release organic compounds that chelate metals and form complexed compounds with these metals. The transport of these complexes into the cell is hampered due to their structure and complexity [37]. This mechanism is vastly demonstrated in the following cases. For example, certain plant species tolerate Al in the rhizosphere with the help of a mechanism, which involves exudation of citric and malic acids [38, 39]. It is also observed that Ni-chelating histidine and citrate, present in root exudates of excluders, play an important role in Ni-tolerance strategy (Hall, 2002). Since the range of compounds exuded could be rather wide, other exudates could also contribute to metal tolerance [40].

Rhamnolipids produced by various *Pseudomonas aeruginosa* strains, are capable of forming complexes with such metals as cadmium (Cd^{2+}), lead (Pb^{2+}) and zinc (Zn^{2+}) and some others [41]. This is caused by the fact that anionic biosurfactants, that include rhamnolipids, carry negatively charged sites (COO^{-}). So, when a rhamnolipid molecule encounters a cationic metal that carries a positive charge, an ionic bond is formed [34].

Therefore, the metal bound into a complex could not penetrate into the cell. Moreover, many studies have demonstrated that rhamnolipids are also able to change the charge of cell walls, which leads to the overall reduction of the negative charge on the cell surface. This reduction entails a drop in the positively charged metals' ability to interact with the cell surface.

Our study has shown that transgenic plants with *rhlA*, *rhlB*, *rhlA+rhlB* genes are able to excrete rhamnolipids from root cells. We hypothesised that our plants apply the same mechanism that was described above. Plant cell membranes are known to be mainly composed of phospholipids and proteins. Since a lipid bilayer confers a considerable negative charge upon the cell surface which favours electrostatic interaction with the cations [41], the removal of molecules containing COO^- groups ($R\text{-}COO^-$) from the cell surface may reduce the negative charge of the cell surface [42]. This alteration of the cell surface could slacken the interaction with metal cations. It is also possible that the released $R\text{-}COO^-$ molecules could bind cationic metals via charged functional groups [41].

To find out if the presence of rhamnolipids alone in media was sufficient to provide plant resistance to heavy metals, the tobacco plants were separated in vessels in the following way: 1st flask with control plants; 2nd flask with both control and transgenic plants; 3rd flask with transgenic plants. Then plants were grown in the liquid media. Since transgenic plants are able to excrete a large amount of rhamnolipids, we presumed that they would produce sufficient amount of rhamnolipids within one week. A week later we added 1800 mgL^{-1} Cu^{2+} to the liquid medium. All control plants wilted after 18 hours and died a few hours later. In contrast, transgenic plants in all flasks did not show any symptoms of wilting. We also grew control and rhlA-containing transgenic plants in the soil contaminated with 1700 mg kg^{-1} Cu^{2+} using the same scheme as for the previous experiment. In two months, we found that all control plants were suppressed in heavy metal presence in contrast with transgenic plants (Figure 7A).

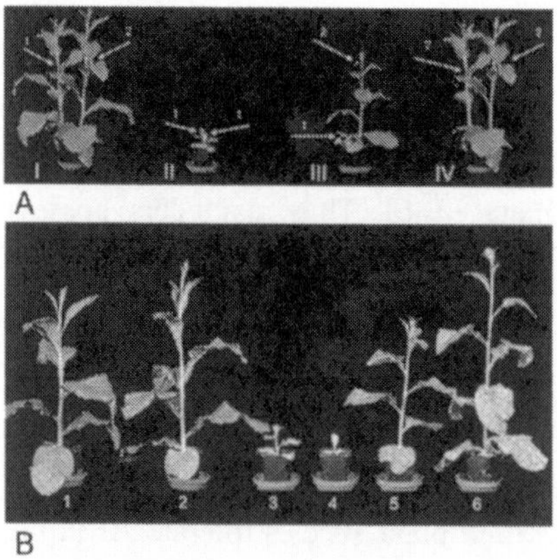

Figure 7. The influence of rhamnolipids on the tolerance to heavy metals. A. Two plants in one pot on soil contaminated with 1600 mg kg^{-1} of copper. I, wild type plants on control soil. II, wild type plants on contaminated soil. III, wild type and transgenic plants with *rhlA* gene on contaminated soil. IV, transgenic plants with *rhlA* gene on contaminated soil. B. The addition of rhamnolipid biosurfactants JBR425 (Jeneil Biosurfactant Co., LLC) in soil contaminated with 1600 mg kg^{-1} of copper. 1, wild type plant on control soil; 2, transgenic plant with *rhlA* gene on control soil; 3, wild type plant on contaminated soil; 4, wild type plant on soil supplemented with rhamnolipid mixture; 5, transgenic plant with rhlA gene on contaminated soil; 6, transgenic plant with *rhlA* gene on contaminated soil supplemented with rhamnolipid mixture (1M Cu per 20M rhamnolipid mixture).

To confirm the result we added rhamnolipid biosurfactants JBR425 (Jeneil Biosurfactant Co., LLC) to the soil contaminated with 1700mg kg^{-1} of copper. We added 20 M rhamnolipid biosurfactants per 1 M Cu. All control tobacco plants on the contaminated soil (with and without rhamnolipids) were suppressed. On the contrary, the growth of the transgenic tobacco plants was considerably facilitated by the addition of rhamnolipids to the contaminated soil (Figure 7B). Consequently, we conclude that exogenous rhamnolipids by themselves cannot provide tolerance to heavy metals. It is also required to alter the cell wall membrane. Thus, we suggest that the interaction of rhamnolipids with metals as well as the alteration of the cell membrane contribute to the prevention of hazardous substance ingress into the root cells. The obtained data are similar to those obtained previously [27].

In conclusion, the introduction of *rhlA* and/or *rhlB* genes from *Pseudomonas aeroginosa* into *Arabidopsis thaliana* and *Nicotiana tabacum* enhanced tolerance of transgenic plants to high concentrations of crude oil and heavy metals in soil. The transgenic plants were not only tolerant to the pollutants, but could effectively promote biodegradation of hydrocarbons in contaminated soil, and could be used as a universal technique for reclamation of soils.

Acknowledgements. The authors are especially grateful to Dr. Peter Kurmann, Dr. Sergey Zui and Dr. Lidia Krasovskaya (Ecomir Research Centre, BLR); Dr. Lionell Hills (John Innes Centre, UK). The *Pseudomonas* genes used were kindly provided by Dr. U. Ochsner (USA).

References

[1] Cunningham, S.D., Anderson, T.A., Schwab, A.P. and Hsu, F.S. (1996). Phytoremediation of soils contaminated with organic pollutants. Adv. Agron. 56, 55-114.

[2] Cunningham, S.D., Berti, W.R. and Huang, J.W. (1995). Phytoremediation of contaminated soils. Trends Biotechnol. 13, 393-397.

[3] Malawska, M. and Wilkomirski, B. (2001). An analysis of soil and plants (*Taraxacum officinale*) contamination with heavy metals and polycyclic aromatic hydrocarbons (PAHs) in the area of the railway junction Ilawa Glowna, Poland. Water Air Soil Poll. 127, 339-349.

[4] Bizily, S.P., Rugh, C.L. and Meagher, R.B. (2000). Phytodetoxification of hazardous organomercurials by genetically engineered plants. Nat. Biotechnol. 18, 213-217.

[5] Guerinot, M.L. and Salt, D.E. (2001). Fortified foods and phytoremediation. two sides of the same coin. Plant Physiol. 125, 164-167.

[6] Semple, K.T., Reid, B.J. and Fermor, T.R. (2001). Impact of composting strategies on the treatment of soils contaminated with organic pollutants. Environ. Poll. 112, 269-283.

[7] Douglas, G.S., McCarthy, K.J., Dahlen, D.T., Seavey, J.A., Steinhauer, W.G., Prince, R.C. et al. (1992). The use of hydrocarbon analysis for environmental assessment and remediation. J. Soil Cont. 1, 197-216.

[8] Mejare, M. and Bullow, L. (2001). Metal-binding proteins and peptides in bioremediation and phytoremediation of heavy metals. Trends Biotechnol. 19, 67-72.

[9] Dahmani-Muller, H., van Oort, F., Gelie, B. and Balabane, M. (2000). Strategies of heavy metal uptake by tree plant spesies growing near a metal smelter. Environ. Poll. 109, 231-238.

[10] Swannel, R.J. and Head, I.M. (1994). Bioremediation comes of ages. Nature 368, 396-397.

[11] Kärenlampi, S., Schat, H., Vangronsveld, J., Verkleij, J.A.C., van der Leie, D., Mergeay, M. et al. (2000). Genetic engineering in the improvement of plants for phytoremediation of metal polluted soils. Environ. Poll. 107, 225-231.

[12] Chayabutra, C. and Ju, L.-K. (2000). Degradation of n-hexadecane and its metabolites by *Pseudomonas aeruginosa* under microaerobic denitrifying conditions. Appl. Environ. Microbiol. 66, 493-498.

[13] Ochsner, U.A., Fiechter, A. and Reiser, J. (1994). Isolation, characterization, and expression in *Escherichia coli* of the *Pseudomonas aeruginosa rhlAB* genes encoding a rhamnosyltransferase involved in rhamnolipid biosurfactant synthesis. J. Biol. Chem. 269, 19787-19795.

[14] Desai, J.D. and Banat, I.M. (1997). Microbial production of surfactants and their commercial potential. Microbiol. Mol. Biol. Rev. 61, 47-64.

[15] Jarvis, F.G. and Johnson, M.J. (1949). A glycolipid produced by Pseudomonas aeruginosa. J. Am. Chem. Soc. 71, 4124-4126.

[16] Ochsner, U.A., Koch, A.K., Fiechter, A. and Reiser, J. (1994). Isolation and characterization of a regulatory gene affecting rhamnolipid biosurfactant synthesis in *Pseudomonas aeruginosa*. J. Bact. 176, 2044-2054.

[17] Torrens, J.L., Hermann, D.C. and Miller-Maier, R.M. (1998). Biosurfactant (rhamnolipid) sorption and the impact on the rhamnolipid-facilitated removal of cadmium from various soils under saturated flow conditions. Environ. Sci. Technol. 32, 776-781.

[18] Mulligan, C.N., Yong, R.N. and Gibbs, B.F. (2001). Surfactant-enchanced remediation of contaminated soil: a review. Engin. Geol. 60, 371-380.

[19] Zhang, Y. and Miller, R.M. (1992). Enchancement of octadecane dispersion and biodegradation by a *Pseudomonas* rhamnolipid surfactant (biosurfactant). Appl. Environ. Microbiol. 58, 3276-3282.

[20] Poirier, Y., Ventre, G. and Caldelari, D. (1999). Increased flow of fatty acids toward ß-oxidation in developing seeds of Arabidopsis deficient in diacylglycerol acyltransferase activity or synthesizing medium-chain-length fatty acids. Plant Physiol. 121, 1359–1366.
[21] Pezeshki, S.R., Hester, M.W., Lin, Q. and Nyman, J.A. (2000). The effect of oil spill and clean-up on dominant US Gulf coast marsh macrophytes: a review. Environ. Poll. 108, 129-139.
[22] Chang, Y.Y. and Corapcioglu, M.Y. (1998). Plant-enhanced subsurface bioremediation of non-volatile hydrocarbons. J. Environ. Engin. 124, 162-168.
[23] Shann, J.R. and Boyle, J.J. (1994). Influence of plant species on *in situ* rhizosphere degradation. In: Bioremediation through Rhizosphere Technology, Vol. 563 (Anderson, T.A. and Coats, J.R. Eds), Washington, DC: American Chemical Society, pp. 69-81.
[24] Gunther, T., Dornberger, U. and Fritsche, W. (1996). Effects of ryegrass on biodegradation of hydrocarbons in soil. Chemosphere 33, 203-215.
[25] Liste, H.H. and Alexander, M. (2000). Plant-promoted pyrene degradation in soil. Chemosphere 40, 7-10.
[26] Heinonsalo. J., Jorgensen, K.S., Haahtela, K. and Sen, R. (2000). Effects of *Pinus sylvestris* root growth and mycorrhizosphere development on bacterial carbon source utilization and hydrocarbon oxidation in forest and petroleum-contaminated soils. Can. J. Microbiol. 46, 451-464.
[27] Nyer, E.K. and Gatliff, E.G. (1996). Phytoremediation. Ground Water Monitor. Remed. 16, 58-62.
[28] Noordman, W.H. and Janssen, D.B. (2002). Rhamnolipid stimulates uptake of hydrophobic compounds by Pseudomonas aeruginosa. Appl. Environ. Microbiol. 68, 4502-4508.
[29] Noordman, W.H., Ji, W., Brusseau, M.L. and Janssen, D.B. (1998). Effects of rhamnolipid biosurfactants on removal of phenanthrene from soil. Environ. Sci. Technol. 32, 1806-1812.
[30] Al-Tahhan, R.A., Sandrin, T.R., Bodour, A.A. and Maier, R.M. (2000). Rhamnolipid-induced removal of lipipolysaccharide from *Pseudomonas aeruginosa*: Effect of cell surface properties and interaction with hydrophobic substrates. Appl. Envir. Microbiol. 66, 3262-3268.
[31] Schere, G.S., Inguva, S. and Gunnam, S. (1995). Rhamnolipid biosurfactant enhancement of hexadecane biodegradation by *Pseudomonas aeruginosa*. Mol. Marine Biol. Biotechnol. 4, 331-337.
[32] Zhang, Y. and Miller, R.M. (1994). Effect of a Pseudomonas rhamnolipid biosurfactant on cell hydrophobicity and biodegradation of octadecane. Appl. Envir. Microbiol. 60, 2101-2106.
[33] Kopponen, P., Utriainen, M., Lukkari, K., Suntionen, S., Kärenlampi, M. and Kärenlampi, S. (2001). Clonal differences in copper and zinc tolerance of birch in metal-supplemented soils. Environ. Poll. 112, 89-97.
[34] Tan, H., Champion, J.T., Artiola, J.F., Brusseau, M.L. and Maier, R.M. (1994). Complexation of cadmium by a rhamnolipid biosurfactant. Environ. Sci. Technol. 28, 2402-2406.
[35] Ochoa-Loza, F.J., Artiola, J.F. and Maier, R.M. (2001). Stability constants for the complexation of various metals with a rhamnolipid biosurfactant. J. Environ. Qual. 30, 479-485.
[36] Raskin, I., Nanda Kumar, P.B.A., Dushenkov, S. and Salt, D.E. (1994). Bioconcentration of heavy metals by plants. Curr. Opin. Biotechnol. 5, 285-290.
[37] Ybarra, G.R. and Webb, R. (1998). Differential responses of *Gro*el and metallothionein genes to divalent metal cations and the oxyanions of arsenic in the *Cyanobacterium synechococcus* sp. strain PCC7942. Proc. Conf. Hazard. Waste Res. 1, 76-86.
[38] Pellet, M.D., Grunes, D.L. and Kochian, L.V. (1995). Organic acid exudation as an aluminum tolerance mechanism in maize (*Zea mays* L.). Planta 196, 788-795.
[39] Larsen, P.B., Degenhardt, J., Tai, C.Y., Stenzler, L.M., Howell, K. and Kochian, L.V. (1998). Aluminum-resistance Arabidopsis mutants that exhibit altered patterns of aluminum accumulation and organic acid release from roots. Plant Physiol. 117, 19-27.
[40] Hall, J.L. (2002). Cellular mechanisms for heavy metal detoxification and tolerance. J. Exp. Bot. 53, 1-11.
[41] Sandrin, T.R., Chech, A.M. and Maier, R.M. (2000). A rhamnolipid biosurfactant reduces cadmium toxicity during naphthalene degradation. Appl. Environ. Microbiol. 66, 4585-4588.
[42] Lasat, M.M., Pence, N.S., Garvin, D.F., Ebbs, S.D. and Kochian, L.V. (2000). Molecular physiology of zinc transport in the Zn hyperaccumulator *Thlaspi caerulescens*. J. Exp. Bot. 51, 71-79.

Genomics for Biosafety in Plant Biotechnology
J.P.H. Nap, A. Atanassov and W.J. Stiekema (Eds.)
IOS Press, 2004

Molecular Responses to Drought in Rice and Maize: Towards Genetic Engineering for Stress Tolerance

Ana Paula FARINHA[1,2], Victoria LUMBRERAS[1] and Montserrat PAGÈS[1*]
[1]*IBMB-CSIC, Jordi Girona 18-26, 08034 Barcelona, Spain;*
[2]*ITQB/IBET,Oeiras, Portugal;*
**email: mptgmm@cid.csic.es*

Abstract. Rice and maize are the most important cereals for human consumption. Given the anticipated future increase of the human population, their production must increase to cover the increasing demand for these crops. A major factor that limits the world's cereal production is abiotic stress caused by drought, high salinity or low temperatures. Plants respond to these environmental changes through a number of homeostatic mechanisms that maintain the water balance and the integrity of tissues. This includes several regulatory mechanisms that activate the expression of tolerance effector genes. Therefore, the study of the molecular mechanisms by which plants tolerate osmotic stress is fundamental to improve crop production. Here, we review recent studies on molecular responses to drought and how this information suggests opportunities in genetic engineering for stress tolerance in crops.

Introduction

Rice (*Oryza sativa*) and maize (*Zea mays*) are two major cereal crops for food consumption worldwide. Total production (2000) was about 600 million tons each. These crops, together with wheat, will in the future account for nearly 80% of the cereal demand in developing countries [1,2]. Rice constitutes the staple food for more than half of the world's population. In Asia alone, more than two billion people obtain 60% to 70% of their daily calorie intake from rice [3]. It is the most rapidly growing food resource in Africa and its consumption is also increasing in Europe and North America [4]. In developing countries, maize is used for human consumption, while in the developed world it is mainly for industrial use and animal feed. The high yields in North and Central America outproduce countries such as Mexico, where maize is the staple cereal grain [5]. In the Near East and North Africa, wheat is the dominant staple and its demand for food supply is increasing in a wide number of developing countries [2].

Besides their importance for food security, rice and maize are model plants due to the advantages of rice genomics and maize genetics. Rice has the smaller genome (430 Mbp) compared with other monocots (3000 Mbp in maize and 16000 Mbp in wheat). The overall colinearity with other cereals make it an excellent model for cereal genomics [6]. Maize is the best-studied and most tractable genetic system among cereals, making it the most important system for cereal genetics [7]. Moreover, both genomes are being sequenced. Due to these advantages, in this review we will only focus on developments in rice and maize systems, despite the obvious importance of wheat.

1. Rice and Maize Sequencing Projects

December 2002, the public consortium "International Rice Genome Sequencing Project" (IRGSP) announced the completion of a high quality draft (at least phase-2 sequences) of the 12 chromosomes of the *japonica* cultivar "Nipponbare". The sequencing IRGSP progress presents chromosomes 7 and 10 as being completed, as well as the centromeres on chromosomes 4 and 8, besides chromosomes 1 [8] and 4 [9]. Monsanto [10] and Syngenta [11] have also focused on sequencing the *japonica* subspecies genome, sharing their sequencing data with the public project, whereas a Chinese genome research project has developed a draft sequence of the genome of the subspecies *indica,* cultivar *93-1.1,* the main subspecies grown in China and Southeast Asia [12]. A collection of over 28,000 full-length cDNA clones has been mapped to the genome of these two subspecies [13], in addition to expressed sequence tag (EST) mapping, containing 6591 entries [14].

September 2002, the National Science Foundation of America announced the launch of the 'Maize Genome Sequencing Project', representing a new challenge for sequencing given its size (about six times larger than that of rice) and complexity [7]. The maize genes comprise only about 20% of the whole genome and they are scattered throughout highly conserved and repetitive sequences [15]. Two deep-coverage bacterial artificial chromosome (BAC) libraries have already been generated [16,17] and an integrated genetic/ physical map is also being produced. Considerable maize sequence data sets include ESTs, genome survey sequences from transposon-tagged sites, random clone insert sequences, among others [7].

Although the existence of so many genome-sequencing projects seems redundant, they are necessary to understand the differences in the mechanisms of species and subspecies-specific phenotypes and adaptations. Despite the conservation in gene order and content among the cereals' genome observed by both genetic and physical mapping [18], there is local rearrangements, such as tandem gene duplication, small inversions and translocations of one or a few genes between chromosomes [7]. Even the comparison between the genomes of rice subspecies *indica* and *japonica,* demonstrates the existence of significant numbers of rearrangements and polymorphisms [19]. All these sequencing data opens the way for the next steps in functional genomics and at the same time it will allow to engineer new crops more productive and hardier.

2. Importance of Genetic Engineering for Stress Tolerance Improvement

Breeding approaches for stress improvement were successful in the development of cultivars or hybrids more tolerant in the field. Rice varieties with tolerance to abiotic stresses, namely problem soils, adverse temperatures, and moisture stress, as well as resistance to diseases and insects, constitute 80 to 90% of the planted rice area in Asia. Wide-scale adoption of improved varieties contributed to increases in rice production and yield potential and stability [1]. However, despite the cultivation of improved varieties, rice production is facing serious constraints, including declining yield growth rates.

To face this problem and promote an increase in the production of rice, the United Nations launched last October the International Year of Rice 2004 under the motto "Rice is Life" (http://fao. org/rice2004/). The maize production in the United States has also shown a net increase in grain yield during the last century related with genetic improvements such as hybrid development and exploitation of heterosis. The selection strategies for improved yield in parental lines and improved yield stability in their hybrid progenies were ultimately related to an improved tolerance to abiotic stresses, especially to water-deficit [20].

Drought, like many other environmental stresses, has adverse effects on crop yield and water stress is one of the major causes for crop yield reductions. On the other hand, as water resources for agronomic uses become more limiting, the development of drought-tolerant lines becomes increasingly more important. Based on population projections (8,300 million people are expected in 2030) and improved consumption patterns in developing countries, world food grain production must increase by almost 50% in the next 30 years. The challenge for food security in the next decades, is thus to achieve an even greater cereal's production with less land, less labour, fewer chemicals and less water [1].
Genetic engineering or modification (GM) tools will contribute to meet this challenge, by introducing foreign genes conferring stress tolerance, in a much faster and efficient manner compared to that of conventional genetic methods. On the other hand, transgenic approaches are powerful tools to gain valuable information towards a better understanding of the mechanisms beyond the stress tolerance, testing how single or sets of genes confer stress tolerance in GM plants.

The ideal GM crop should possess a highly regulated stress-response capability that doesn't affect crop performance under normal conditions. Until date most of the genetically modified stress-tolerant plants still are non-agronomic plants and only few of them have been tested in field conditions [21]. The introduction of foreign genes into crop plants raises questions about the possible post-transcriptional and post-translational modifications that might occur and its impact on human health and on the environment.

In this review, we will discuss how transgenic approaches serve as powerful means to elucidate stress-response mechanisms. We summarise recent cases where stress-responsive genes proved to confer some kind of tolerance, with special emphasis on rice and maize in the framework of considerations about the biosafety of transgenic plants.

3. Plant Molecular Responses to Drought

The responses to abiotic stress are complex, involving many genes and biochemical-molecular mechanisms [22-29]. These stress-related genes can be grouped into two major categories: those that are involved in signalling transduction cascades and in the control of gene expression, and those that encode proteins that directly protect cellular structures against stresses [22,23]. The first group includes protein kinases e.g., calcium-dependent (CDPK), mitogen-activated (MAPK), SNF1-related protein kinases type 2 and 3 (SnRK2- and SnRK3); putative osmosensors, such as the transmembrane histidine kinase *AtHK1;* enzymes involved in phosphoinositide metabolism; and stress-inducible transcription factors belonging to a range of protein families, e.g. DREBs, ERFs, MYBs, bZIPs, NAC and others [22-26,28]. The second group of gene products includes heat shock proteins, chaperones, late embryogenesis abundant (LEA) proteins, osmoprotectants (proline, glycine betaine, trehalose) and antioxidants (e.g. catalase, ascorbate, carotenoids) [24,25,29].

3.1 Signal Perception and Transduction in Drought Responses

The perception is the first step in stress molecular responses, triggering a cellular signal transduction pathway that will translate a physical stress into a biochemical response. For example, there are several aspects of cellular water loss that can be sensed by the plant, such as the decrease on turgor, change in cell volume or membrane area and alterations in cell wall - plasma membrane connections [29]. In the case of osmotic stress, two osmosensors have been identified in yeast: one is a transmembrane protein that is activated under conditions of high osmolarity, and another is a complex of three proteins that is inhibited in the same conditions [30]. In plants, the putative osmosensor *AtHK1* [31], a transmembrane histidine kinase, is thought to relay changes in osmotic potential outside the cell to the transduction pathway(s) inside the cell, regulating drought-inducible gene expression [28].

Following stress perception, a signalling mechanism must be activated to induce genes specifically involved in the stress tolerance acquirement. It is becoming clear that the linear pathways identified in this context are only part of a more complex signalling network with multiple functional interconnections [22,23,28]. For instance, many genes that are induced by cold are also induced by drought or ABA [22,32], probably because many cold-inducible genes encode proteins to protect the plant from the consequences of freezing stress, which includes dehydration. In nature, plants suffer multiple stresses either at the same time or separated temporally and they must present an integrated and co-ordinated response to them. This might explain the existence of crosstalk between different signalling pathways in response to stress [28].

Molecular and genetic approaches have greatly contributed to start elucidating about the components of signal transduction pathways induced by drought and other abiotic stresses. The gene *RD29A* from *Arabidopsis* (responsive to dehydration) [33], also known as COR78 (cold-regulated) or LTI78 (low-temperature-induced), is a cold and drought-regulated gene that has been used to examine the convergence of these two pathways. The promoter analysis of this gene has been found to contain two major *cis*-acting elements: the dehydration-responsive element/C-repeat (DRE/CRT) and the ABA-responsive element (ABRE), that function in ABA-independent and ABA-dependent gene expression in response to abiotic stress, respectively [23].

In *Arabidopsis* two groups of transcription factors were found to bind to the DREB/CBF *cis*-acting element and were designed as *CBF/DREB1* and *DREB2* [34,35]. The *CBF/DREB1* genes are quickly and transiently induced by cold, whereas the *DREB2* genes are induced by dehydration or salt (osmotic stress), leading to the expression of target genes involved in stress tolerance acquirement. In rice, it was isolated four genes homologues to *CBF/DREB1* and one homologue to *DREB2* [36]. Gene expression of *OsDREB1A* and *OsDREB1B* was induced by cold, whereas the expression of *OsDREB2A* was induced by dehydration and high salinity, in a similar way to what was described for *Arabidopsis*. The overexpression of *OsDREB1A* in transgenic *Arabidopsis* resulted in plants with a higher tolerance to drought, freezing and high salinity, indicating that is similar functional to *DREB1A* [37,38]. This is, therefore, a potentially useful gene for genetic engineering to generate transgenic cereal crop plants hat show tolerance to multiple abiotic stresses.

Detailed analyses of ABRE and DRE/CBF elements in the expression of the *RD29A* gene revealed that DRE/CBF acts co-operatively with ABRE as a coupling element in ABA-responsive gene expression, in response to drought stress [39]. This is indicative that crosstalk can occur at the level of different *cis*-DNA regulatory elements that mediate stress responses [22]. Other example comes from the *CBF/DREB1* genes that are mainly induced by cold, despite *CBF4/DREB1D* gene induction by osmotic stress [40], suggesting that *CBF4* can mediate the crosstalk between DRE2 and CBF/DREB1 regulatory systems. In maize, the DBF1 transcription factor binds to DRE2 element activating the *rab17* promoter in an ABA-dependent pathway, pointing to a differential regulation of stress-inducible genes in maize and *Arabidopsis* [41]. Such interactions constitute real advantages to the plant since they provide the crosstalk between different stress signalling pathways and the plasticity of the responses.

3.2 Importance of ABA in Drought Responses

ABA is synthesised *de novo* in response to drought and high salinity, but not in response to cold. More than half of the drought-inducible genes identified by cDNA microarrays are also induced by salt and/or by ABA, while only 10% of the drought inducible genes were also induced by cold [22]. Some genes that codify for enzymes involved in ABA biosynthesis [42] were strongly induced by drought. Furthermore, the overexpression of bZIP transcription factors that bind to the ABRE *cis*-element e.g., *ABF3* and *AREB2/ABF4,* caused ABA hypersensitivity, reduced transpiration rate and enhanced drought tolerance in transgenics [43]. All these data indicates the existence of significant crosstalk among the drought and ABA pathways that seems greater than that of ABA and cold [22].

4. Gene Function Analyses

4.1 Transgenic Approaches

The use of transgenic model plants is one of the most powerful tools to ascribe a function to a particular gene. The insertional mutagenesis either through heterologous maize transposons or *Agrobacterium*-mediated T-DNA insertions, have been quite valuable for the identification and isolation of genes displaying a mutant phenotype. However, many genes don't display a phenotype when mutated, being necessary to engineer the insertion sequences to monitor or change the expression pattern of adjacent genes. Also the employment of various forms of gene silencing technology (*e.g.* RNAi mediated gene silencing) allows recovering of knockout phenotypes for genes with redundant function. These transgenic approaches can be addressed by high throughput reverse genetics methods that help to provide functions to the genes discovered by genome sequencing [44].Rice can provide powerful insertion mutagenesis and reverse genetics tools for functional genomics. Rice can now be transformed in a more efficient and routine way than any other cereal [45]. The efficient rice transformation procedure for generating large numbers of T-DNA plants recently published facilitates routine integration of these genomics technologies [46]. With respect to maize, collections with a knockout in each gene of the genome (not including lethal mutations) derived by insertional mutagenesis will be made available in databases.

4.2 Transcriptomics and Proteomics

Microarray technology using cDNAs or oligonucleotides, among other transcript-profiling techniques, is extremely important to analyse the patterns of gene expression in response to abiotic stress [47,48,49,50,51]. Despite the fact that knowing when and where a gene product (RNA and/or protein) is expressed thus providing important clues about its biological function, it doesn't necessary prove a causal relationship between gene sequences and their function. Other factors than the mRNA alone, such as protein modifications, determine the activity of a gene product at cellular level. Therefore, the true picture of gene expression can be offered by proteomics that addresses the protein expression of a cell type and the post-translational modifications. The recent developments in mass spectrometry and the creation of algorithms for automatic spot quantification, along with a better resolution and reproducibility of two-dimensional polyacrylamide gelelectrophoresis, have simplified protein analysis and their characterisation. On the other hand, the availability of genome sequences has also contributed to simplify and speed up the identification of proteins.

Proteomics is a valuable tool to analyse drought stress response. Studies on the regulation of gene expression during maize embryogenesis through two-dimensional PAGE analysis, revealed the expression of a set of polypeptides that were rapidly induced by ABA in young embryos [52]. These polypeptides accumulated along with maturation, coinciding with the period when the endogenous level of ABA attained a maximum [53]. These polypeptides corresponded to the maize LEA proteins, Rab17 and Rab28 [54,55]. Furthermore, proteomics offers an interesting new approach to analyse the modifications that occur on gene products when overexpressed in transgenic plants.

5. Genes Conferring Drought Tolerance in Transgenic Plants

Most genetic engineering involving abiotic stress-tolerance has been made in non-agronomic plants, like *Arabidopsis* and tobacco. This fact is related to their greater facility in transforming and reduced life cycles compared with crop plants. Following below, we report some examples of rice transgenic plants that were engineered for drought tolerance, based on different mechanisms of stress response. All these experiments make reference to genes encoding a single specific stress-protective protein, although other approaches, such as the use of transcription factors, confer improvement to several abiotic stresses simultaneously [38]. For a detailed review of this subject, see [56].

5.1 Osmoprotectants

Organic compounds of low molecular weight are known as compatible solutes, osmolytes or osmoprotectants. They accumulate in plants in response to water stress and high salinity. The primary function of compatible solutes is to maintain cell turgor and consequently the adjustment of the osmotic potential in the cytoplasm [57]. Recent studies indicate that they can also act as free-radical scavengers or chemical chaperones by directly stabilising membranes and/or proteins (for review, see [24]). Compatible solutes are grouped into three major classes: amino acids (*e.g.* proline), quaternary amines (*e.g.* glycine betaine) and polyol/sugars (*e.g.* mannitol, trehalose).

5.1.1 Proline

The enzyme P5CS (Δ^1–pyrroline-5-carboxylate synthetase) is involved in the biosynthesis of proline. In rice, a mothbean (*Vigna aconitifolia*) gene codifying for this enzyme was overexpressed under a stress and ABA-inducible promoter complex (e.g. AIPC-ABA), leading to the accumulation of proline. The transgenic plants accumulated up 1.6 to 2.5-fold more proline under drought conditions, as compared to control plants. The second generation of transgenic plants showed an increase in biomass under drought and salt stress in respect to the non-transformed plants, reflected as higher levels of fresh shoot and root weight. Thus, the extent of drought and salt stress tolerance in transgenic plants were correlated to the levels of proline accumulated [58].

5.1.2 Glycine Betaine

The compatible solute glycine betaine is biosynthesised by different pathways. The choline oxidase (*codA*) gene from the bacterium soil *Arthrobacter globiformis* is involved in a single-step pathway and was transferred to rice to enhance salt and cold tolerance [57]. In transgenic plants, the levels of glycine betaine were as high as 1-5 µm per gram fresh weight of leaves. Although treatments with 150 mM NaCl inhibited the growth of both wild type and transgenic plants, the last recovered earlier after the elimination of the stress. The authors suggested that this tolerance could be related with the stabilisation of the structure/function of proteins, since the levels of concentration of this solute were rather low for osmotic adjustment.

5.1.3 Trehalose

It is known that trehalose has a high water retention activity, which maintains the fluidity of membranes under dry conditions [59]. Dicotyledonous transgenic plants expressing the trehalose-6-phosphate (T-6-P) synthase (TPS) and/or T-6-P phosphatase (TPP) genes from bacteria, exhibited an increased drought tolerance, but also strong developmental alterations. However, two recent works on rice expressing the TPS and TPP bacterial genes under constitutive [59] and tissue/stress inducible promoters [60], respectively, showed that the overproduction of trehalose resulted in the improvement of abiotic stress tolerance, without stunting growth. Maize transgenic plants overexpressing an *Arabidopsis* gene e.g., *AtTPS* codifying for TPS, under the CaMV35S promoter were obtained and are tested for drought stress improve-ment (A.M. Almeida, personal communication).

5.2 LEA Proteins

Plants activate the expression of specific stress-associated proteins, such as the LEA proteins, in addition to the accumulation of compatible solutes. These proteins accumulate in high concentrations in mature seeds and are induced by ABA in immature embryos and in vegetative tissues, by abiotic stress and/or ABA, [61]. Although their function still remains unknown, they have been proposed several roles in protecting cellular structures from dehydration e.g., hydration buffer, sequestering ions, protection of other proteins and renaturation of unfolded proteins [62]. In rice, the overexpression of a LEA protein from barley (*HVA1*) under a constitutive promoter, conferred drought and salt tolerance that was correlated with the high levels of accumulation of this protein. [63]. Transgenic plants expressing LEA proteins from maize under a constitutive promoter, also presented a higher tolerance to drought and salinity in respect to control plants. These studies support the role of LEA proteins in protecting plants against drought and/or high salinity stress.

6. Biosafety considerations

The Green Revolution in the 1960s allowed the introduction of higher-yielding varieties of rice, maize and wheat, leading to an increase in food production to keep pace with population growth. Nowadays, a “Doubly Green” Revolution is necessary. Future agriculture must be both more productive and more environmentally friendly in terms of the conservation of natural resources and the avoidance of pollution, than was achieved in the first Green Revolution [64]. As discussed above, genetic engineering tools may help to meet this challenge.

In 1999, 40% of all cotton, 35% of soy and 25% of maize grown in the United States, were genetically modified (GM) [65]. In 2001, world production of GM crops occurred mainly in the United States, Brazil, Argentina, Canada and China [66]. In Europe, almost none GM crops were cultivated probably due to the highly restricted release of GM crops and their controversy. However, European Union (EU) generally permits the release and commercialisation of GMOs [67]. Many research studies about biosafety have been conducted in the last 15 years. The most important environmental concerns include the possibility of: (1) the novel traits increasing the weediness or invasiveness of crop plants or related species; (2) the spread of the novel genes and traits through hybridisation with wild, sexually compatible relatives, thus potentially increasing the fitness levels of these wild species and causing the disturbance or loss of rare populations; (3) novel insect resistance traits, and (4) the evolution of pests with increased resistance that could necessitate the use of higher doses of pesticides [68]. Moreover, environmental safety assessment should exist also with respect to the novel traits that are achieved through mutagenesis and selective breeding, which also introduces genetic modifications. The need of a constant assessment of the risks associated with their environmental release is important to control biosafety.

Another type of concern is related with the use of antibiotics and herbicides as selectable markers in transformation procedures. These markers raise questions about the possible antibiotic resistance on humans and herbicide resistance into wild types. However, these kind of problems associated with the presence of marker genes in plants have been known for quite some time and various studies have been carried in order to remove these genes [69]. Therefore, the future transgenic crops will have this problem obviated.

Establishing the biosafety of the drought-resistant GM crops discussed above will be particularly challenging in terms of risks and benefits. On the one hand, such plants will have an extended habitat with possibly undesired ecological consequences. Also, the changes brought about in overall plant physiology will have to be assessed for their consequences for food, feed and ecology. There may be unexpected or unintended changes accompanying GM drought tolerance. Novel developments in ecological genomics and all other 'omics' approaches will prove helpful, if not crucial, for the acceptable assessment of the biosafety of GM drought tolerance in crops. On the other hand, however, drought resistance in crops clearly aims to help solving or alleviating a pressing problem in current-day and future agriculture. At the end of the day, it should therefore be a proper risk-benefit analysis that will allow a balanced and informed decision on the acceptability and desirability of the use of GM drought tolerance in crops in a given area. This way, genetic modification should help to meet the challenge of the "Doubly Green" Revolution. We believe that the future and foreseeable problems of notably the lack of proper water in agriculture should be an important issue in such a risk-benefit analysis.

Acknowledgements. The authors wish to acknowledge FCT-Portugal (Ministry for Education and Science) (grant SFRH/BD/1185/2000) for financial support (APF). This work was also supported by grants BIO2003-01133 and GEN2001-4890-CO7-02 from the Spanish MCYT (MP).

References

[1] Khush, G. (2003). Productivity improvements in rice. Nutr. Rev. 61, S114-116.

[2] Hoisington, D., Khairallah, M., Reeves T., Ribaut J.M., Skovmand, B., Taba, S. et al. (1999). Plant Genetic Resources: What can they contribute toward increased crop productivity? Proc. Natl Acad. Sci. USA 96, 5937-5943.

[3] FAO (2003). The International Year of Rice 2004. Concept Paper. Rome, Italy: FAO (www.fao.org/rice2004/en/f-sheet/concept.pdf).

[4] Nguyen, V.N. and Tran, D.V (2002). Rice in Producing Countries. FAO Rice Information, Vol. 3, Chapter III, Rome, Italy: FAO (www.fao.org/DOCREP/005/Y4347E/y4347e04.htm#bm04).

[5] FAO (1992). Maize in Human Nutrition. Rome, Italy: FAO (www.fao.org/docrep/T0395E/T0395E00.htm).

[6] Goff, S.A. (1999). Rice as a model for cereal genomics. Curr. Opin. Plant Biol. 2, 86-89.

[7] Chandler, V.L. and Brendel, V. (2002). The maize genome sequencing project. Plant Physiol. 130, 1594-1597.

[8] Sasaki, T., Matsumoto, T., Yamamoto, Sakata, K. K., Baba, T., Katayose, Y. et al. (2002). The genome sequence and structure of rice chromosome 1. Nature 420, 312-316.

[9] Feng, Q., Zhang Y, Hao, P., Wang, S., Fu, G., Huang Y. et al. (2002). Sequence and analysis of rice chromosome 4. Nature 420, 316-320.
[10] Barry, G. (2001). The use of the Monsanto draft rice genome sequence in research. Plant Physiol. 125, 1161-1165.
[11] Goff , S.A., Ricke, D., Lan, T.H., Presting, G., Wang, R. Dunn, M. et al. (2002). A draft sequence of the rice genome (*Oryza sativa* L. ssp *japonica*). Science 296, 92-100.
[12] Yu, J., Hu, S., Wang, J, Wong, G.K.S., Li, S. Liu, B. et al. (2002). A draft sequence of the rice genome (*Oryza sativa* L. ssp *indica*). Science 296, 79-91.
[13] Kikuchi, S., Satoh, K., Nagata, T., Kawagashira, N., Doi, K., Kishimoto, N. et al. (2003). Collection, mapping, and annotation of over 28,000 cDNA clones from *japonica* rice. Science 301, 376-378.
[14] Wu, J., Maehara, T., Shimokawa, T., Yamamoto, S., Harada, C., Takazaki, Y. et al. (2002). A comprehensive rice transcript map containing 6591 expressed sequence tag sites. Plant Cell 14, 525-535.
[15] San Miguel, P., Tikhonov A., Jin, Y.K., Motchoulskaia N., Zakharov, D., Melake-Berhan, A. et al. (1996). Nested retrotransposons in the intergenic regions of the maize genome. Science 274, 765-768.
[16] Cone, K., McMullen, M., Bi, I.V., Davis, G. Yim, Y.S., Gardiner, J. et al. (2002). Genetic, physical and informatics resources on maize: on road to an integrated map. Plant Physiol. 130, 1598-1605.
[17] Tomkins, J.P., Davis, G., Main, D., Yim, Y., Duru, N. Musket, T. et al. (2002). Construction and characterization of a deep-coverage bacterial artificial chromosome library for maize. Crop Sci. 42, 928-933.
[18] Gale, M.D. and Devos, K.M. (1998). Comparative genetics in the grasses. Proc. Natl Acad. Sci. USA 95, 1971-1974.
[19] Sasaki, T. and Sederoff, R. (2003). Genome studies and molecular genetics. The rice genome and comparative genomics of higher plants. Curr. Opin. Plant Biol. 6, 97-100.
[20] Bruce, W.B., Edmeades, G.O. and Barker, T.C. (2002). Molecular and physiological approaches to maize improvement for drought tolerance. J. Exp. Bot. 53, 13-25.
[21] Dunwell, J.M. (2000). Transgenic approaches to crop improvement. J. Exp. Bot. 51, 487-496.
[22] Shinosaki, K, Yamaguchi-Shinosaki, K. and Seki, M. (2003). Regulatory network of gene expression in the drought and cold stress responses. Curr. Opin. Plant Biol. 6, 410-417.
[23] Seki, M., Kamei, A., Yamaguchi-Shinosaki, K. and Shinosaki. K. (2003). Molecular responses to drought, salinity and frost: common and different paths for plant protection. Curr. Opin. Biotechnol. 14, 194-199.
[24] Wang, W., Vinocur, B. and Altman, A. (2003). Plant responses to drought, salinity and extreme temperatures: towards genetic engineering for stress tolerance. Planta 218, 1-14.
[25] Sung, D.Y., Kaplan, F., Lee, K.J. and Guy, C.L. (2003). Acquired tolerance to temperature extremes. Trends Plant Sci. 8, 179-187.
[26] Hazen, S.P., Wu, Y. and Kreps, J.A. (2003). Gene expression profiling of plant responses to abiotic stress. Funct. Integr. Genom. 3, 105-111.
[27] Zhu, J.K. (2001). Plant salt tolerance. Trends Plant Sci. 6, 66-71.
[28] Knight, H. and Knight, M.R. (2001). Abiotic stress signalling pathways: specificity and cross-talk. Trends Plant Sci. 6, 262-267.
[29] Bray, E.A. (1997). Plant responses to water deficit. Trends Plant Sci. 2, 48-54.
[30] Posas, F., Wurgler-Murphy, S.M., Maeda, T., Witten, E.A., Thai, T.C. and Saito, H. (1996). Yeast HOG1 MAP kinase cascade is regulated by a multistep phosphorelay mechanism in the SLN1-YPD1-SSK1 "two component" osmosensor. Cell 86, 865-875.
[31] Urao, T., Yakubov, B., Satoh, R., Yamaguchi-Shinozaki, K., Seki, M., Hirayama, T. et al. (1999). A transmembrane hybrid-type histidine kinase in Arabidopsis functions as an osmosensor. Plant Cell 11, 1743-1754.
[32] Shinosaki, K. and Yamaguchi-Shinosaki, K. (2000). Molecular responses to dehydration and low temperature: differences and cross-talk between two stress signalling pathways. Curr. Opin. Plant Biol. 3, 217-223.
[33] Yamaguchi-Shinosaki, K. and Shinosaki, K. (1994). A novel cis-acting element in *Arabidopsis* gene is involved in responsiveness to drought, low temperature, or high-salt stress. Plant Cell 6, 251-264.
[34] Liu, X., Kasuga, M. Sakuma, Y. Abe, H., Miura, S., Yamaguchi-Shinosaki, K. et al. (1998). Two transcription factors, DREB1 and DREB2, with an EREBP/AP2 binding domain separate two cellular signal transduction pathways in drought- and low temperature-responsive gene expression, respectively, in *Arabidopsis*. Plant Cell 10, 1391-1406.
[35] Sakuma, Y., Liu Q., Dubouzet, J.G., Abe, H., Shinosaki, K. and Yamaguchi-Shinosaki, K. (2002). DNA binding specifity of the ERF/AP2 domain of *Arabidopsis* DREBs, transcription factors involved in dehydration- and cold-inducible gene expression. Biochem. Biophys. Res. Comm. 290, 998-1009.
[36] Dubouzet, J.G., Sakuma, Y., Ito, Y., Kasuga, M., Dubouzet, E.G. Miura, S. et al. (2003). *OsDREB* genes in rice, *Oryza sativa* L., encode transcription activators that function in drought-, high-salt and cold-responsive gene expression. Plant J. 33, 751-763.

[37] Gilmour, S.J., Sebolt, A.M., Salazar, M.P., Everard, J.D. and Thomashow, M.F. (2000). Over-expression of the *Arabidopsis* CBF3 transcriptional activator mimics multiple biochemical changes associated with cold acclimatation. Plant Physiol. 124, 1854-1865.
[38] Kasuga, M., Liu, Q., Miura, S., Yamaguchi-Shinosaki, K. and Shinosaki, K. (1999). Improving plant drought, salt, and freezing tolerance by gene transfer of a single stress-inducible transcription factor. Nat. Biotechnol. 17, 287-291.
[39] Narusaka, Y., Nakashima, K, Shinwari, Z.K. Sakuma, Y., Furihata, T. Abe, H. et al. (2003). Interaction between two *cis*-acting elements, ABRE and DRE, in ABA-dependent expression of *Arabidopsis rd29A* gene in response to dehydration and high salinity stresses. Plant J. 34, 137-148.
[40] Haarke, V. Cook, D., Riechmann, J.L, Omaira, P., Thomashow, M.F. and Zang, J.Z. (2002) Transcription factor CBF4 is a regulator of drought adaptation in *Arabidopsis*. Plant Physiol. 130, 639-648.
[41] Kizis, D. and Pagès. M. (2002). Maize DRE-binding proteins DBF1 and DBF2 are involved in *rab17* regulation through the drought responsive element in an ABA-dependent pathway. Plant J. 30, 679-689.
[42] Iuchi, S., Kobayashi, M., Taji, T., Naramoto, M,, Seki, M., Kato, T. et al. (2001). Regulation of drought tolerance by gene manipulation of 9-cis epoxycarotenoid dioxygenase, a key enzyme in abscisic acid biosynthesis in *Arabidopsis*. Plant J. 27, 325-333.
[43] Kang, J.Y., Choi., H.I., Im, M.Y. and Kim, S.Y. (2002). *Arabidopsis* basic leucine zipper proteins that mediate stress-responsive abscisic acid signalling. Plant Cell 14, 343-357.
[44] Pereira, A. (2000). A transgenic perspective on plant functional genomics. Transg. Res. 9, 245-260.
[45] Delseny, M. (2003). Towards an accurate sequence of the rice genome. Curr. Opin. Plant Biol. 6, 101-105.
[46] Sallaud, C., Meynard, D., Boxtel, J., Gay, C., Bès, M., Brizard, J.P. et al. (2003). Highly efficient production and characterisation of T-DNA plants for rice (*Oryza sativa* L.) functional genomics. Theor. Appl. Genet. 106, 1396-1408.
[47] Matsumura, H., Nirasawa, S. and Terauchi, R. (1999). Transcript profiling in rice (*Oryza sativa* L.) seedlings using serial analysis of gene expression (SAGE). Plant J. 20, 719-726.
[48] Seki, M., Narusaka, M., Abe, H., Kasuga, M., Yamaguchi-Shinosaki, K., Carninci, P. et al. (2001). Monitoring the expression pattern of 1300 *Arabidopsis* genes under drought and cold stresses by using a full-length cDNA microarray. Plant Cell 13, 61-72.
[49] Kawasaki, S., Borchert, C., Deyholos, M., Wang, H., Brazille, S., Kawai, K. et al. (2001). Gene expression profiles during the initial phase of salt stress in rice. Plant Cell 13, 889-905.
[50] Yu, L.X. and Setter, T.L. (2003). Comparative transcriptional profiling of placenta and endosperm in developing maize kernels in response to water deficit. Plant Physiol .131, 568-582.
[51] Cooper, B., Clarke, J.D., Budworth, P., Kreps, J., Hutchison, D., Park, S. et al. (2003). A network of rice genes associated with stress response and seed development. Proc. Natl Acad. Sci. USA 100, 4945-4950.
[52] Sanchez-Martinez, D., Puigdomenech, P. and Pagès, M. (1986). Regulation of gene expression in developing *Zea mays* embryos. Plant Physiol. 82, 543-549.
[53] Goday, A., Sánchez-Martinez, D., Gómez, J., Puigdomenech, P. and Pagès, M. (1988). Gene expression in developing *Zea mays* embryos: regulation by abscisic acid of a highly phosphorylated 23-25-kD group of proteins. Plant Physiol. 88, 564-569.
[54] Vilardell, J., Goday, A., Freire, M.A., Torrent, M., Martinez, M.C., Torné, J.M. et al. (1990). Gene sequence, developmental expression and protein phosphorylation of RAB-17 in maize. Plant Mol. Biol. 14, 423-432.
[55] Pla, M., Gómez, J., Goday, A. and Pagès, M. (1991). Regulation of the abscisic acid-responsive gene rab28 in maize viviparous mutants. Mol. Gen. Genet. 230, 394-400.
[56] Zhang, J.Z. (2003). Overexpression analysis of plant transcription factors. Curr Opin. Plant Biol. 6, 430-440.
[57] Sakamoto, A., Murata, A. and Murata, N. (1998). Metabolic engineering of rice leading to biosynthesis of glycine betaine and tolerance to salt and cold. Plant Mol. Biol. 38, 1011-1019.
[58] Zhu, B., Su, J., Chang, M., Verma, D.P., Fan, Y.L. and Wu, R. (1998). Overexpression of a Δ^1–pyrroline-5-carboxylate synthetase gene and analysis of tolerance to water-and salt-stress in transgenic rice. Plant Sci. 139, 41-48.
[59] Jang, I.C., Oh, S.J., Seo, J.S., Choi, W.B., Song, S.I.K., Kim, C.H. et al. (2003). Expression of a bifunctional fusion of the *Escherichia coli* genes for trehalose-6-phsphate synthase and trehalose-6-phosphate phosphatase in transgenic rice plants increases trehalose accumulation and abiotic stress tolerance without stunting growth. Plant Physiol. 131, 516-524.
[60] Garg, A.K., Kim, J.K., Owens, T.G., Ranwala, A.P., Choi, Y.D., Kochian, L.V. et al. (2002). Trehalose accumulation in rice plants confers high tolerance levels to different abiotic stresses. Proc. Natl Acad. Sci. USA 99, 15898-15903.

[61] Dure III, L. (1997). LEA proteins and the desiccation tolerance of seeds. In Cellular and Molecular Biology of Plant Seed Development. (Larkins, B.A. and Vasil, I.K. Eds). The Netherlands: Kluwer Academic Publishers, pp. 525-543.
[62] Wise, M.J. and Tunnacliffe, A. (2003). POPP the question: what do LEA proteins do? Trends Plant Sci. 124, 1-5.
[63] Xu, D., Duan, X., Wang, B., Hong, B., Ho, T.D. and Wu, R. (1996). Expression of a late embryogenesis abundant protein gene, *HVA1*, from barley confers tolerance to water deficit and salt stress in transgenic rice. Plant Physiol. 10, 249-257.
[64] Abelson, P.H. and Hines, P.J. (1999). The plant revolution. Science 285, 367-368.
[65] Borlaug, N.E. (2000). Ending world hunger, the promise of biotechnology and the threat of antiscience zealotry. Plant Physiol. 124, 487-490.
[66] Wisniewski, J.P., Frangne, N., Massonneau, A. and Dumas, D. (2002). Between myth and reality: genetically modified maize, an example of sizeable scientific controversy. Biochim. 84, 1095-1103.
[67] Saeglitz, C. and Bartsch, D. (2003). Regulatory and associated political issues with respect to Bt transgenic maize in the European Union. J. Invertebr. Pathol. 83, 107-109.
[68] Macdonald, P. and Yarrow, S. (2003). Regulation of Bt crops in Canada. J. Invertebr. Pathol. 83, 93-99.
[69] Scutt, C.P., Zubko, E. and Meyer P. (2002). Techniques for the removal of marker genes from transgenic plants. Biochimie 84, 1119-1126.

4. Biosafety Research in Central and Eastern Europe: A Call for Genomics?

Genomics for Biosafety in Plant Biotechnology
J.P.H. Nap, A. Atanassov and W.J. Stiekema (Eds.)
IOS Press, 2004

Transgenic Plant Products and their Introduction into the Environment and Crop Protection Systems, a Risk Assessment

Slavomír RAKOUSKÝ[1*], Miloš ONDŘEJ[2], František SEHNAL[3], Oxana HABUŠTOVÁ[3], Hany Mohamed HUSSEIN[3], Jaroslava OVESNÁ[4], Ladislav KUČERA[4], František KOCOUREK[4], Karel ŘÍHA[4], Radmila DOSTÁLOVÁ[5], Marek SEIDENGLANZ[5], Eva TEJKLOVÁ[5] and Miroslav GRIGA[5]

*[1]Institute of Plant Molecular Biology of the Czech Academy of Sciences (IPMB), Branišovská 31, 370 05 České Budějovice, Czech Republic (CR); [2]Faculty of Biological Sciences, University of South Bohemia, same address; [3]Institute of Entomology of the Czech Academy of Sciences (IEAS), sane address; [4]Research Institute of Crop Production (RICP), Prague, CR; [5]AGRITEC Ltd., Šumperk, CR; *email: ray@umbr.cas.cz*

Abstract. Various issues associated with the development and use of genetically modified (GM) crops in the Czech Republic (CR) prompted the evaluation of the concerns to generate data to develop proper legislation. Investigations are directed to the types of GM crops and traits that may be considered for future cultivation in CR, notably winter oilseed rape (*Brassica napus*), maize (*Zea mays*), wheat (*Triticum aestivum*), pea (*Pisum sativum*) and flax (*Linum usitatissimum*) and the traits herbicide tolerance and insect resistance. Evaluations include potential changes in crop productivity and their impact on field management, as well as the extent and impact of pollen flow between GM plants and traditional cultivars or related wild species. When relevant, non-GM plant material with dominant traits that could be easily assayed (e.g. flower colour) was used for research, rather than actual GM crop material. Research attention was particularly devoted to the influence of insect-resistant GM crops on non-target organisms that are considered beneficial for agroecosystems. This paper gives an overview of approaches, results and conclusions.

Introduction

Like many other European countries, the Czech Republic (CR) is considering the cultivation of genetically modified (GM) crops, but it also faces many concerns and problems regarding the social and agronomic desirability of such crops. Agriculture in CR underwent major changes, including changes in the cropping system. The issue is whether GM crops could play a role in CR agriculture. Various government- and privately funded research projects into the environmental impact of GM crops have been or are now being undertaken. These studies focus on GM crops that may be developed and grown in CR. At the forefront with respect to field release and commercialisation is herbicide-tolerant (HT) winter oilseed rape. Also GM insect-resistant maize (*Zea mays*), carrying a *Bacillus thuringiensis* cry gene, is being evaluated in the field, notably for unintended effects on non-target organisms. GM wheat (*Triticum aestivum*) is available for release and risk assessment studies are carried out for pea (*Pisum sativum*) and flax (*Linum usitatissimum*).

1. Oilseed Rape (*Brassica napus*)

Oilseed rape (*B. napus*; AACC genome, 2n=38) is an allotetraploid species derived from the hybridisation of turnip (*B. rapa*; AA genome, 2n=20) and *B. oleracea* (CC genome, 2n=18). In contrast to both parental species, in which a sporophytic incompatibility mechanism prevents self-fertilisation, oilseed rape is a predominantly self-pollinating crop. Average outcrossing ranges between 15 and 45 % and is largely determined by environmental factors [1]. There are two types of oilseed rape, winter and summer oilseed rape. Winter oilseed rape needs a period of cold before it flowers and can survive winter.

During the last decade oilseed rape became one of the dominating crops in the fields of CR. The cultivated area varies between 250 and 350 thousands hectares annually. Rape is now an important item for export, especially to the countries of the European Union. In CR mostly winter oilseed rape is grown. Because oilseed rape is at the forefront of field releases and commercialisation, both companies and the government were engaged in specific research programs aimed at biosafety assessment of HT oilseed rape cultivation.

Oilseed rape can be crossed to other species in the family *Cruciferae* resulting in inter-specific and inter-generic hybrids. With *B. oleracea*, *B. rapa* and *Raphanus sativus* such crosses even resulted in new crops such as *Raphanobrassica* or Radicole. The success of hybridisation between GM crops and its wild relatives depends on the relationship between species involved [2,3]. GM winter oilseed rape tolerant to phosphinothricin (glufosinate-ammonium) "Liberty-link" of Aventis Crop Science was widely tested in field experiments for its performance, herbicide registration and possible market release of a crop for a given period of time. Hybridisation of transgenic, herbicide tolerant oilseed rape was intensively studied [e.g. 4, 5, 6]. Recently, great attention has been paid in many European and national programs to evaluation of possible impact of HT oilseed rape on both non-GM oilseed rape and its wild relatives.

Transgenic oilseed rape (*B. napus*) was released into the environment and cultivated experimentally on fields in CR in accordance with national laws. Transgenic oilseed rape carrying the *bar* transgene conferring phosphinothricin tolerance was grown by Aventis Crop Science on several small fields belonging to different breeding and/or experimental stations located in different places of CR. Used were the breeding lines pHW01-439, pHW01-440 and pHW99-424. The assessments had as major goals monitoring of (a) transgene transfer by pollen to neighbouring oilseed rape fields and (b) survival of HT rape seeds in fields during succeeding years. Related *Brassica* species or wild relatives are grown rather seldom in CR. As a result, the possibility of the transgene transfer to related species or wild relatives is limited. Volunteers or wild type *B. rapa* plants, which hybridise to *B. napus* and are quite common e.g. in the UK, do not persist cold winter or grow in small populations in CR only, respectively.

Assessments performed in 2001 shows that transfer of transgene by pollen to neighbouring rape fields occurs in a low frequency. Outcrossed transgenic events were detected after spraying seedlings by phosphinothricin solution and evaluating the frequency of phosphinothricin tolerant seedlings. Altogether, 85,000 seedlings were sown. The presence of *bar* gene in the phosphinothricin-tolerant seedlings was confirmed by PCR and by a segregation ratio of 3 : 1 of phosphinothricin tolerant : sensitive plants in 25 offspring tested. The frequency of admixture of transgenic seeds in non-GM rape cultivars, when cultivated in the vicinity of transgenic cultivars on a distance of 5 - 20 m did not exceed 0.4%. This is below the legally tolerated frequency (0.9 %). At a distance of 500 m, no admixtures of transgenic seed were detected in the offspring.

Transgenic seeds unintentionally disseminated in the fields of transgenic rape have been followed in two successive years. In experiments with transgenic rape (year zero), transgenic plants as well as non-GM controls were cultivated together. Small proportion of plants from disseminated seeds has been grown the next and the second year, when usually wheat or barley was the crop cultivated. In altogether 4 fields followed on different sites, rare single transgenic as well as non-GM rape plants appeared for at least one of two years after cultivation of HT oilseed rape. Their incidence was in the range of a single plant in 20 m^2 (as maximum) to a few plants in a whole field. The incidence of HT oilseed rape plants in the second year, however, depended on cultivation conditions. In two out of the four fields, the incidence of transgenic rape plants decreased the second year an order of magnitude. In the third field, the incidence did not change considerably. In the fourth field, no transgenic oilseed rape plants were found the first year but a few of them were observed in the second year.

Based on the results obtained in a period of two succeeding years it was concluded that the probability of transgene escape to non-GM oilseed rape is relatively low, even if the separating distance is short. Under usual cultivation conditions the admixture of transgenic seeds in conventional oilseed rape grown in the vicinity of HT rape field is below the accepted legal limit. The frequency of HT volunteers is influenced by post-harvest treatments. Therefore, attention is warranted for precautionary measures to avoid the risk of unintended contamination of conventional or organic oilseed rape production.

2. Maize (*Zea mays*)

The maize hybrid MON 810 (Monsanto; Bt maize) carries the *Bacillus thuringiensis* gene encoding the Cry1Ab insecticidal toxin. This hybrid was used in field experiments jointly performed from 2002 to 2004 by the Research Institute of Crop Production (RICP; Praha-Ruzyně) and the Institute of Entomology of the Czech Academy of Sciences (IEAS; České Budějovice). RICP fields were located in Praha-Ruzyně and Ivanovice na Hané (Central Moravia). The field experiments concerned one-hectare plots and ran from 2002 to 2004.

The trials of IEAS were performed in Southern Bohemia, in what is a rather cold region of CR, where maize is grown mostly for silage. Plot sizes were 7.6 ha in 2003 and 14 ha in 2004. GM maize was grown on five 0.5 ha plots, and the parental, non-GM cultivar on alternating 0.5 ha plots. Plots were separated by ca. 10-m stripes of barren land. Field margins were sown with the parental, non-GM cultivar. Standard agronomics, including one (in 2002) or two (2003) herbicide treatments, was applied. Experiments were terminated at late milk-ripening stage when the plants were shredded to 1 cm pieces and ploughed about 25 cm deep into soil. Regular sampling done by IEAS revealed that Bt maize expressed 5-10 PPM Cry1Ab toxin (measured with ELISA) in the leaves, stem, and roots throughout the vegetation period. The Bt toxin could not be detected in the flowers and grain [7]. In the other parts of the plant it persisted for several months after their full desiccation and working into the soil. Next season (March), however, the toxin could neither be detected in the field, nor in experimentally ensilaged plant tests.

2.1 Plant Resistance against the European Corn Borer

The European corn borer (ECB; *Ostrinia nubilalis*) is the major insect pest of maize in CR. Resistance of Bt-maize to ECB was verified both by RICP and IEAS. It was observed that adult ECB larvae do (or can) not distinguish between Bt maize and non-Bt maize. They deposit eggs on both, but on Bt maize hatched larvae die before they bore into the plant tissue [8]. ECB Infestation of the non-Bt maize in the fields of IEAS amounted to 10% of plants in 2002 and 15% of plants in 2003.

A new method for the evaluation of Bt maize resistance to ECB was developed with a laboratory strain of *O. nubilalis* [9]. Resistance was evaluated on 9 hybrids of maize. Four were genetically modified and contained the Cry1Ab insecticidal protein (MON 810). The method is based on feeding larvae with semi-synthetic diet containing fresh maize tissue material. The optimal age and start weight of larvae for such feeding experiments was experimentally determined as the weight at which both sex of larvae show the same growth rate. The presence of tissues of the Bt maize in semi-synthetic diet became manifested by retardation of larval growth whereas the larvae on the standard diet grew at normal rates. Three days after starting consumption of Cry1Ab protein the weight increase of larvae was minimal, and in next days their body mass decreased.

Differences in resistance levels between maize hybrids were manifested by different times of survival of larvae on each semi-synthetic diet. The average time of survival of Cry1Ab-fed larvae was not identical in all hybrids. With 12 percent of maize tissue in the diet, mean time of survival was 7.14 days for hybrid X0897AT and 10.73 days for hybrids X0957GT, X0829ZT, X0768MT. When 6 percent of maize tissue was added to the diet, the mean time of survival was 9.63 days for the hybrids X0897AT and X0768MT but 13.04 days for hybrids X0957GT and X0829ZT. Ultimately, all larvae reared on semi-synthetic diet with Bt maize tissue died (final mortality was always 100%).

Differences in survival rates of larvae correlated well with the content of Cry1Ab protein in leaves of maize hybrids, as established by ELISA. The amount of Cry1Ab protein in the hybrid maize varied from 2.1 to 7.7 μg/g of fresh maize leaves. Larvae exposed to Cry1Ab in their diet were able to recover and exhibit normal gut function after the transfer to the standard diet without Bt toxin. These results indicate that for a lethal effect of Bt toxin on larvae of *O. nubilalis* not only the concentration of Bt toxin is important, but also the time of exposure to the toxin. The laboratory bioassay may be used in breeding programs for testing hybrids combining different transgenes and genetic backgrounds. In insect resistant management (IRM) this method enables to test the Bt resistance of local populations of larvae, comparing larvae from field populations with susceptible laboratory populations.

2.2 Efficacy of Bt Maize against ECB, Compared to Biological Control

In the fight against ECB, biological control, for example with the parasitoid *Trichogramma*, is proposed as alternative to genetic modification with Bt genes. The efficacy of Bt maize in regulating *O. nubilalis* populations was compared to biological control of ECB with the parasitoid *Trichogramma* and to non-GM, non-treated plants [10]. Three hybrids of maize were tested: transgenic MON810 (Bt maize), Monumental (non-GM, isogenic to MON810) and Raissa (a commonly grown local maize variety). *Trichogramma* was applied once using the commercial preparation Trichocarp at the time of occurrence of ECB eggs. The application dose was 300 capsules per hectare. This is equivalent to at least 208 thousand *Trichogramma* individuals per hectare.

In 2002, plant injury due to ECB larvae tunnelling on untreated plots before harvest ranged from 2.1 to 2.5 in Ivanovice na Hané and from 0.75 to 0.95 in Ruzyně. In 2003, the same trend was observed, although overall tunnelling was lower. In Ivanovice na Hané, it ranged from 1.3 to 1.4 and in Ruzyně from 0.2 to 0.6. Judged on the basis of plant injury the efficacy of transgenic Bt in comparison with untreated control was 100%, the efficacy of *Trichogramma* ranged from 35 to 69% in 2002 and from 0 to 56% in 2003. The transgenic strategy seems therefore more potent in reducing plant injury than the biological control with the parasitoid *Trichogramma*.

In 2002, no mycotoxins were found in any Bt maize (by ELISA), while mycotoxins were detected in many samples of the other non-GM maize varieties. The mycotoxins deoxynivalenon (DON), T 2– toxin, fumonisin (FUM) and zearalenon (ZEA) were detected in non-GM maize samples from Ivanovice na Hané. The frequency of occurrence of DON varied from 20 to 100%. The average amount ranged from 26 to 122 PPM and the maximum was 397 PPM. The frequency of occurrence of ZEA ranged from 20 to 40%, the average content ranged from 112 to 156 PPM and maximum content was 1535 PPM. The frequency of occurrence of T-2 toxin was up to 20% and its content ranged from 164 to 183 PPM. The frequency of FUM was also up to 20% with a maximum content of 20 PPM. In Ruzyně, only DON and ZEA mycotoxins were detected in non-Bt maize. The frequency of occurrence ranged from 20 to 60% and the average amount varied from 22 to 333 PPM.

The above data is corroborated by the yield data. In 2002, the yield per hectare of Bt maize was 120% in comparison with untreated control in Ivanovice na Hané and 128% in Ruzyně. In 2003, figures were 115% in Ivanovice na Hané and 107% in Ruzyně. Plots treated with *Trichogramma* showed yields increases varying from 100-120 % to the same control in both years. In combination, the data collected confirm the efficacy of Bt in maize protection and production.

2.3 Non-target Effects of Bt Maize

Toxin production in Bt-maize could affect other organisms than the target pest. There could be adverse effects on the predators and parasitoids of the current and potential pests, on the natural enemies of weeds, and on the decomposing (micro)organisms in the soil. Such effects would present a threat to sustainable agriculture. The occurrence and extent of such undesired and unintended side effects was investigated by recording the occurrence of main organisms on the maize plants and the major components of the epigeal fauna in the maize fields. No statistically significant differences were found between Bt maize and control maize in the occurrence of three aphid species (*Rhopalosiphum padi*, *Metopolophium dirhodum*, *Aphis maidi*), as well as the occurrence of common thrips (*Franklinella occidentalis*). This indicates that these sap-sucking insects are not affected by the presence of Bt and expression of the Bt gene in maize.

The most abundant natural enemy of these herbivores is a predatory bug (*Orius* sp.). This species was also found in similar numbers on Bt and non-Bt maize. The abundance of other beneficial insects, such as ladybirds, was relatively low and differences between individual plots could not be evaluated reliably. Ground beetles (*Carabidae*), rove beetles (*Staphylinidae*), and spiders (*Araneae*) dominated among the epigeal arthropods collected in pitfall traps. Biodiversity and abundance were generally higher in the fields studied in 2002 than in that examined in 2003, possibly due to differences in weather conditions or field position. Despite considerable variation within each field, it could be established that the occurrence of arthropods in individual plots was determined by plot position and not by the type of maize [11].

As expected, the fields were dominated by a few species. Nine carabid species made up 92% of all ground beetle individuals, and about half of the carabid species was represented in the catches by just one or a few specimens. In case of spiders, the juveniles of *Oedothorax apicatus*, possibly carried into the field by wind, represented 90% of all spider species caught in 2002, while the second most common species, *Erigone dentipalpis*, made up only about 2.6% of the total spider catch. All differences detected between the plots of Bt-maize and control maize in species composition and abundance were statistically insignificant.

3. Wheat (*Triticum aestivum*)

Wheat belongs to the most important crops in the CR as a source of food and feed. Wheat is grown on large areas. Its breeding and cultivation have had a long tradition. Transgenic HT wheat was introduced by Monsanto in the environment of CR for a short period of time. In view of missing and/or unclear biosafety data, this wheat variety was quickly withdrawn from the market. A major issue is the possible pollen flow between the traditional and GM cultivars, especially in relationship to the desired coexistence of traditional agriculture, organic agriculture and GM agriculture [12]. The European Parliament has adopted a Commission Recommendation 2003/557/EC "On coexistence between genetically modified crops and conventional and organic crops". This calls for a Europe-wide regulation for coexistence, liability and labelling. Several investigations were set up to determine the possible impact of pollen-mediated gene flow into other bread-wheat cultivars, into annual wild wheat relatives and into perennial wild wheat relatives.

3.1 Pollen-Mediated Gene Flow to Non-GM Wheat Cultivars

Wheat is primarily self-pollinating, but a large number of biological, agronomic and environmental factors can interact to cause a small frequency of cross-pollination. We investigated the extent of pollen-mediated gene flow in wheat as a function of distance from the pollinator source. We used genetic markers to measure gene flow [13]. A marker gene (red colour) was used in our study, assuming that any transgene will behave as any other chromosomal gene. A 100 x 100 m central plot of red spiked wheat (*T. aestivum* var. *ferrugineum* (ALEF.) MANSF.) was used as pollinator. Two white-spiked wheat cultivars (*Triticum aestivum L.* var. *lutescens* (ALEF.) MANSF.) were used as acceptors of the pollen. The gametocide GENESIS (Monsanto) was applied to half of the acceptor plots to detect long distance pollen flow. The number of red spiked wheat plants was scored in the progenies of plants from strips of the recipient cultivars harvested at different distances from the pollinator. An area of nine hectare was used.

At the moment of writing, the results of the first two-year period are available. Donor pollen is able to produce viable seed at a distance of 100 m, but at relatively low frequencies. This confirms the findings of other investigators. Several factors, notably weather conditions affect the outcome. Trials will continue to develop reliable data for multiple environments under multiple climatic conditions. Such data will help establishing a science-based legal framework for risk assessment in CR.

3.2 Pollen-Mediated Gene Flow to Wild Relatives of Wheat

Wheat is a hexaploid species with three genomes from its ancestors: the A genome from *T. urartu*, and the B and D genomes from *Aegilops* species. Wide hybridisation occurs in nature and has been used to introduce some genes from wild relatives to cultivated wheat [14]. Similar wide hybridisation with GM wheat may be undesirable. Especially jointed goatgrass (*Aegilops cylindrica*) is a problematic case. It is a weed with a very close genetic relationship with wheat having C and D genomes [15]. IN the USA, *A. cylindrica* is considered an important weedy species. Gene flow from wheat to jointed goatgrass was documented in Europe [16]. Hybrids of durum wheat and bread wheat with *A. cylindrica* have been obtained without embryo rescue techniques [17]. Hybrid output (of pollinated flower number) in the field was 1%, 15.3% and 10.0% in the crosses *T. durum* × *A. cylindrica*, *A. cylindrica* × *T. durum* and *T. aestivum* × *A. cylindrica*, respectively. A high level of meiotic chromosome pairing between homologous D genomes of bread wheat and *Aegilops* has been observed (c = 80.0–83.7%). Herbicide-tolerant wheat will increase the probability of producing herbicide resistant jointed goatgrass [18].

Hybridisation between some more distant perennial species and wheat is seldom and embryo rescue had to be used to develop hybrids between perennial *Elytrigia* and wheat. Both annual and perennial weedy species do occur in the territory of CR. While *A. cylindrica* is found only in Southern Moravia, *Elytrigia* sp. are important weeds in the whole CR. Two main species occur widely, *E. intermedia* and *E. repens*. *E. intermedia* can form hybrids with wheat [19]. We have investigated (1) the dispersion of *E. intermedia*, (2) the genetic diversity of E. *intermedia* (genomic constitution $E^eE^eS^t$) and *E. repens* (genomic constitution S^tS^tH) populations and (3) the ability of *Elytrigia* ecotypes to form viable hybrids with wheat. Stands of *E. intermedia* were found across CR. Such areas should be GM wheat-free to prevent any transgene flow into this species. Diversity was investigated using gliadine patterns of clones obtained at different places. The diversity of the natural populations is high in comparison with cultivated bread-wheat. More stands and clones should be investigated for their ability to form hybrids.

Hexaploid wheat was pollinated by *E. intermedia* and *E. repens* pollen and *in vitro* cultures were used to produce hybrid seed. Over 2000 flowers were pollinated, caryopsis developed in some extend. However, no embryos were obtained (Figure 1). Haploid and double haploid *E. intermedia* and *E. repens* were developed as tool for further investigation.

Figure 1. Developing caryopses after pollination of recipient wheat flowers with pollen of *E. repens*

Available data indicate that the probability of transgene transfer into non-GM cultivars is low, but possible when both species are grown in close proximity. Data indicating possibility of a transgene escape into wild related species have to be developed for different regions. We are developing data suitable for science-based risk assessment before the introduction of GM wheat into the environment of CR. We are investigating long-distance pollen flow, the occurrence of wheat related species in the territory and the possibility of wide hybridisation of these species with durum and bread wheat. Current data indicate that the likelihood of transgene transfer into wild related species is extremely low in CR. Further investigation of hybrid formation with *E. intermedium* is warranted. More attention should be paid to transgenic wheat with respect to the desired coexistence of traditional agriculture, organic farming and modern technologies.

4. Pea (*Pisum sativum*)

In Europe, there are no wild *Pisum* species that are crossable with culture pea (*Pisum sativum* L.). Wild pea forms / species (*P. fulvum*, *P. elatius*, *P. jomardi*, *P. humile*) occur in the Near East, in the gene centre of the genus *Pisum* evolution and domestication. They are only partially crossable with culture pea. The progeny is sterile or semi-sterile [20]. These wild forms usually do not survive the climatic conditions of Central or Western Europe in case of occasional introgression. Culture pea also does not cross successfully with other cultivated legumes (genera *Phaseolus*, *Vicia*, *Glycine*, *Lathyrus*) or the wild European *Fabaceae* family. This is documented by many (generally unpublished) breeders´ attempts.

Pea is a cleistogamous species, *i.e.* pollination is completed in closed flower buds (before flower opening). So, cross-pollination mediated by honey or pollen-collecting insects is extremely low. Bumblebees, bees and similar insects usually visit open flowers. Cross-pollination may be realised in exceptional case. It may require, for example, higher affinity of foreign pollen to the host stigma. Cross-pollination may also be caused by small pests or other insect species, which visit still undeveloped flowers, such as aphids (*Acyrthosiphon pisum, Homoptera, Aphididae*), or thrips (*Kakothrips robustus, Thysanoptera, Thripidae*), and especially by the pea weevil (*Bruchus pisorum* L.). Sexually immature females of the beetle feed on pollen before they can lay eggs. While seeking for pollen, they injure other parts of the pea flower, notably petals and the female organs (gynaecium). They take nectar too [21]. Pollen on the insect body may this way transfer between different pea flowers and may establish a route for cross-pollination. The pea weevil appears in pea fields in the beginning of the flowering period. Beetles come into contact with already pollinated/fertilised flowers, but also with flower buds containing mature pollen and stigmas ready to pollination.

Based on studies of the mutual relationships between pea host and phytophagous insects, notably the pea weevil [22], there is an -at least theoretical- possibility for cross-pollination. Due to the biology of pea, it makes little sense to follow pollen transfer from transgenic peas to wild leguminous relatives. In contrast, it is necessary to monitor pollen transfer between pea individuals or to monitor the survival of plants from seeds of shattered pods during harvest. The last possibility may be influenced by global warming and the occurrence of winter pea varieties in Western and Southern Europe.

To investigate possible cross-pollination within culture pea, field experiments were carried out in 2001-2003 at Šumperk-Petrov, CR with non-GM plant material. Single gene markers for flower colour (dominant violet versus recessive white) and seed coat colour (dominant coloured versus recessive colourless) were used for rapid identification of hybrids. Dry-seed, white-flowering pea cv. Zecon (*Pisum sativum* ssp. *sativum* L.) and fodder, violet-flowering pea cv. Arvika (*Pisum sativum* ssp. *arvense* L.) were used as parents. Both cultivars are of Czech origin. The flowering period of both peas overlap and offer the hypothetical possibility for pollen transfer.

The experimental design is outlined in Figure 2. Seed harvest was carried out separately. All harvested seeds of Zecon were sown next year (Figure 2) and the occurrence of violet-flowering hybrids was recorded. In the case of cross-pollination mediated by insect pollinators, the dominant trait (= violet flower) should already be seen in F1 generation. The incidence of potential insect pollinators was monitored by four methods: pitfall traps, beating down, sweeping and glue strips. Insect monitoring was carried out both in the cross-pollination trials and in neighbouring pea fields. The experiment was carried out twice during the period 2001-2003. The arrangement illustrated in Figure 2 was established in 2001 and 2002 and monitoring of hybrids in the F1 progeny (Figure 3) was carried out in 2002 and 2003.

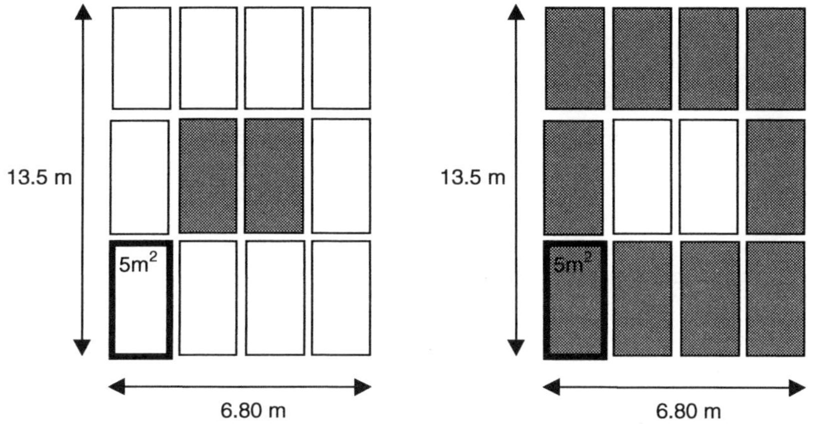

Figure 2. Experimental design of the 1st-year field experiment monitoring the potential for pollen transfer between white- and violet-flowering peas; open boxes, white colour (cv. Zecon), closed boxes, grey colour (cv. Arvika).

The growing seasons in the years 2001-2002 gave relatively high temperatures compared to long-term temperature means. Adults of *Sitona lineatus* (*Coleoptera*) occurred in the field practically from emergence (April to June). In the flowering period, a high abundance of the pea aphid (*Acyrthosiphon pisum, Homoptera*) and *Cydia nigricana* (*Lepidoptera*) was recorded. Besides the pea aphid, sporadically also *Aphis fabae* occurred, especially on tendrils. Pea weevil adults (*Coleoptera*), the most important candidate for pollen transfer, occurred from the beginning of the flowering period, but with a low population density (on he average 1 adult per 25 sweeps). This may be due in part to the adverse weather conditions for this insect. The frequency of seeds affected by the pea weevil was below 1% of the total number of seeds inspected.

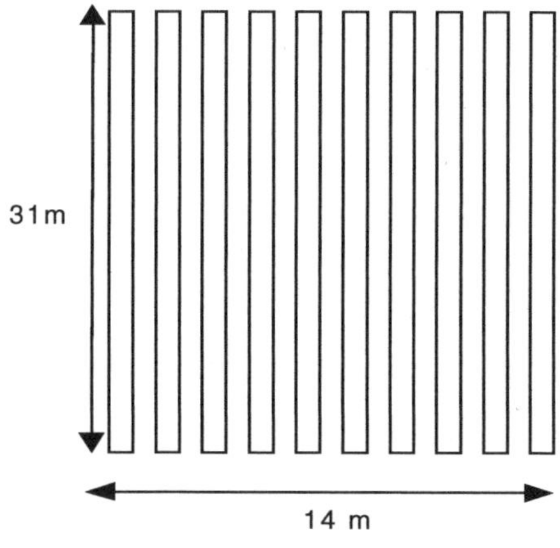

Figure 3. Experimental design for the 2nd-year field experiment monitoring the potential pollen transfer between white- and violet-flowering peas; boxed areas, white-flowering cv. Zecon.

Analysis of potential hybrids was carried out in approximately 40 thousand F1 plants (self-pollinated cv. Zecon) in both 2002 and 2003. No violet-flowering plants were recorded in either year, indicating that no cross-pollination had occurred. Although cross-pollination may increase by a higher incidence of potential insect pollinators in more suitable weather conditions, based on these results we conclude that the probability of cross-pollination mediated by insects in recent commercial pea cultivars is extremely low. Transgenic peas are therefore practically without environmental risk caused by pollen-mediated gene flow.

5. Flax (*Linum usitatissimum*)

In many countries, flax (*Linum usitatissimum* L.) is considered a marginal crop, despite its intense cultivation in the past. Flax, however, may be on the way back. Also in CR its cultivation area sharply decreased in the early nineties, but the production area increased again and stabilised on approximately 10,000 ha/year. This increase can be attributed to the introduction of oil flax (linseed) into agricultural practice, the restoration of agroeconomic subsidies to flax growers and the successful market introduction of new flax varieties (for example, the cultivars Merkur, Jordan, Bonet, Jitka and Venica) that ensured high yield of stems, fine quality fibre and good resistance to *Fusarium.*

Breeding of new flax and linseed varieties is now supported by research and development of new breeding methods. New approaches include *in vitro* techniques (doubled haploids), plant transformation and molecular markers [23]. In Canada and the USA, a transgenic linseed variety was approved and released [24]. In CR, successful application of plant transformation has produced new sources for genetic variability in flax, notably through the use of insertion mutagenesis and anther culture for double-haploid production [25, 26]. IPMB in co-operation with Agritec has generated as much as 1100 transgenic lines. Some of these were evaluated in field trials over five subsequent generations for their agronomic characteristics and performance.

In the context of the evaluation of GM flax lines, biosafety studies with flax have been initiated. These concentrate on the occurrence and prevention of gene transfer to genomes of cultivated conventional (non-GM) flax, wild type flax and related species, and on the persistence of viable seeds following the harvest in fields and composting. Cultivated flax is a predominantly self-pollinating species but various -low- levels of outcrossing are reported. Its relatively heavy pollen could be transferred to nearby flax or linseed plants via insect vectors (bees, bumblebees or thrips) or by direct plant contact.

Experiments were performed to determine the occurrence of outcrossing relevant for the climate and field conditions of CR. Two double marked linseed lines that differed in stem and flower colours were chosen to mimic the transgenic situation. One line was characterised by a yellow-green stem (recessive) and blue flowers (dominant). The second line carried a green stem (dominant) and white flowers (recessive). Both traits were determined by a single gene and segregated in a Mendelian way. The contrasting lines were sown every other row in a regular array and offspring was evaluated. The maximum frequency of outcrossing at a plant distance of 10 cm was 0.4% in 2002. It was somewhat higher in 2003, but still below the limit of 0.9% outcrossing. Future experiments will be performed to determine the minimal distance of separation that fully prevents outcrossing in the field.

Some wild type flax species are endemic in CR, but only *Linum flavum* L. (a perrenial species) has the same chromosome number (2n=30) as flax. It was investigated if fertile hybrids could be formed. In all experiments where *L. flavum* was grown in close distance to flax, no spontaneous hybridisation was found. Attempts to hybridise the two species artificially by forced hand pollination were not successful. Pollen-mediated gene transfer to *L. flavum* may therefore not have a high probability.

In general breeding, flax lines that do not perform sufficiently well should be eliminated efficiently. The viable seeds from such plants, as well as from GM flax has to be inactivated efficiently to avoid of undesirable cross-contamination of conventional production. We have studied the possibility to clean plant stems and husks containing viable flax seeds by composting. Flax seeds do not need any period of dormancy to start germination. Seeds germinate within 3-7 days if the humidity is high enough. Experimental compost was established in the autumn of 2001. All flax seeds occurring on the surface of the compost surface germinated and died in the winter cold plants died. Seeds inside the compost decayed. In subsequent seasons, no viable seeds were ever observed. The same results were obtained when the seeds with husks were left freely in a 20-cm compost layer over the soil surface. Composting therefore seems a good method to eliminate viable flax seeds. These experiments will continue to extend our knowledge on potential risks and possible precautionary measures required for sufficiently safe cultivation of transgenic flax.

6. Conclusions and Future Outlook

A variety of approaches is used to establish a proper baseline of data and knowledge for informed decisions on the cultivation of selected GM crops in CR. Most attention now focuses on the potential ecological and environmental impact of such crops on agriculture and nature. The use of traditional plants lines that are appropriately marked for analysis prior to the actual cultivation of GM crops is a suitable approach to mimic that cultivation. Current plant breeding is advanced substantially by the use of modern genomics technologies, such as the extensive use of markers, the increase in variability by high-throughput mutagenesis and the increased knowledge of plant genomes and plant development by 'omics' approaches. Recent developments in ecological genomics may also help understanding the role and functioning of ecosystems. However, ecological genomics may not be able offer suitably cost-effective approaches for informed decisions about given GM crops. With the continued detailing of questions and issues, however, this may change in the not-too-distant future. The environmental biosafety assessment of GM crops may then profit from the advances in genomics.

Acknowledgements. This work was supported by project grants of the Czech Ministry of Agriculture (NAZV QD 1360, QC 1362 and QC 0056), the Grant Agency of the Czech Academy of Sciences (Z5007907-I011 and 521/97/1135), the Ministry of Environment (MZP/340/1/12/GMO/02, MZP/340/1/2/GMO/03) and the Ministry of Education, Youth and Sports (ME 210 and ME 434).

References

[1] Becker, H.C, Damgaard, C. and Karlson, B. (1992). Environmental variation for outcrossing rate in rapeseed (*Brassica napus*). Theor. Appl. Genet. 84, 303-306.

[2] Dale, P.J. (1994). The impact of hybrids between genetically modified crop plants and their related species: general considerations. Mol. Ecol. 3, 31-36.

[3] Messeguer, J. (2003). Gene flow assessment in transgenic plants. Plant Cell Tiss. Org. Cult. 73, 201-212.

[4] Scheffler, J.A. and Dale, P.J. (1994). Opportunities for gene transfer from transgenic oilseed rape (*Brassica napus*) to related species. Transg. Res. 3, 263-278.

[5] Metz, P.L.J., Jacobsen, E. and Stiekema, W.J. (1997). Aspects of the biosafety of transgenic oilseed rape (*Brassica napus* L.). Acta Bot. Neerl. 46, 51-67.

[6] Chèvre, A.M., Eber, F., Darmency, H., Fleury, A., Picault, H., Letanneur, L.C. and Rennard, M. (2000). Assessment of interspecific hybridisation betweeen transgenic oilseed rape and wild radish under normal agronomic conditions. Theor. Appl. Genet. 100, 1233-1239.

[7] Sehnal, F., Habuštová, O., Spitzer, L., Hussein, H.M. and Růžička, V. (2004). A biannual study on the environmental impact of Bt-maize. IOBC/WPRS Bull., in press.

[8] Habuštová, O., Hussein, H. and Sehnal, F. (2004). Insect communities on maize expressing a Bt-toxin. Acta Phytotech. Zootech., in press.
[9] Říha, K., Kocourek, F. and Habuštová, O. (2003). Using of semi-synthetic diet with content of Bt-maize tissue for evaluation of resistance of Bt-maize to European corn borer. In Book of Abstracts, IOBC/WPRS working group meeting Ecological Impact of Genetically Modified Organisms, November 26-29, Prague, Czech Republic, p. 42.
[10] Kocourek, F, and Říha, K. (2003). Evaluation of efficacy of *Bt*-maize and Trichogramma wasp against European corn borer (*Ostrinia nubilalis*) in field trial, In Book of Abstracts, IOBC/WPRS working group meeting Ecological Impact of Genetically Modified Organisms, November 26-29, Prague, Czech Republic, p. 34.
[11] Spitzer, L., Růžička, V., Hussein, H.M, Habuštová, O., and Sehnal, F. (2004). Expression of a *Bacillus thuringiensis* toxin in maize does not affect epigeal communities of carabid beetles and spiders. Acta Phytotech. Zootech., in press.
[12] Bock, A.K., Lheureux, K., Libeau-Dulos, M., Nilsgard, H., and Rodriguez Cereo, E. (2002). Scenarios for Co-existence of Genetically Modified, Convential and Organic Crops in European Agriculture. EC JRC Report, IPST Sevilla, Spain: IPST.
[13] Latta, R.G., Linhart, Y.B., Fleck, D. and Elliot, M. (1998). Direct and indirect estimates of seed versus pollen movement within a population of ponderosa pine. Evolution 52, 61-67.
[14] Van Aken, J. (1999). Centres of Diversity: Global Heritage of Crop Varieties Threatened by Genetic Pollution. Greenpeace International, Berlin, Germany: Greenpeace International.
[15] Kimber, G. and Sears, E.R. (1987). Evolution in the genus *Triticum* and the origin of cultivated wheat. In Wheat and Wheat Improvement (Heyne E.G., Ed.). Agronomy Monograph No. 13, Madison, Wisconsin, USA: ASA-CSSA-SSSA.
[16] Guadagnuolo, R., Savova-Bianchi, D. and Felber, F. (2001). Gene flow from wheat (*Triticum aestivum* L.) to jointed goatgrass (*Aegilops cylindrica* Host.), as revealed by RAPD and microsatellite markers. Theor. Appl. Genet. 103, 1-8.
[17] Avsenin, V.I., Motsnyy, I.I., Rybalka, A.I. and Fayt, V.I. (2003). Hybrids of *Aegilops cylindrica* Host. with *Triticum durum* Desf. and *T. aestivum* L. Cytol. Genet. 37, 11-17.
[18] Hanson, D.E., Ball D.A. and Mallory-Smith, C.A.. (2002). Herbicide resistance in jointed goatgrass (*Aegilops cylindrica*): simulated responses to agronomic practices. Weed Tech. 16, 156-163.
[19] Friebe, B., Mukai, Y., Gill, B.S. and Cauderon, Y. (1992). C-banding and in situ hybridisation analyses of *Agropyron intermedium*, a partial wheat x *Ag. intermedium* amphiploid, six derived chromosome addition lines. Theor. Appl. Genet. 84, 899-905.
[20] Griga, M. and Novák, F.J. (1990). Pea (*Pisum sativum* L.). In Biotechnology in Agriculture and Forestry, Vol. 10, Legumes and Oil Seed Crops (Bajaj, Y.P.S., Ed.). Berlin, Germany: Springer-Verlag, pp. 65-99.
[21] Clement, S. L. (1992). On the function of pea flower feeding by *Bruchus pisorum*. Entomol. Exp. Appl. 63, 115-121.
[22] Horne, J. and Bailey, P. (1991). *Bruchus pisorum* L. (Coleoptera; Bruchidae) control by a knockdown pyrethroid in field peas. Crop Prot. 10, 53-56.
[23] Rowland, G.G. and Wilen, R. (1998). New trends in linseed breeding. Nat. Fibres 2, 22-31.
[24] McHughen, A., Rowland, G.G., Holm, F.A., Bhatty, R.S. and Kenaschuk, E.O. (1997). CDC Triffid transgenic flax. Can. J. Plant Sci. 77, 641-643.
[25] Tejklová, E. and Pavelek, M. (2000). Comparison of variability in flax lines derived from traditional breeding methods and anther culture. Vortr. Pflanzenzücht. 47, 57.
[26] Rakouský, S., Tejklová, E. and Wiesner, I. (2001). Recent advances in flax biotechnology in the Czech Republic. Nat. Fibres 1, 151-155.

Genomics for Biosafety in Plant Biotechnology
J.P.H. Nap, A. Atanassov and W.J. Stiekema (Eds.)
IOS Press, 2004

Evaluation of Gene Flow from Transgenic Tobacco in Bulgaria

Tsvetan POPOV[1], Rossitza BATCHVAROVA[1], Slavtcho SLAVOV[1], Petya CHRISTOVA[1], Milena ALEXANDROVA[1], Atanas ATANASSOV[1*], Isamu YAMAGUCHI[2] and Hiroyuki ANZAI[3]
[1]AgroBioInstitute, 8 Dragan Tsankov blv., Sofia 1164, Bulgaria; [2]RIKEN Plant Science Center, Wako, Saitama,, Japan; [3]Gene Research Center, Ibaraki University, Ibaraki, Japan;
**email: atanas_atanassov@abi.bg*

Abstract. Gene flow in transgenic tobacco was studied for transgenic tobacco lines carrying the *ttr* (tabtoxin resistance) gene from *Pseudomonas syringae* pv. *tabaci*. Homozygous transgenic lines from transformed cultivars Nevrokop 1146 and Plovdiv 50 were characterised with PCR and Southern blot analyses and for complete resistance to wildfire. Field trials with both transgenic lines were carried out to evaluate the direction and distance of gene flow in tobacco. Transgenic plants were surrounded by non-transgenic controls from parental cultivars. Seed of the recipient controls was screened for the presence of the *ttr* gene by PCR. At the largest distance evaluated, 13,8 m, the *ttr* gene was already found present in recipient seed batches in both trials. This demonstrates the rather unexpected occurrence of gene flow at such a distance in a self-pollinating plant species as tobacco. Additional trials are required to determine the ecologically safe cultivation distance for transgenic tobacco.

Introduction

One of the more frequently raised concerns accompanying the introduction of genetically modified (GM) crops in the environment is the potential 'escape' of transgenes. Transgenes may be transferred by pollen to non-transgenic individuals or wild relatives. This could result in what is sometimes considered as a 'disruption of natural diversity' or as 'pollution of gene pools'. Undesired pollen flow from a GM crop to wild relatives or non-GM plants of the same crop is obviously a major issue in cross-pollinating crops such as maize (*Zea mays*) [1], sugar beet (*Beta vulgaris*) [2], carrot (*Daucus carota*) [3], oil seed rape (*Brassica napus*) [4] and many others. Many pollen dispersal studies have been performed and most studies are carried out with such cross-pollinating crops.

In contrast, relatively few gene flow studies are performed with self-pollinating crops such as tobacco (*Nicotiana tabacum*) [5]. The risk of gene flow is considered to be negligible for such crops. In Europe, there are no wild relatives of tobacco that are able to outcross with tobacco. Therefore, the issue is pollen gene flow is between GM and non-GM tobacco. Also in case of tobacco, there is the need for co-existence of conventional, organic and GM agriculture. For such co-existence, an ecologically safe distance, has to be determined experimentally to prevent undesired gene flow from GM tobacco.

1. Experimental Approach

1.1 Plant Material

In this study, homozygous transgenic tobacco lines (*N. tabacum*) Nevrokop 1146-*ttr* and Plovdiv 50-*ttr* were used. As controls, the parental, non-transformed cultivars Nevrokop 1146 and Plovdiv 50 were used. Transgenic lines were obtained via *Agrobacterium*-mediated transfer of the *ttr* (tabtoxin resistance) gene [6] as previously described [7]. These lines were grown for more than 10 generations and multiplied by continued selfing. Every generation, transgenic plants expressing the *ttr* gene and showing resistance to wildfire caused by *Pseudomonas syringae* pv. *tabaci* were selected and propagated. Seeds from resistant plants of the 11th generation (R_{11}) of line Nevrokop 1146-*ttr* and the 10th generation (R_{10}) of line Plovdiv 50-*ttr* were used in separate field trials to determine the distance and extent of gene flow in tobacco.

1.2 PCR Analysis

Genomic DNA was isolated from tobacco leaf tissue or seed as previously described [8]. Polymerase chain reaction (PCR) amplification was carried out using Ready-to-Go PCR beads (Amersham Pharmacia). PCR reactions were performed in a total volume 25 μl that was assembled as follows: 1 μl from each primer (15 pmol), 21 μl sterile water and 2 μl plant genomic DNA. The *ttr* gene used for transformation has a length of 531 nulceotides [9]. The specific primers used to amplify this gene were 5‘-CTT CGC CCA TTA TCG CCA TGG TC-3‘ and 5‘-CAG CCC GCG TTT GTG TTT TAC TG-3‘. PCR was carried out at 94°C for 4 min, followed by 30 cycles at 94°C for 45 sec, 64°C for 45 sec, 72°C for 45 sec, and 72°C for 5 min.

1.3 Southern Blot Analysis

Total DNA was extracted from leaf material as described [8]. Genomic DNA from *ttr* positive plants was isolated, digested with *Xba*I and hybridised with DIG-labelled DNA probes. DIG labelling was performed using the DIG Probe Synthesis Kit (Roche Diagnostics, GmbH) according to the recommendations of the manufacturer. Detection of hybridisation signals was performed with the DIG Luminescent Detection Kit for Nucleic Acids (Roche Diagnostics, GmbH) following the manufacturer's recommendations. In all cases, *Xba*I-digested genomic DNA from non-transgenic parental plants was used as negative control.

1.4 Test for Resistance to Wildfire

Resistance of plants to wildfire caused by *P. syringae* pv. *tabaci* was evaluated with a detached leaf bioassay. Fully expanded tobacco leaves were cut from the transgenic lines Nevrokop 1146-*ttr* and Plovdiv 50-*ttr* and from the parental cultivars Nevrokop 1146 and Plovdiv 50 as susceptible controls. These were inoculated with a bacterial suspension of *P. syringae* pv. *tabaci* with a concentration of 1×10^9 cfu/ml and incubated in Petri dishes at 28^0C in high humidity and an alternating light/dark regime of 12/12 hrs and 4000 lux. The bacterial inoculum was prepared from the most severe strain of *P. syringae* pv. *tabaci* known, strain ATCC 17914. The bacterium was grown on Nutrient agar pH 6.0 (Difco) at 28°C in the dark for 24 hrs. Bacterial material was washed from the medium with distilled water and the concentration was brought to $1x10^9$ cfu/ml using Specol 11. Leaves without any visible symptoms after ten days of incubation were considered resistant [7].

1.5 Field Trial Set-Up

For each transgenic line, transgenic tobacco plants were placed in a central plot of 8 by 9 meters surrounded by non-transgenic control plants in eight directions on a total plot size of 25 by 30 meters (Figure 1). The largest distance therefore is 13.8 meters from the pollen source. After flowering and ripening, seeds were collected at every meter in all directions from the central plot. Seed samples were analysed separately. For each seed sample, PCR analysis was performed to detect the presence of the *ttr* gene.

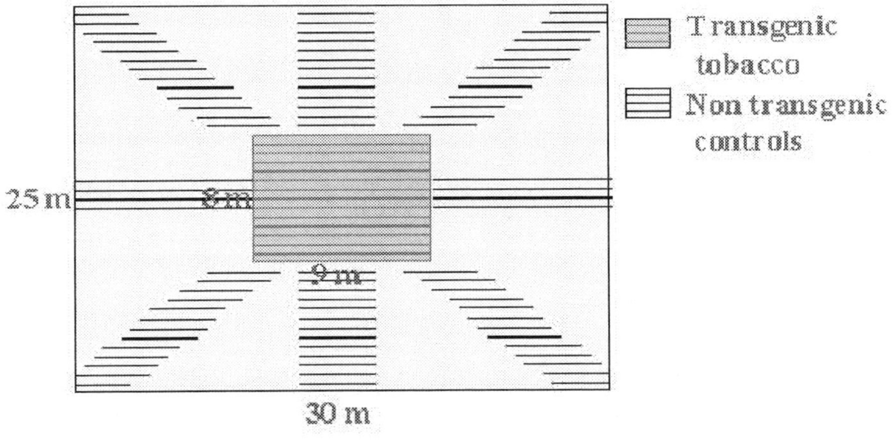

Figure 1. Set-up of the field trial to evaluate gene flow in transgenic tobacco. The same field set-up was used for both transgenic tobacco lines Nevrokop 1146-*ttr* and Plovdiv 50-*ttr*.

2. Molecular Analyses of Transgenic Lines

Transgenic nature of the tobacco lines Nevrokop 1146-*ttr* and Plovdiv 50-*ttr* was confirmed by PCR using *ttr* primers (Figure 2) and Southern blot analysis with a *ttr* gene probe (Figure 3). Proper transgene expression was analysed with a detached leaf bioassay to determine resistance of plants to wildfire caused by *P. syringae* pv. *tabaci* (Figure 4).

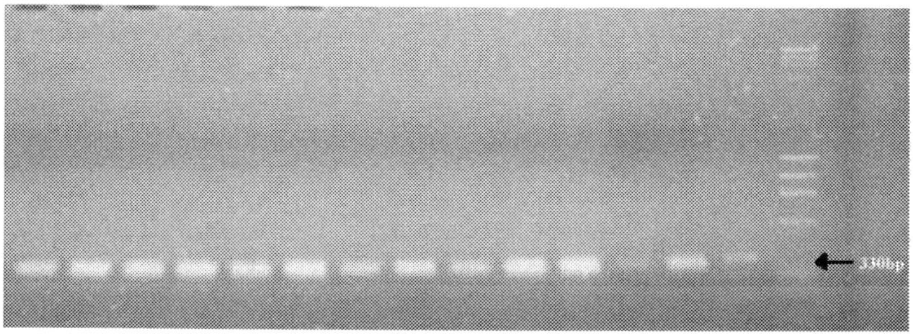

Figure 2. PCR analyses show the presence of the *ttr* gene in the genome of transgenic tobacco lines.

Both PCR and Southern blot analyses showed the proper integration of the ttr gene in the genome of the tobacco lines analysed. Southern blot hybridisation (Figure 3) indicates a difference in copy number of the *ttr* gene between the lines Nevrokop 1146-*ttr* and Plovdiv 50-*ttr.* All plants of plant line Nevrokop 1146-*ttr* have two copies of the *ttr* gene.

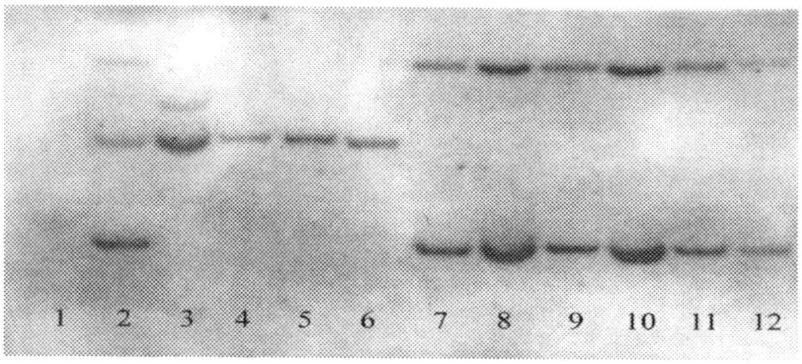

Figure 3. Southern blot analysis using a DIG-labelled *ttr* probe shows the presence of the *ttr* gene in the genome of transgenic tobacco. Lane 1, Negative control (DNA from a non-transformed plant), Lanes 2-6, DNA of transgenic plants from tobacco cv. Plovdiv 50-*ttr*, Lanes 7-12, DNA of transgenic plants from tobacco cv. Nevrokop 1146.

Full resistance to wild fire in a detached leaf assay (Figure 4) demonstrated proper expression of the ttr gene. For both transgenic lines, transgenic plants with a complete resistance to wild fire were grown in a greenhouse to obtain seeds for the field trial.

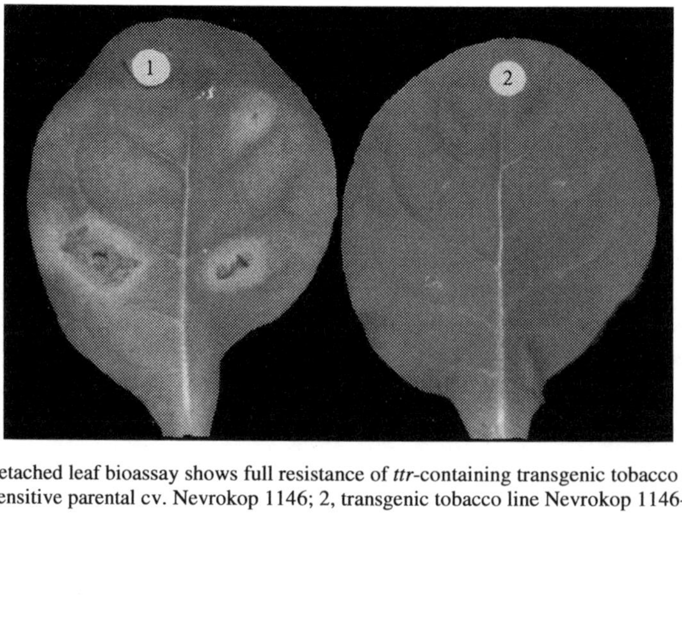

Figure 4. Detached leaf bioassay shows full resistance of *ttr*-containing transgenic tobacco to wild fire. 1, sensitive parental cv. Nevrokop 1146; 2, transgenic tobacco line Nevrokop 1146-*ttr*.

3. Gene Flow in Transgenic Tobacco

Seed samples from the recipient wild-type tobacco were analysed for the presence of the *ttr* gene by gene-specific PCR. Results indicated that the *ttr* gene could be detected in many of the samples from both fields. Rather unexpectedly, the presence of the *ttr* gene could also be detected in samples collected from the recipient plants most far away from the transgenic pollen source, at 13.8 meters. Even at 13.8 meters gene flow apparently occurs in tobacco. The results of all PCR analyses are summarised in maps of the trial site for tobacco line Nevrokop 1146-*ttr* (Figure 5) and for tobacco line Plovdiv 50-*ttr* (Figure 6). There are no obvious differences in the pattern or distance of gene flow in both trials.

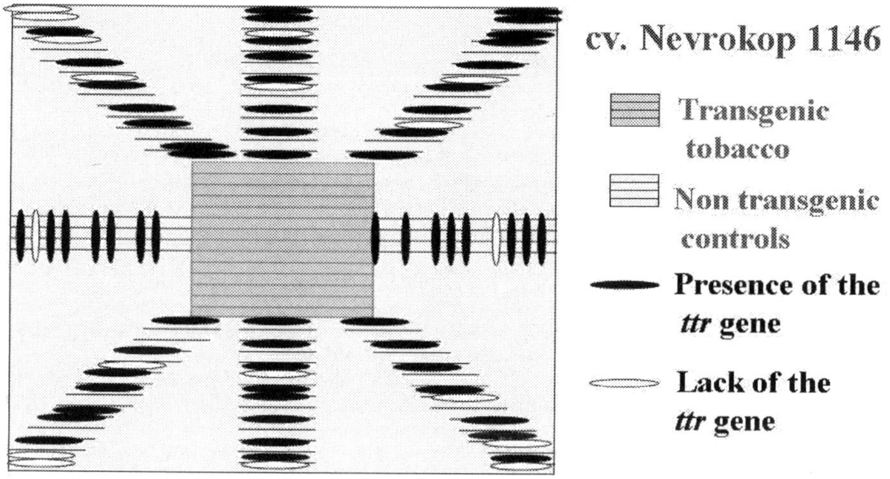

Figure 5. Summary of the dispersal of the *ttr* gene in the plot of tobacco cv. Nevrokop 1146

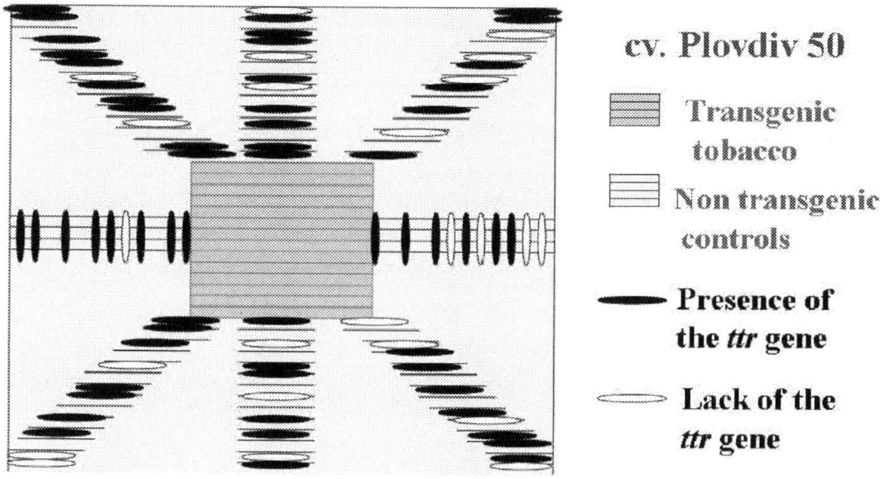

Figure 6. Summary of the dispersal of the *ttr* gene in the plot of tobacco cv. Plovdiv 50

Our analysis of gene flow in tobacco demonstrates that there is a considerable gene flow in tobacco, up to 13.8 m in the experimental field trials performed. This was unexpected for a self pollinating species as tobacco. It was also unfortunate, because 13.8 meters is also the largest distance evaluated. Therefore, further experiments and more analyses will have to be performed to judge this result and to determine an ecologically safe cultivation distance for preventing gene flow from transgenic tobacco to non-transgenic tobacco. The results indicate that also for self-pollinating species the phenomenon of outcrossing may require more attention than previously assumed.

References

[1] Jarosz, N., Loubet, B., Durand, B., McCartney, A., Foueillassar, X. and Huber, L. (2003). Field measurements of airborne concentration and deposition rate of maize pollen. Agric. Forest Meteorol. 119, 37-51.

[2] Bartsch, D., Lehnen, M., Clegg, J., Pohl-Orf, M., Shuphan, I. and Ellstrand, N. (1999). Impact of gene flow from cultivated beet on genetic diversity of wild sea beet populations. Mol. Ecol. 8, 1733-1741.

[3] Hauser, T.P. and Bjorn, G.K. (2001). Hybrids between wild and cultivated carrots in Danish carrot fields. Genet. Res. Crop Evol. 48, 499-506.

[4] Paul, E.M., Thompson, C. and Dunwell, J.M. (1995). Gene dispersal from genetically modified oil seed rape in the field. Euphytica 81, 283-289.

[5] Hudson, L., Chamberlain, D. and Steward, C. jr. (2001). GFP-tagged pollen to monitor pollen flow of transgenic plants. Mol. Ecol. Not. 1, 321-324.

[6] Anzai, H., Yoneyama, K. and Yamaguchi, I. (1989). Transgenic tobacco resistant to a bacterial disease by the detoxification of a pathogenic toxin. Mol. Gen. Genet. 219, 492-494.

[7] Batchvarova, R., Nikolaeva, V., Slavov, S., Bossolova, S., Valkov, V., Atanassova, S. et al. (1998). Transgenic tobacco cultivars resistant to *Pseudomonas syringae* pv. *tabaci.* Theor. Appl. Genet. 97, 986-989.

[8] Dellaporta, S.L., Wood, J. and Hicks, J.B. (1983). A plant DNA minipreparation: version II. Plant Mol. Biol. Rep. 1, 19-21.

[9] Anzai, H., Yoneyama, K. and Yamaguchi, I. (1990). The nucleotide sequence of tabtoxin resistance gene (*ttr*) of *Pseudomonas syringae* pv. *tabaci.* Nucl. Acids Res. 18, 1890 (EMBL accession x17150).

Genomics for Biosafety in Plant Biotechnology
J.P.H. Nap, A. Atanassov and W.J. Stiekema (Eds.)
IOS Press, 2004

Romanian Biosafety Research: Current Approaches and Developments

Elena Marcela BADEA[1*], Ion ROSCA[2], Emilian MADOŞĂ[3] and Ion CIOCAZANU[4]
*[1]Institute of Biology, Splaiul Independentei 296, 060031 6 Bucharest, Romania; [2]University of Agriculture and Veterinary Medicine, Bucharest, Romania; [3]University of Agriculture and Veterinary Medicine, Timişoara, Romania; [4]Pioneer Hi-Bred Seeds Agro SRL Romania; *email: elena.badea@ibiol.ro*

Abstract. Roundup Ready (RR) soybean and Superior New Leaf (Bt) potato are approved for commercialization in Romania, whereas RR and Bt corn are being evaluated for registration. The biosafety research is carried out by universities and institutes, in good collaboration with the private sectors seeking approval. The research deals with the major environmental concerns for each crop-trait combination. No obvious adverse effects on the biodiversity of weed populations were identified upon cultivation of RR soybean. The risk of cultivating RR soybean with use of Roundup is negligible. Cultivation of Bt potato did not show harmful effects on non-target organisms; it tended to increase insect biodiversity. Adverse environmental effects of Bt corn on non-target invertebrates are negligible. A study into pollen flow in corn indicates that the conditions for co-existence of GM corn with non-GM corn could be achieved without major changes of existing farming practices, if a distance of 200 m isolation between GM and non-GM fields is realized.

Introduction

In Romania, four genetically modified (GM) crop-trait combinations are currently being evaluated for environmental release and marketing. It concerns herbicide-tolerant soybean (*Glycine max*) and corn (*Zea mays*) as well as insect-resistant potato (*Solanum tuberosum*) and corn (Table 1). Roundup ready (RR) soybean (Monsanto) is approved for commercialization till 2005. Bt potato (Monsanto) is approved for market introduction, but is not actually grown. RR corn and Bt corn are approved for field testing for registration.

Developments with GM crops in Romania were accelerated in 2000 by private seed companies starting to apply for approval for deliberate release and commercialization. The Romanian Biosafety Commission (RBC) is overseeing all risk assessments of the deliberate release of GM crops. The RBC also is the main advisor of the Romanian Competent Authority that has the power of law to decide over deliberate release and market introductions. The relevant biotechnology law requires the monitoring of direct and indirect, immediate as well as long-term impacts of the commercial release of any GM crop. RBC defined the main issues and concerns associated with the applications for deliberate release in Romania (Table 1). It subsequently approached the applying industries to help performing the research necessary to address these issues. The research was carried out by research groups in Romanian universities and research institutes with appropriate experience in GM plants and transgenic technology [1-3]. The main environmental concerns differ per crop-trait combination (Table 1) and these will be discussed for each individual case below.

Table 1. Status of release and commercialization of GM crops in Romania

GM crop[1]	Type of release		Main environmental concerns	Partners/ Research project[3]
	Trials (for registration)	Market[2]		
RR Soybean	+	+	Survival of weed populations	Institute of Biology-B./ Monsanto: "Surveillance of RR soybean - 2002/2003"
Bt Potato	+	(+)	Effect on non-target species	Agricultural University-B./ Monsanto: "Impact on non-target insects - 2001 "
Bt Corn	+	–	Gene flow	Agricultural University-T./ Monsanto/Pioneer: "Study of gene flow in corn in Western Romania - 2003"
			Effect on non-target species	Agricultural University-B./ Monsanto: "Impact on non-target insects - 2001"
RR Corn	+	–	Gene flow	Agricultural University-T./ Monsanto/Pioneer: "Study of gene flow in corn in Western Romania - 2003"
			Effect on non-target species	Institute of Biology-B./ Monsanto: "Impact on earthworms - 2001"

[1]Abbreviations used: RR, Roundup ready, Bt, *Bacillus thuringiensis* toxin-containing.
[2]Although Bt potato (Superior New Leaf) is approved for market introduction, it has not yet been grown commercially in Romania.
[3]Abbreviations used: B., Bucharest; T., Timişoara.

Romanian legislation is being adjusted to be in accordance with EU legislation. Therefore, a major issue is co-existence, the need and/or desire to have GM- and non-GM-crops being grown and kept separately throughout the whole production chain. The economic and market implications of GM and non-GM crops being grown in close proximity should be investigated and proper agronomic measures should be defined to guarantee proper co-existence. Adventitious presence of one crop in another can arise by cross-pollination. Therefore, a study was initiated to estimate the potential level of the adventitious presence of GM crops in non-GM crops depending on agronomic practice by estimating pollen (gene) flow in corn. The economic implications in terms of costs and the agronomic consequences in terms of cultivation practices, are studied in relationship to adventitious presence not exceeding the legal threshold in a project that is currently financed by the Biotech program of the Romanian government.

1. Herbicide-Tolerant (RR) GM Soybean

Since 2000, RR soybean is grown in Romania on thousands of hectares. The GM crop carries the CP4 5-enolpyruvylshikimate-3-phosphate synthase (EPSPS) gene resulting in tolerance to the herbicide Roundup. RR soybean is approved for commercialization till 2005. To monitor indirect and long-term impacts of the commercial release of this crop, a post-market surveillance program was initiated to investigate potentially unexpected effects of this introduction on the evolution of new or endogenous weed populations.

At four different locations in Romania, an inventory was made of the most frequent plant (weed) species in RR soybean fields of more than 50 hectare immediately before each herbicide treatment of a total of two, and prior to harvest. The same cultivation area was followed for one to two years after RR soybean cultivation in crop rotation schemes.

Table 2. Summary of the field surveillance program (2002-2003) of the potentially adverse effects of RR soybean cultivation on weed populations

Location	Description	Results
Viziru, Urleasca, Ianca (Braila) Ograda (Ialomita) Galati (Galati) Timişoara (Timis)	Inventory of the most frequent species in RR soybean fields of more than 50 ha before each herbicide treatment and at harvest	Most frequent species • Before first treatment: *Sorghum halepense* *Descurainia sophia* *Capsella bursa-pastoris* *Convolvulus arvensis* *Xanthium strumarium* *Senecio vernalis* *Matricaria perforata* *Cardaria draba* *Lactuca seriolla* • Before second treatment: *Xanthium strumarium* *Phragmites australis* *Cirsium arvense* *Solanum nigrum* *Conyza canadensis* *Thlaspi arvense* *Convolvulus arvensis* *Polygonum aviculare* *Hordeum murinum* *Agropyron repens* • At harvest: *Solanum nigrum* *Matricaria perforata* *Chenopodium album* *Sorghum halepense* *Hibiscus trionum* *Convolvulus arvensis* *Abutilon teophrasti* *Brassica rapa* *Cannabis ruderalis* *Diplotaxis muralis* *Datura stramonium* *Sinapis alba* *Panicum miliaceum* *Eragrostis minor* *Arctium lappa* *Rumex sanguineus* *Setaria glauca* *Amaranthus retroflexus* *Xanthium italicum*
	Inventory of the most frequent species in crops fields 1-2 years after RR soybean cultivation	Weed species normally grow in fields after RR soybean, although with a relatively lower frequency. RR soybean is not a source of volunteers in rotational crops
	Surveillance of plant populations in aside fields and ruderal areas	- Feral soybean populations were never identified - Common weeds grow and develop normally and set seed - On out-of-crop agricultural fields and particularly at the cropped field edges, proliferating species characteristic for the pastures from covered areas include: *Poa angustifolia, Galium humifusum, reseda lutea, Agropyron repens, Taraxacum officinale, Melilotus officinalis* and others.

Soybean itself never appeared as volunteer crop. The diversity of weed species was similar compared to fields were non-GM soybean was grown, although the total amount of plants is somewhat less. In ruderal areas, feral soybean plants were not observed and common weed species proliferated. The main results of this research program are summarized in Table 2.

Weed species are eliminated from RR soybean crops by herbicide applications around emergence of the crop. Several weeks after the treatment, annual weed species had germinated from remaining seeds, developed rapidly and flowered. This way the annual weed seed reserve recovered. Perennial weed species are also drastically eliminated from RR soybean crops but they proliferate in the field edges, the ruderal areas, and along the access pathways. In the regions where RR soybean is grown, in addition there are out-of-crop areas, where in a similar way to the field edges the characteristic zonal pasture vegetation develops and maintains structure and size. The large quantity of seeds produced by the perennial species is sufficient for their perpetuation. Their existence is not being jeopardized.

Biosafety Networks in the Region of Central and Eastern Europe and the Role of the AgroBioInstitute

Atanas ATANASSOV* and Kristina GEORGIEVA
AgroBioInstitute, 8 Dragan Tsankov blv., Sofia 1164, Bulgaria;
**email: atanas_atanassov@abi.bg*

Abstract. Organisation and regulation of GM crops is a major challenge for the economies of the countries of Central and Eastern Europe (CEE). How future genomics technologies will contribute to the regulation of biosafety may take lessons from the way biosafety capacity is now organised. Three ongoing initiatives in biosafety in the AgroBioInstitute in Bulgaria are used to illustrate this point: the UNEP-GEF project, the CEE Steering Committee and the Bulgarian Biotechnology Information Center (BgBIC). Joining forces, sharing and exchanging expertise and transparent legislative systems are just as important aspects of biosafety capacity building as are scientific excellence, appropriate and accredited laboratory protocols and research facilities.

Introduction

In plant science, genomics is one of the faster growing branches that promises considerably to impact future sustainable agroproduction. Agrogenomics will point to genes and phenotypes that will be of use in further improving plants or plant agronomy for specific human uses, either with or without the use of genetic modified (GM) crops or living modified organisms.

As is the case for each new technology, future products from agrogenomics will face all the ethical, moral, social, technical and organisational issues associated with agrobiotechnology in general and the safety of GM crops in particular. At the same time, agrogenomics may also generate the technology and knowledge for the appropriate evaluation of the safety of GM plants.

Major issues are how the developments in genomics technologies can contribute to the regulation of biosafety and what kinds of regulation agrogenomics itself requires. The outcome of these discussions will shape the future of agrogenomics and biosafety. Organisation and regulation of GM crops is a major challenge for the economies of the countries of Central and Eastern Europe (CEE). Obviously these countries are eager to make use of the new technologies for biosafety assessments, provided that these are properly implemented in the national and regional legislation and can be afforded. Applications may be found in national regulation, as well as in the formulation of future research and new collaborative research projects.

There is a need to investigate the appropriateness of current regulations and assessments of biosafety. As example of current developments, we will here give an overview of recent initiatives in biosafety regulations in Bulgaria and the CEE region.

1. The UNEP-GEF Biosafety Project

In 2000, at a Ministerial Round Table, the United Nations Environment Progamme (UNEP) acknowledged the need for capacity-building in biosafety at the national level, in order to allow 'the safe use of modern biotechnology, in particular the safe transfer of living modified organisms (LMOs) resulting from modern biotechnology that may have adverse effects on the conservation and sustainable use of biological diversity between countries which may have very different climatic, social and economic condition'. A decision was taken by the Council of the Global Environment Facility (GEF) to support a second phase of the UNEP/GEF Pilot Biosafety Enabling Activity Projects in which Bulgaria participates, along with China, Cuba, Cameroon, Kenya, Uganda, Namibia and Poland. The UNEP-GEF project started in September 2002. It provides funds and assistance to Bulgaria to further develop and implement its national biosafety framework. The aim of the project is that by 2005 Bulgaria has a workable and transparent national biosafety framework that is in line with its international obligations and desires to join the European Union.

A national biosafety framework is a combination of policy, legal, administrative and technical instruments that should be developed to address safety for the environment and human health in the context of developing and applying modern biotechnology. Therefore, the project is focused on the development of the five distinguishable elements: (1) a policy on biosafety, (2) a regulatory regime for biosafety-related activities, (3) a system to handle notifications and/or requests for authorisations for certain activities, (4) a regulatory system for monitoring and enforcement and (5) an information system for public communication.

The Government of Bulgaria has indicated that its general approach with respect to modern biotechnology and biosafety will be to actively stimulate the technology, because of its potential to contribute significantly to sustainable food production, improved health care and environmental protection. In addition, the Government of Bulgaria also announced that, for the country to benefit maximally from this new technology, it would establish an adequate, workable and transparent national biosafety framework, which will be implemented in consultation with all relevant stakeholders. In the framework of the UNEP-GEF project, the draft law on GMOs was finalised. This included an independent review by international experts from international organisations, governments, academia, non-governmental organisations and the private sector. The various ordinances for the implementation of the law are expected to be prepared and finalised in the course of 2004. These regulations will specify the information requirements, the requirements for containment levels and risk assessments.

In addition, a regulatory system for handling notifications and requests for permits will be developed and implemented. While the scientific knowledge and practice to carry out proper risk assessments has been accumulated in Bulgaria over the last few years, additional developments are required for the implementation of the actual administrative system to handle notifications as well as the regulatory system to carry out risk assessments, the procedures for public communication and participation, as well as decision making. In the draft Law on GMOs, plans for monitoring are part of the notification procedure or requests for permits. Monitoring can consist of general surveillance or case specific monitoring. This will depend on the results of the risk assessment. Part of the UNEP-GEF project is the establishment of local 'botanical files'. This is the assembled information that includes all available data about plants grown commercially in Bulgaria, their economic and ecological importance, their wild relatives in Bulgarian flora and preliminary assessments on actual hybridisation and/or outcrossing. Such information will be essential to estimate the likelihood of gene transfer by pollen in Bulgarian agronomy.

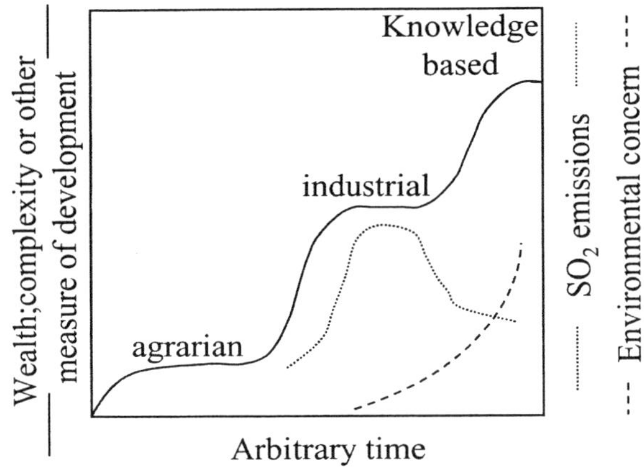

Figure 1. Relationship between economic development and environmental concern. The three primary economic systems of agrarian-, industrial-, and knowledge-based service are indicated with arbitrary indications of wealth and development. SO_2 emission is used merely as an indicator of industrial development and the subsequent environmental concern [2].

1. Innovation through Genomics in Crops

Present-day GM crops on the market contain only a few transgenes, such as the genes conferring insect resistance (Bt) or herbicide tolerance (CP4). Analysis of the research literature opens up a much more diverse picture for the future: it will be possible to enhance natural resistance systems. Functional genomics will reveal the true richness of hundreds of other resistance genes. Ecologically more relevant characteristics such as resistance to cold and drought are based on interactions of numerous genes and these are still poorly understood. However, due to functional genomics approaches, knowledge about such genetic properties is increasing rapidly. It may soon become possible to considerably enhance the drought and cold resistance in crops. Looking back in years from now, it will be recognised that the present day period of agrobiotechnology should be considered a first phase of development, an era of "palaeogenetics". The future elegant breeding methods based on genome-wide profiling and analyses offer the possibility of an ecological future for agriculture.

New research goals are being formulated using the approaches of genomics and model plants [3-15]. Plants are used as factories in the molecular farming of antibodies [9]. New possibilities of weed management are established through enhanced allelopathy [10]. Resistance to biotic and abiotic stress [11] and new approaches to improved nutrient content [12] and biological nitrogen fixation [13] will be future topics for genomic plant breeding research. Genomics research is proceeding at an amazing and accelerating pace, with numerous tools and approaches being documented [14-20]. A typical example is the approach to gain insight in the collaboration of genes in a concerted way [21]. Comprehensive genetic screens of mouse, plant and human transcriptomes consider gene expression as a quantitative trait, combining genetics, genomics and statistics to identify networks of regulation for traits-of-interest. The dramatic acceleration of the speed of analysis of genomes, traits, and genes is currently outperforming Moore's law (Figure 2). This is triggering feelings uttered as "Prometheus unbound".

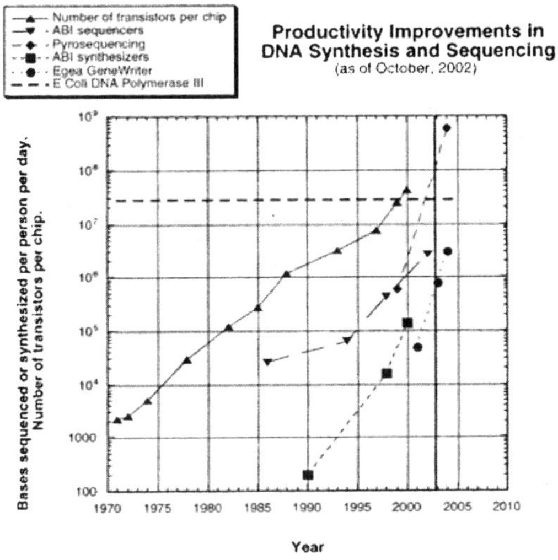

Figure 2. Prometheus unbound: revolutionary advances in biological technologies. Rapid acceleration of genomic analysis [40]. This plot shows that the productivity of DNA synthesis and sequencing are both increasing at least as fast as Moore's Law (upwards triangles). Each of the remaining points is the amount of DNA that can be processed by one person running multiple machines for one eight-hour day, defined by the time required for preprocessing and sample handling on each instrument. Not included in these estimates is the time required for sequence analysis. For comparison, the approximate rate at which a single molecule of *E. coli* DNA polymerase III replicates DNA is shown (dashed horizontal line), referenced to an eight-hour day.

Obviously, the genomics-based crops of the future will be the future objects of risk assessment and regulation. Crops of future generations produced with modern breeding methods of all kinds need a thoughtful and new approach in risk assessment, since the complexity of genomic changes will add to the complexity of risk assessment. Thought should be given to new approaches in assessing the risk and the likelihood of hazards upon the introduction of such new crops. There is no reason to be particularly pessimistic and the Chinese risk assessment scheme could be considered (Figure 3), rather than the usual scheme, in which risk is defined as hazard x likelihood. The latter is focusing solely on the negative side. A recent report concludes that current agricultural biotechnology has not had a permanently detrimental effect on biodiversity, despite many claims it would [22].

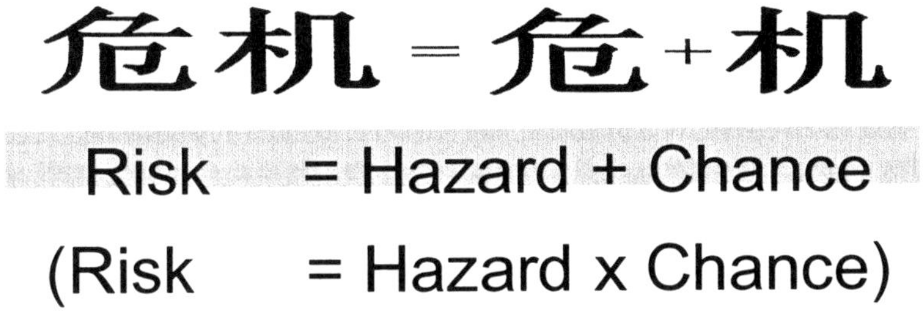

Figure 3. Chinese and European ways of interpreting risk assessment differ.

The case of the monarch butterflies shows that one paper in *Nature,* with the information that Bt corn could kill larvae of Monarch butterflies, very quickly resulted in the freezing by the EU Commission of the approval of new Bt corn varieties. Greenpeace neglected the caution expressed by John Losey of Cornell, and called for a ban on all GMO's. When field research in American Midwest showed that Monarch butterfly populations are hardly affected by large scale Bt corn cultivation, the (political) damage was already done.

How could this have happened so fast? The Monarch butterfly case is only one among many cases in which the so-called Precautionary Principle (PP) has been invoked to advocate preventative action to forestall possible harm even before the likelihood or the possible extent of the latter has been scientifically well established. This principle is highly contested because it may hamper innovation. Every new technology or innovation, like plant genomics, may eventually share the same fate as Bt corn. The aim of this paper is to reflect on the consequences of the PP, and to develop it into a principle of innovation that stimulates safe innovations. Before we describe the Innovation Principle and apply it to an example of plant genomics, we will first describe the PP and its problems.

1. The Precautionary Principle

Precaution as a principle may be derived from the idea of sustainability and it provides an intuitively simple idea that any action likely to result in serious environmental harm is morally wrong. The PP states that if there is reason to suspect that a particular substance or practice is endangering the environment then action should be taken immediately even if the evidence is not ironclad. The rationale behind this view is that it will be too late to respond effectively if we wait for a final scientific resolution years down the road [4]. According to Milne the origins of the PP can be found in the early 60s in West Germany. The PP was invented around 1965 by a committee of German civil servants who called it the *Vorsorgeprinzip* - literally, the 'foresight principle' [5]. In the 70s the German State tried to avoid environmental damage by careful forward planning. The PP was used to implement vigorous policies to tackle acid rain, global warming and pollution of the North Sea in the mid- to late-eighties [6].

In German Federal Law [7], the Principle of Precautionary Action requires that:

(1) harm should be avoided before it occurs;

(2) high quality scientific research should act to provide early warnings of the potential for damage;

(3) action not merely can, but must be taken on the basis of available information as necessary to avoid threats, even when evidence of causality remains limited;

(4) all technical, social and economic developments should tend towards a progressive reduction in overall environmental burden.

In 1992, the first international example of a law that is an implementation of the PP can be found in European Union Guideline 90/220, about the introduction of genetically modified organisms in the environment. In the case of genetically modified organisms there is indeed no clear proof of any special dangers. The principle is operationalised as a case-by-case assessment of every request to modify a certain crop. A scientific committee evaluates the acceptability of thinkable environmental risks. They use norms that depend on the latest views of science and definitions of 'damage' which are suggested by scientists. They do not determine risks, but the acceptability of uncertainties of possible environmental effects. The PP was further enshrined in Agenda 21, the Declaration of Rio Conference in June 1992. The core of the 'Rio' version is that when there are threats of serious or irreversible damage, lack of full scientific certainty shall not be used as a reason for postponing cost-effective measures that will prevent the environmental degradation [8].

Definitions such as these raise many questions. Is there ever full scientific certainty? Do we need a minimal threshold of scientific certainty or plausibility before we may (or should) undertake preventative action? And do we really know how to prevent harm if we are so much ignorant about the underlying cause-effect relationships? There is neither a generally agreed definition nor a set of criteria to guide its implementation. At the core of the PP is the intuitively simple idea that decision-makers should act in advance of scientific certainty to protect the environment from incurring harm [6]. The definitions that are currently on offer fail to spell out the precise conditions that have to be fulfilled before the PP may be invoked or the nature of the preventative action that has to be taken. The types of action suggested range from implementing a ban, imposing a moratorium while further research is conducted, allowing the potentially harmful activity to proceed while closely monitoring its effects, to just conducting more research. The PP does not have a very precise meaning as long as such crucial aspects are left largely unanswered.

In practice, however, the PP is often given a more definite meaning by reducing it to what we will call the 'dogmatic' version of the PP. In this case, no minimal threshold of plausibility is specified as a 'triggering' condition, so that even the slightest indication that a particular product or activity might possibly produce some harm to human health or the environment will suffice to invoke the principle. And just as often no other preventative action is contemplated than an outright ban on the incriminated product or activity. The intervention of Greenpeace in the monarch butterfly case seems to fit this pattern. This 'strong' version of the PP amounts to a complete reversal of the burden of proof on the technological innovators and to forbid the use of any innovation until there is proof that they are harmless. The 'strong' version would be the end of innovation because every new technology would be excluded as long as it is made plausible that there may be as yet unknown dangers connected to it. It also is, in principle, always possible to point at hypothetical causalities that could lead to a disaster.

Is the 'strong' version tenable? First, we must examine more closely the underlying justification for the 'strong' version of the PP. Why should the prospect of harmful effects of a new technology take precedence over the prospect of beneficial effects, quite apart from the inherent likelihood of each of these possibilities? The obvious answer seems to be that such a priority is defensible only when the harmful effects are of such magnitude that they carry catastrophic potential. The infinite costs of a possible catastrophic outcome necessarily outweigh even the slightest probability of its occurrence. Manson [9], therefore, dubs the PP the 'catastrophe principle'.

This type of reasoning exhibits a remarkable resemblance to a well-known example of a 'zero-infinity dilemma', namely Pascal's famous 'wager.' When it comes to wagering on the existence of God, the 17th century French philosopher argued incisively in his *Pensées* that it is better to be safe than sorry [10]. Given an unknown but nonzero probability of God's existence and the infinity of the reward of an eternal life, the rational option would be to conduct one's earthly life as if God exists. However, Pascal's reasoning contains a fatal flaw. His argument is vulnerable to the 'many gods' objection [9]. Consider the possible existence of another deity than God, for example Odin. If Odin is jealous, he will resent our worship of God, and we will have to pay an infinite price for our mistake. Never mind that Odin's existence may not seem likely or plausible. It is sufficient that we cannot exclude the possibility that he exists with absolute certainty. Therefore, the very same logic of Pascal's wager would lead us to adopt the opposite conclusion not to worship God. Pascal's argument, then, cannot be valid.

If the wager argument is not valid, the 'strong' version of the PP cannot be valid either. Take the application of this principle to the problem of global warming. Environmentalists often argue that even if it is not conclusively established that the emission of carbon dioxide and other gases causes an enhanced greenhouse effect, the mere prospect of an ecological catastrophe due to such a scenario should lead us to drastically curb our emissions of greenhouse gases now. By the same logic, however, one could conjure up the possibility of a coming ice age. The mere prospect of this equally catastrophic scenario should then induce us to avert this outcome by stepping up the emission of greenhouse gases. Thus, the strong version of the PP would lead to contradictory recommendations [11]. In a similar way, it could be argued that this principle commits us to either of two contradictory and mutually exclusive policies: (1) we must not develop GM crops, and (2) we must develop GM crops.

The first alternative is argued vehemently by many environmentalists who appeal to the PP. To support the second possibility, Comstock [12] conjures up a dramatic scenario in which people are forced to seize upon the remaining reserves of nature in a desperate effort to overcome food shortages resulting from global warming. He then argues, in the style of the environmentalists, that lack of full scientific certainty that GM crops will prevent environmental degradation shall not be used as a reason for postponing this potentially cost-effective measure. In conclusion, the 'strong' version of the PP is untenable.

Closely linked to the 'strong' version of the PP, but also to many of the 'milder' versions of the PP, is the idea of reversing the onus of proof. Proponents of the Wingspread Statement declare that the applicant or proponent of an activity or process or chemical needs to demonstrate that the environment and public health will be safe. The proof must shift to the party or entity that will benefit from the activity and that is most likely to have the information [13]. Greenpeace also holds that effective implementation of the PP requires a shift in the burden of proof [14]. Shifting the burden of proof seems a fairly straightforward way to ensure that greater weight will be given to the negative aspects than to positive aspects.

Is the proposed shifting of the onus of proof tenable? Reversing the burden of proof would amount to substituting the maxim 'guilty until proven innocent' for the age-old legal principle 'innocent until proven guilty'. Biotech enthusiasts and antiregulationists resent this departure from what they consider time-honoured legal sanity [15]. They are prone to counter the frequent invocation of the PP with an equally insistent demand of 'sound science.' The critics of the PP assert that the burden that environmentalists and regulators want to impose on the proponents of new technologies tends to be unbearable [15]. In the name of absolute safety, the latter are asked nothing less than to demonstrate conclusively that the new technologies they advocate offer no possible harm. This is a formidable, perhaps even logically impossible, task. One cannot prove a negative [16]. Moreover, a risk-free world is not a real option. Thus, a consistent application of the PP would in a final analysis stifle all innovation.

A closer analysis of what is involved in applying the classical principle 'innocent until proven guilty,' however, reveals that the situation needs not be as black and white as it seems at first sight. Take the paradigm case of criminal justice. There are two main ways in which a miscarriage of justice can come about. Either the suspect did not commit the crime, but the verdict found him guilty; or the suspect did commit the crime, but the verdict found him not guilty. In a civilised system of justice, the risks of the first type of error are minimised as far as possible. That is what is meant by the phrase 'innocent until proven guilty'. The system contains safeguards and precautions in the form of high standards of proof so as to ensure that a suspect will be condemned for a certain criminal offence only if it has been established 'beyond reasonable doubt' that he in fact committed the alleged offence. There is a price for this cautious and civilised approach, namely the possibly large number of wrongdoers who have to be acquitted due to 'lack of sufficient proof'.

To a certain extent, the risks of the two types of error are inversely related. We may try to reduce the risk of condemning an innocent person by demanding ever more exacting standards of proof but only at the expense of increasing the risk of acquitting culpable offenders. Therefore, we must recognise that there is an inevitable trade-off involved in the design of our system of criminal justice. We may attempt to set our standards as high as we can, but somewhere a balance must be struck, lest the system will become unworkable by making it too difficult to pass sentence on the majority of wrongful offenders. In statistical testing, there is a similar trade-off to be made between the chances of committing a type I or a type II error, i.e. rejecting the null hypothesis of 'no effect' when it is in fact true or failing to reject the null hypothesis when in fact it is false. By selecting a significance level, we implicitly strike a particular balance. Ideally, this balance should depend on our estimation of the costs, economic and otherwise, associated with either of the two types of error.

The above considerations show that the matter at issue is not just where to place 'the' burden of proof. As soon as we allow for more or less exacting standards of proof, an extra dimension of variation immediately becomes visible. In other words, the burden we want to put on the shoulders of one or the other party becomes more or less heavy depending on whether we set our standards of proof more or less highly. This consideration may help us to escape from the unduly polarised opposition of PP versus sound science and develop the PP into an Innovation Principle.

2. Towards an Innovation Principle (IP)

So far we have endorsed the two main points of criticism against the 'strong' version of the PP. First, it is used to examine only one side of the risk equation, the risk of using new technologies, and not the other side: their benefits or the `opportunity costs' in the form of foregone possibilities of mitigating known risks. Second, the principle tends to put a well-nigh-impossible onus of proof on the shoulders of the proponents of new technologies. It all boils down to the fact that the 'strong' version of the PP in its interpretation precludes any balancing between the risks of harm with the potential benefits of innovation. We think that to preclude such balancing as a matter of principle is highly unreasonable. That is not to say that the risks and benefits of new technologies are all on an equal par and that their balancing is a simple and straightforward technical task that can be safely delegated to the practitioners of CBA (cost-benefit analysis), RBA (risk-benefit analysis) and kindred methodologies. Indeed, there may be good ethical reasons to assign unequal weight to various considerations and concerns.

Most critics of the PP swear by 'sound science' and their ideal is a fully quantified risk assessment. However, it is only possible to meet this objective in more limited contexts, where direct and short-term hazards such as toxicity or pathogenicity are at issue. Even then the expression 'sound science' is disingenuous because it obscures the extra-scientific value judgements that necessarily enter into the whole exercise (e.g. identification of hazard types, baselines of acceptability, and trade-offs between type I and type II errors). In other contexts, where indirect, cumulative, or subtler ecological effects are at issue, the format of the fully quantified risk assessment is unattainable. Adherents of 'sound science' will be tempted to downplay such less straightforward hazards as purely hypothetical, conjectural, or theoretical risks that can safely be ignored. However, as the proponents of the PP are never tired in pointing out, lack of evidence of harm is not evidence of lack of harm. If we are really concerned about such hazards, we can put in additional investigative effort to learn more about their plausibility or likelihood. It would be absurd to halt our inquiries with an appeal to 'sound science.'

The inability of science to give absolutely unequivocal evidence of safety, opens the door to the creation and contestation of environmental problems [17]. All scientists can give are estimates of probability which often vary widely from one to another. This lack of certainty allows people to assert that the situation is alarming and the risk is too high. There is a shift from prevention of clear and manifest dangers towards preventing hypothetical dangers. Now we can interfere when there is a possible danger. Even when it is not clear if there is any risk involved. A large measure of the disagreement here resolves around how science should be done: reductionistic or holistic. Causation may be indirect or multiple, making it all but invisible to the reductionistic perspective. Conventional science handles ambiguity by claiming that further research is needed, but when we use the PP it is, in principle, possible to hide behind hypothetical causalities, of which the existence cannot be proven empirically, nor theoretically excluded. Also, the PP could generate new problems or is even not applicable in some situations. Our position can be compared to the view expressed in the report of the British Nuffield Council on Bioethics [18]. The Working Party on GM crops that prepared the report noted that on a stringent definition of the PP there could be no balancing of the risk of harms with the benefits of innovation, since even a suspicion of possible harm, no matter how ill-founded, would be sufficient to prohibit a new technology [18].

According to the advocates of minimal regulation, the only grounds providing a sufficient case for regulating new technologies are valid reasons based on 'sound science'. In their view, considerations derived from some interpretation of the PP or from worst-case analyses are inherently biased, unscientific, and therefore inadmissible as grounds for regulation. The British investigators of the ESRC Global Environmental Change Programme take precisely the opposite view. They hold that due to the inevitable selectivity in framing assumptions science is inherently incapable of dealing adequately with the profound uncertainties that are raised by biotechnology and other new technologies. They suggest that science cannot provide definitive answers in these cases, so the policy of relying on claims of 'sound' science may, ironically, itself be unsound. Ethical issues are central [19].

In view of the criticism on the 'strong' version of the PP we propose to replace it by what can be called a broad Innovation Principle (IP). A first definition could be: from the very start of an innovative trajectory, all societal stakeholders should co-operate in balancing the risks of harm and the benefits of innovation. The six basic assumptions of the IP are:

- the PP is evoked when an innovation is already introduced in society, we think it is better to start from the beginning of an innovative trajectory;
- the IP is about all possible harm, and not only about environmental degradation, like the 'Rio' version;
- a situation of scientific uncertainty is the main reason to evoke the PP, we think that there is always a degree of scientific uncertainty involved, especially at the start of an innovative trajectory;
- we embrace the rationale behind some of the 'mild' versions of the PP, like the 'Rio' version, where lack of full scientific evidence is not used as a reason to postpone action;
- in time, different versions of the PP may replace each other. For example, the 'Rio' version may be replaced by the so-called 'institutional' version when the scientific uncertainty diminishes. It allows for some scientific criteria with a degree of uncertainty that define 'harmlessness'. Without any empirical obviousness there would be no selectivity;
- the IP combines technical, economic, social, and ethical considerations.

The switch, during the development and application of an innovation, from the 'Rio' version to the 'institutional' version of the PP is the core of the IP. This requires that some institutional tools are made operational, for example a committee that develops and use standards to control the quality of food. For this reason, during the development and application of an innovation in a broad societal context, the following activities are compulsory: (1) pre-market screening; (2) the development of monitoring devices; (3) the development of safety standards; (4) post-market screening. This way the IP contributes to an early societal acceptance of innovations, because risks, in society, are more acceptable the easier they are identified, personally detected, appreciated and avoided.

3. Outline of an Example of the Innovation Principle

In this paragraph we will discuss the Centre for BioSystems Genomics (CBSG) as an example of an attempt to apply a kind of IP [20]. The CBSG is part of the Netherlands Genomics Initiative. On July 18th 2002 the Dutch Cabinet approved this initiative and allocated a budget of M€ 195 for a period of five years. In January 2003, four Centres' of Excellence started: the Cancer Genomics Centre, the Centre for BioSystems Genomics, the Centre for Medical Systems Biology, and the Kluyver Centre for Genomics of Industrial Fermentation. This year the Centre for Society and Genomics started, and also the Netherlands Bioinformatics Centre, the Netherlands Proteomics Centre, and five Innovative Clusters (consortia which are headed by a company). Right from the start of the Genomics Initiative, several societal sectors co-operated: knowledge centres, industry, government and politics. It is important to observe that the societal organisations and the consumers/public are absent.

The Centre for Biosystems Genomics, a consortium comprising four Universities, two research institutes and thirteen industrial partners, will conduct research into the key sustainability issues in agroproduction, with a strong focus on ecology and biodiversity: disease resistance to drastically reduce pesticides and product quality to address changing consumer and industrial demands. This consortium comprises top scientists in the field of plant sciences and genomics, and the major Dutch companies in plant genomics, breeding, cultivation and processing. From the start the CBSG is embedded in a societal context that requires increasingly efficient ways to produce safe and high-quality nutritious food and natural compounds. At the same time, it is necessary to maintain biodiversity and to conserve natural resources. Sustainable agroproduction systems are crucial to meeting both needs, and the CBSG will exploit the potential of biosystems genomics to realise new opportunities for sustainable agroproduction systems, contributing to the quality improvement of food crops and their derived products for both the consumer and the environment.

Biosystems genomics is the study of genes, gene expression and gene functioning at the level of the organism. It encompasses the complex interaction at all levels of biological information: DNA, mRNA, proteins, metabolites and how these informational networks work together. Biosystems genomics helps to address the complex sustainability issues at the level of whole plants and their environment. This provides opportunities for innovative, sustainable improvement of crops and entire agro-food production chains and enables a transition to precision agro-production, minimising use of energy, chemical treatments and wasted product while maximising product quality. The mission of the CBSG is threefold:

- to exploit genomics as basis for the development of sustainable agroproduction systems;
- to contribute to high quality, safe and healthy food crops and derived products for consumers and the environment;
- to contribute to the advancement of science and technology while creating economic and societal value.

This mission is accomplished by a research focus upon unveiling mechanisms contributing to crop quality, environmental quality, the reduction agrochemicals, resistance to diseases, availability of soils and water, and cold/drought/salt/metal tolerance. The CBSG tries to influence the food and non-food quality by changing consumer demands (the flavour, fragrance, and health aspects of products), by improving the food supply chain (changing the freshness, storability, and transport of the products), reducing post-harvest chemicals, waste, and processing energy.

The Centre focuses on two related crop species of national and global importance: potato and tomato, in addition to the model plant *Arabidopsis thaliana*. This choice is driven by economic, social, environmental and scientific considerations. Both crops are strong economic drivers of the Dutch agro-industry, internationally renowned for its breeding, propagation, cultivation, processing and export. The industrial partners are convinced that a joint innovative research programme is critical to maintaining and expanding their market position. Growing consumer demand creates a clear *social* need for sustainable production of high-quality products. The programme aims to enhance the sustainability of potato and tomato production and processing, important for all systems of agriculture, including organic farming. Devastating diseases, such as late blight in potato culture, are one of the main hindrances for sustainable agro-production systems. Potato cultivation alone requires over 70% of all chemical pesticides used in the agro-industry today. Disease-resistant crop varieties are needed to drastically reduce the use of pesticides, thus relieving pressure on the environment.

The overall research programme of the CBSG will identify the complex genetic and biochemical mechanisms in potato and tomato underlying pathogen resistances and consumer and industrial quality traits. It will develop a research platform for genome-based approaches in the crop species potato and tomato and the model plant Arabidopsis. For this, it will explore the rich biodiversity of Solanaceous species to identify new traits and valuable genetic alleles and fully exploit the emerging field of bioinformatics to integrate genomics data from sequence analysis, transcriptomics, proteomics and metabolomics. In addition, it will conduct multidisciplinary research into the various societal implications of genomics. Knowledge transfer to society is one of the main overall objectives of the Centre. The commitment of researchers and partners is key to successful knowledge transfer and to the Centre reaping the financial benefits of the commercialisation of intellectual property.

A way to assure the societal imbedding of the CBSG is the societal research programme. This research generates knowledge about the links between genetic information and the interaction of organisms with their environment to develop new plant cultivars for food and non-food applications. The aim is to influence people's view on sustainable agro-industrial production systems (they can choose for a safe environment and healthy food), and their responsibility to implement these novel opportunities for society.

The results of the CBSG genomics research need to be implemented and embedded, responsibly, in society. Continuous reflection on the societal components of genomics is an integral part of the total programme. The CBSG researchers in the life sciences will participate in deliberation about the meaning and implications of their work for society. Therefore, the CBSG will perform multidisciplinary research into the societal implications of genomics and encourage its life science researchers to become actively involved. The Centre will co-operate in the Genomics & Society programme of the Centre for Bioethics and Health Law (University Utrecht) and Centre for Methodical Ethics and Technology Assessment (META; Wageningen UR). This join t programme aims to develop links between genomics research and research and deliberation on its societal aspects. The co-operation will include:

- Participation in essays, publications and research projects on the societal component of genomics,
- Organisation of joint colloquia examining ethical questions on the implications of genomics research in general and our Biosystems research in particular,
- Introduction of modules on implications of genomics applications in plants in courses for BSc, MSc, and PhD students in the life sciences and in other disciplines.
- Organisation of workshops with the stakeholders in society.

Could the CBSG be considered as an example of the IP? Plant genomics is in its infancy. In contrast to GMO's, the public in general is still unaware of its existence. The Netherlands Genomics Initiative provided the opportunity to combine different research groups and the potato and tomato industry into a coherent centre. We may conclude that almost from the start plant genomics is placed in a societal context with a specific attention for risks and benefits. This is also caused by earlier negative experiences of many researchers in the field of GMO's [21]. Getting societal stakeholders and consumers/public involved, remains difficult. Since the programme of the CBSG is already in place, there are almost no possibilities to change it. However, there is still four years left to co-operate with the stakeholders, and to develop new projects.

The information gap between society and CBSG is an important issue. Some participants within the CBSG are reluctant to inform society, because this could bring a lot of misunderstanding and trouble. Others are in favour of a more transparent information policy. One of the first priorities of the societal research is to develop an agenda for joint activities with the stakeholders. How and when to involve what kind of people outside the CBSG? The CBSG has still a long way to go. It has to contribute to the development of monitoring devices and screening standards, but it is on its way towards the IP.

4. Conclusions and Future Outlook

For centuries, humans have selected plant varieties that best fit their purposes and developed crop plants that have many advantages compared to natural (wild) plants in quality, quantity and farming practises. In the last century modern biotechnology managed to improve mono-factorial traits. Many people reject this technology all over the world, applying the 'strong' version of the PP. However, multi-factorial traits involved in resistance and quality have proven to be extremely difficult to improve, certainly in combination. The revolution in life sciences signalled by genomics dramatically changes the scale and scope of our experimental enq uiry and application in plant breeding. The scale and high resolution power of genomics enables to achieve a broad as well as detailed genetic understanding of plant performance at multiple levels of aggregation. The complex biological processes that make up the mechanisms of pathogen resistance and provide quality to our crops are now open for a systematic functional analysis by means of plant genomics. Although we have shown that the 'strong' version of the PP is untenable, the endeavour of plant genomics, like the GMO technology before, runs the risk of rejection by society. To plant genomics centres like the CBSG, the IP may be helpful in attempts to embed plant genomics into society.

References

[1] van den Belt, H. and Gremmen, B. (2002). Between precautionary principle and "sound science": distributing the burden of proof. J. Agric. Environ. Ethics 15, 103-122.

[2] Losey, J.E., Rayor, L.S. and Carter, M.E. (1999). Transgenic pollen harms monarch butterfly. Nature 399, 214.

[3] Ortman E.E. Barry B.D., Buschman L.L., Calvin D.W., Carpenter J., Dively, G.P. et al. (2001). Transgenic insecticidal corn: the agronomic and ecological rationale for its use. BioScience 51, 900–903.

[4] Hannigan, J. (1995). Environmental Sociology. London, UK: Routledge.

[5] Milne, A. (1993). The perils of green pessimism. New Sci. 138, 34-37.

[6] O'Riordan, T. and Jordan, A. (1995). The precautionary principle in the contemporary environmental politics. Environ.Val. 4, 191-212.

[7] Santillo, D., Belazzi, T. and Johnston, P. (1998). A precautionary approach to the regulation of endocrine disrupting substances. Paper presented at: "Endocrine Disrupters - How to Address the Challenge", Joint Conference of the European Commission, DG XI and the Austrian Presidency, Vienna.

[8] Gremmen, B. and van den Belt, H. (2000). The precautionary principle and pesticides. J. Agric. Environ. Ethics 12, 197-202.

[9] Manson, N. (2002). Formulating the precautionary principle. Environ. Ethics 24, 263–274.

[10] Haller, S. (2000). A prudential argument for precaution under uncertainty and high risk. Environ. Ethics 5, 175–189.
[11] Graham, G. (2002). Genes: a Philosophical Inquiry. London, UK: Routledge.
[12] Comstock, G. (2000). Are the policy implications of the precautionary principle coherent? Talk, Harvard University, 23 September. (www.cid.harvard.edu/cidbiotech/comments/ comments72.htm).
[13] Raffensberger, C. and Tickner, J. (Eds) (1999). Protecting Public Health and the Environment: Implementing the Precautionary Principle. Washington, DC: Island Press.
[14] Greenpeace (1999). Safe Trade in the 21st Century. Amsterdam, The Netherlands: Greenpeace International.
(http://archive.greenpeace.org/politics/wto/ciel.pdf).
[15] Miller, H. and Gonko, G. (2000). The science of biotechnology meets the politics of global regulation. Issues in Science and Technology On Line. (www.nap.edu /issues/17.1/miller.htm).
[16] Wildavsky, A. (1995). But is it True? A citizen's guide to environmental health and safety issues. Cambridge, MA: Harvard University Press.
[17] Hannigan, J. (1995). Environmental Sociology. London and New York: Routledge.
[18] Nuffield Council on Bioethics (1999). Genetically Modified Crops: the ethical and social issues. (http://www.nuffieldbioethics.org/gmcrops/index.asp).
[19] ESRC Global Environmental Change Programme (1999). The Politics of GM food: risk, science & public trust, Special Briefing No 5 (www.susx.ac.uk/Units/gec/gecko/gm-brief.htm).
[20] Stiekema, W.J. and Kok, R. (2001). Letter Of Intent: Centre for BioSystems Genomics, Wageningen: Plant Research International.
[21] Nap, J.P., Jacobs, J., Stiekema, W.J. and Gremmen, B. (2002). Genomics and Sustainability: exploring a societal norm. Note 204, Wageningen: Plant Research International, 37 pp.

Genomics for Biosafety in Plant Biotechnology
J.P.H. Nap, A. Atanassov and W.J. Stiekema (Eds.)
IOS Press, 2004

The Regulation of Genomics: What Did We Learn from the GMO Story?

Mike ADCOCK* and Julian KINDERLERER
Sheffield Institute of Biotechnological Law & Ethics, Law Department, University of Sheffield, Sheffield S10 1FL, United Kingdom;
**email: m.adcock@sheffield.ac.uk*

Abstract. The introduction of a new technology such as agricultural biotechnology may depend on the perceived balance between the benefits of the technology and the potential risks to the environment and human health. It is the responsibility of governments to regulate the technology so as to minimise the risks while maximising the benefits. This paper provides a background into the shaping of policy and regulatory frameworks within the European Union and the United States. It outlines the global framework in which all countries have to operate. The EU and the United States policies on genetically modified organisms are very different in concept, although in practice the formal treatment is very much the same. The paper highlights the ongoing debate in the areas of environmental protection, intellectual property rights and public perception of GM crops.

Introduction

Biology has been used for millennia to provide food and other products for use by humans. It is used to make food palatable or safe, and many mechanisms have been 'discovered' that enable foods to be 'modified' using organisms (usually micro-organisms). Bread, beer, wine, cheese, tofu and yoghurt are amongst the modified food products, many of which use micro-organisms to modify starting ingredients to form the final food product. Less obvious is the modification of all our food crops. The wild, undomesticated relatives of all food crops are almost unrecognisable in comparison to that which we grow in fields today. Genetic modification of crop plants has been practised since the crops were first identified. The first human to choose to keep the best seeds for growing in the following season rather than eating them was practising biotechnology.

Modern biotechnology[1] is seen as different to traditional selection, for it permits the transfer of characteristics that could not be achieved naturally. Scientists often argue that the new techniques are simply an extension of the continuum of selection and genetic modification that has been used over hundreds of years. These 'traditional' techniques have changed markedly during the 20th century as our understanding of the biological processes has improved. Deliberate mutation and many other artificial techniques have allowed selection of characteristics between weakly compatible organisms.

There are other scientists who believe that the transfer of genes between non-compatible organisms is truly different from traditional techniques and constitutes something completely new. Modern biotechnology that permits modifications that cannot happen naturally has raised many concerns and has been regulated from almost the initial experiments that allowed genetic material to be transferred among unrelated organisms.

1. Historical Perspective

It is often the case that safety legislation is brought in because an accident has occurred and systems need to be put into place in order to ensure that it does not reoccur. The possible risks of modern biotechnology were recognised at the very beginning of its use, and steps were taken to ensure that it was used safely. Policy on the safe use of biotechnology set precedents.

It is 50 years since the paper identifying the double helical structure of deoxyribonucleic acid was published by Watson and Crick in *Nature* [1], and over 30 years since the discovery of enzymes that could cut the DNA sequences at specific points in the sequence mechanisms became available for transferring information between organisms in a precise and controlled manner. Initially, these techniques were only effective in micro-organisms, as the insertion of specific molecules into the whole genome of plants is much more complicated.

The discovery of the use *Agrobacterium tumefaciens* allowing the insertion of a sequence of DNA into a random position in the plant genome allowed the generation of GM plants to be realised. The ability to regenerate a complete plant from a single cell made it possible to easily modify many plant species. It was later discovered that if ballistics was used to literally fire DNA into cells a small proportion would have sequences stably integrated into their genome. Selection techniques involving (usually) choosing to have one of the inserted sequences express antibiotic tolerance mean that the tiny proportion of cells modified using the technique can be identified relative to the rest.

The potential uses of genetic modification[2] were obvious from the moment that the techniques that enabled the transfer of genes from one organism to another unrelated organism were first identified. A United Kingdom government-initiated committee reported in 1975 that genetic manipulation techniques would provide "substantial though unpredictable benefits" ... "application of the techniques might enable agricultural scientists to extend the climatic range of crops and to equip plants to secure their nitrogen supply from the air" [2].

A meeting of scientists using the very new recombinant DNA technology in Asilomar in California in February 1975 produced a set of guidelines for the use of biotechnology. The formal goals of the meeting included the need to identify the "possible risks involved for the investigator and or others" and "the measures that can be employed to test for and minimise the biohazards so that the work can go on" [3]. The benefits of the new technology far outweighed the risks *if suitable precautions were put in place* in the view of the Ashby committee in 1975 [2].

Parliaments have not been as reticent, and have recognised both benefits that may arise from its use and the risks that it theoretically poses. In 1993 the Parliamentary Assembly of the Council of Europe passed recommendation 1213 (13th May 1993) on developments in biotechnology, for which there were many wonderful prospects, but also for which there were many concerns. The Council of Europe includes many countries in Central and Eastern Europe as well as those of the European Union. The resolution noted that the gene pool has been widened far beyond the limits of sexual compatibility to encompass the possibility of transferring genes from almost any organism to others. Amongst the many uses of biotechnology it identified were the raising of agricultural outputs (or reducing inputs), the replacement of chemical herbicides and insecticides or more efficient targeting, the use of plants in industry, changes in responses of crop plants to stress. The resolution noted that there might be significant drawbacks resulting from the application of the new biotechnology. The possibility of new diseases was raised, as were the potential environmental effects of transgenic organisms[3].

The need for specific legislation in regard to the use of genetically modified organisms was never a presumption even though it was recognised that regulation was needed from the earliest days of the use of this technology. The UK had regulated the genetic 'manipulation' of micro-organisms starting in 1978, and by 1983 had a full set of legally binding regulations in place. The United States, on the other hand, had specified guidelines (the NIH Guidelines) which identified the manner in which such organisms should be used by those funded by the National Institutes of Health. In 1986 the US Government published its Coordinated Framework for the Regulation of Biotechnology [4] which described the 'comprehensive federal regulatory policy for ensuring the safety of biotechnology research and products" "Existing statutes provide a basic network of agency jurisdiction over both research and products; this network forms the basis of this co-ordinated framework and helps assure reasonable safeguards for the public. This framework is expected to evolve in accord with the experiences of the industry and the agencies". The laws existing in the United States regulated specific product uses, such as foods or pesticides. It was considered that genetically modified organisms posed no new risks that could not be covered using the existing system. "This approach provides the opportunity for similar products to be treated similarly by particular regulatory agencies".

The Cartagena Protocol was agreed in 2000 after years of negotiation and argument, with many misgivings, but in an atmosphere which had changed from that which pertained at the time the negotiations started. Article 19(3) of the Convention on Biological Diversity (June 1992) had required parties to consider the possibility of a Protocol to the convention that addressed the use (and primarily transboundary movement) of living modified organisms that might have an adverse impact on biological diversity. Yet eight years later Europeans were no longer accepting modern biotechnology; products had disappeared from the shops, there was an unofficial moratorium on the approval of new GM crops and there was a gloom and distrust in many countries not observed elsewhere. Few if any products derived using modern biotechnology are now available in Europe [5]. In North America, farmers adopted transgenic organisms with little opposition, and products derived from them have been in the shops for over 6 years.

The Cartagena Protocol requires 50 ratifications to come into force. On 13th June 2003, The Republic of Palau became the 50th country to ratify the Protocol. It was agreed that the Protocol would come into force ninety days after the fiftieth country has ratified it. Therefore, the Protocol came into force on the 11 September 2003. January 2004, the Protocol has received 78 ratifications[4], one of which is by the European Union, which does not count as part of the 50 required. However, not all the European Union countries have so far ratified the Protocol[5]. The main objective of the Protocol is to contribute to ensuring an adequate level of protection in the field of the safe transfer, handling and use of living modified organisms resulting from modern biotechnology that may have adverse effects on the conservation and sustainable use of biological diversity, also taking into account risks to human health, and specifically focusing on transboundary movements.

2. Regulatory Systems

Countries have chosen to use a variety of triggers for regulation of biotechnology. In Europe it is the fact of using modern biotechnology as defined in the Directives that triggers the regulatory process. In the United States, because current law is used, the trigger tends to be the use of organisms that are pests, for example plant pests, in the manufacture of the organism if the Department of Agriculture is to be involved. Canada has chosen to use a concept of novelty to trigger the regulatory process. Many analyses have suggested that once the process is started, the risk assessment and management processes are very similar. In Europe, primary legislation to regulate the deliberate release of GMOs is made by the European Community, either in the form of Directives, which have to implemented in the individual Member States by domestic legislation, or in the form of Regulations, which are directly applicable and implementing legislation is unnecessary for Regulations to become law in Members States.

- Directive 2001/18 on the deliberate release into the environment of genetically modified organisms is the main legislation, which governs experimental releases and placing on the market of GMOs. The Directive repealed the earlier Directive 90/220 that had been in place since the early 1990s. The later Directive tightened up considerable on the manner in which permission for commercial release of GMOs into the environment could be given, including limiting the life of permits and requiring motoring on the impact of the GMO on the environment. The Directive provided a step by step approval process on a case by case assessment of the risks to human health and the environment.
- Directive 98/81 on the contained use of genetically modified micro-organism regulates the contained use of genetically modified micro-organisms for research and industrial purposes.
- Regulation 258/97 on Novel food and novel food ingredients governing the marketing of GM food lays out the rules for authorisation and labelling of novel foods including products containing, consisting or produced using GMOs. This current legislation requires that all food in which there are detectable levels of genetically modified DNA or protein must go through the full authorisation procedure before being placed on the market. Regulation 258/97 requires mandatory labelling to indicate the presence of GMOs. After October 2002, Directive 2001/18 requires that Member States should ensure labelling of GMOs as or in products. Foods produced from GMOs but no longer containing GMO DNA or protein do not have to be labelled if they are "substantially equivalent".
- Council Regulation 1139/98 concerns the labelling of food or food ingredients from one GM soybean and one GM maize variety which were authorised under Directive 90/220, before Regulation 258/97 was in place.
- Regulation 50/2000 lays down the requirements for labelling of food and food ingredients containing additives and flavourings from GM sources.
- Regulation 49/2000 introduces a 1% threshold for the adventitious presence of DNA or protein from GM material in conventional food. There is a proposal for a new Regulation on traceability and labelling of GMOs and products produced from GMOs. The proposed labelling legislation extend the labelling requirements to all food and food ingredients produced from GMOs regardless of the detectable presence of DNA or protein with in the final food product. The proposed Regulation will require business to transmit and retain information about products that contain or are produced from GMOs at all stages of being placed on the market. There is also a proposed Regulation on GM food and feed, extending the labelling provisions to all genetically modified food or feed which consist of, contain or are produced from GMOs. It would not matter whether modified DNA or protein was detectable in the food or feed.

The EU Directives 90/220 and 2001/18 were implemented into UK law via Statutory Instrument 2002 No. 2443 - Genetically Modified Organisms (Deliberate Release) Regulations 2002. These regulations apply to experimental field trials and commercial agriculture and governing the release for research and experimentation and consent for release to market. In addition, Part VI of the Environmental Protection Act 1990 was used for the meaning of certain terms including "genetically modified organisms" and related terms and the meaning of "damage to the environment", "control" and related terms.

The EU Directives 90/219 and 98/81 were implemented into UK law via the Genetically Modified Organisms (Contained Use) Regulations 2000 and Statutory Instrument 2002 No. 63 (Amendment). The Act lays down the regulations governing risk assessment and notification of activities involving genetic modification, (both micro-organisms and non micro-organisms) and the conduct of activities involving genetic modification.

In the UK, there is now a framework of committees to regulate and provide advice on safety for human health and the environment (Figure 1). Three committees provide scientific advice of GM issues that cover food safety, the contained use of GMOs in the laboratory and release into the field. These are ACNFP, ACGM and ACRE respectively.

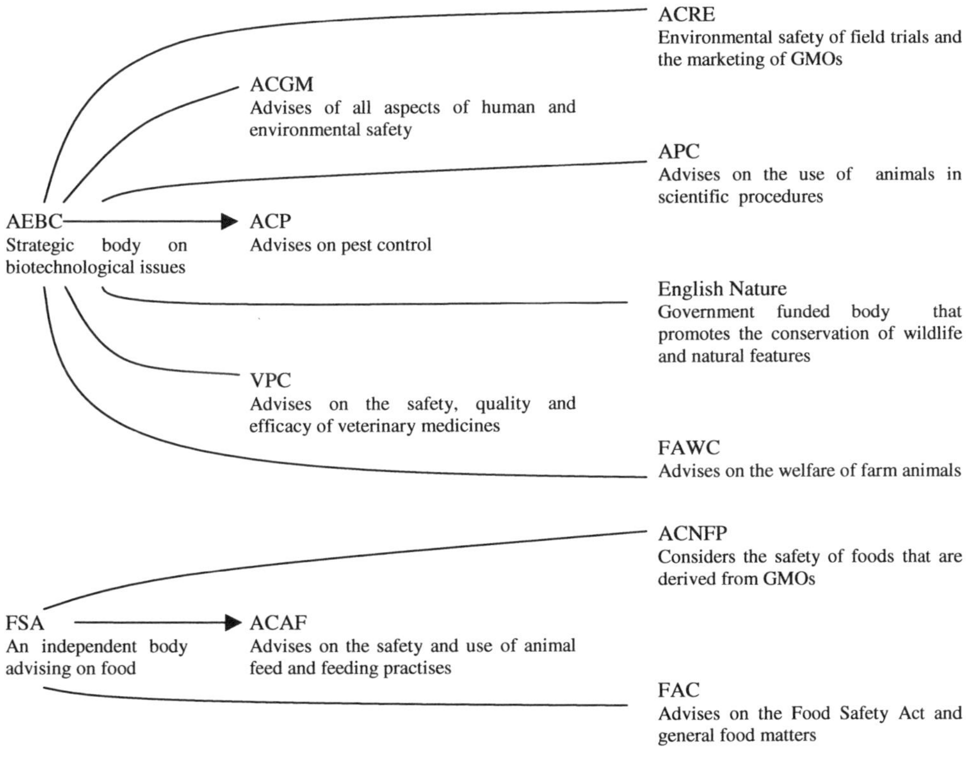

Figure 1. The advisory and regulatory framework for biotechnology in the UK. The Agriculture and Environment Biotechnology Commission (AEBC) interlinks with seven committees with concerns that are relevant to biotechnology issues. The Foods Standards Agency (FSA) is also linked to many other committees — as well as the three with specific biotechnology relevance that are shown — which advise on food issues, including the Advisory Committee on the Microbiological Safety of Food (ACMSF), the Advisory Committee on Research (ACR), the Committee on Carcinogenicity (COC), the Committee on Mutagenicity (COM), the Committee on Nutrition (SACN), the Committee on Toxicity (COT), the Expert Group on Vitamins and Minerals (EVM), the Meat Hygiene Advisory Committee, the Spongiform Encephalopathy Committee (SEAC) and several other working groups and working parties. ACAF, Advisory Committee on Animal Feeding Stuffs; ACGM, Advisory Committee on Genetic Modification; ACNFP, Advisory Committee on Novel Foods and Processes; ACP, Advisory Committee on Pesticides; ACRE, Advisory Committee on Releases into the Environment; APC, Animal Procedures Committee; FAC, Food Advisory Committee; FAWC, Farm Animal Welfare Committee; VPC, Veterinary Products Committee.

In 1999, a review of the framework concluded that socio-economic and ethical issues should be included, the Government introduced a further overarching committee, the AEBC, to provide strategic advice on all aspects of biotechnology as they pertain to agriculture and the environment. Food safety issues are covered by a second set of advisory committees that is overseen by the Foods Standards agency (FSA). A new committee considering animal feed, ACAF, is soon to be established.

In the United States, the framework for the regulation of biotechnology is based on the principle that GMOs do not differ fundamentally from unmodified organisms or conventional products. In essence it is the product rather than the process that should be regulated. In 1986 the Coordinated Framework for Regulation of Biotechnology issued by the US Office of Science and Technology Policy (51 FR 23302), made the decision that existing laws provided adequate authority for regulating biotechnology products. Other countries did not (at the time) have the range of environmental, food, drug and safety legislation in place that permitted effective use of existing legislation. The three federal regulatory agencies that primarily regulate biotech crops: the Environmental Protection Agency (EPA), the Food and Drug Administration (FDA), and the United States Department of Agriculture (USDA), but precisely which agency reviews a particular product will depend on its nature (Figure 2).

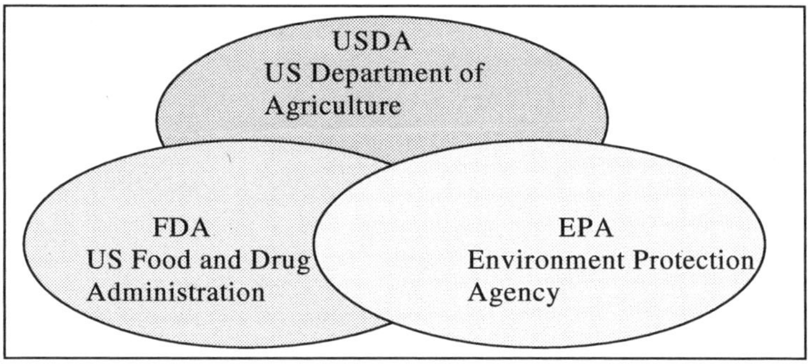

Figure 2. The United States co-ordinated framework for the regulation of biotechnology. For example, Bt Corn comes under the auspices of the United States Department of Agriculture (USDA). Viral resistance in a food crop might have environmental and food-safety implications, and so is considered by all three regulatory agencies. Modified oil content is not expected to have any environmental implications, and so the Environment Protection Agency (EPA) is not involved. Modified flower colour is not anticipated to have any environmental implications and ornamental crops are not food, so neither the Food and Drug Administration (FDA) nor the EPA are involved.

Using a voluntary consultation process, FDA determines whether foods and animal feeds developed from GM crops are as safe as their conventional counterparts. It does this by determining whether the companies producing them have answered all the appropriate questions about the new plant varieties, such as whether new allergens are present and whether there are increased levels of natural toxicants or perhaps reductions of important nutrients. Any changes in nutritional properties or crop processing or the presence of new allergens could require labelling to inform consumers of important changes to food or feed.

The USDA is responsible for protecting US agriculture against pests and diseases. All GM crops with for example, pesticidal traits are considered plant pests until USDA concludes that the crop is not a plant pest and makes a determination of non-regulated status—that is, decides that the plant will no longer be regulated by USDA as a plant pest. Until that determination is made, the plants are subject to USDA oversight for importation, interstate movement and environmental release. The EPA's oversight focuses on the pesticidal substance produced (such as Bt protein) and the genetic material necessary for its production in the plant (such as cry genes). The EPA calls this unique class of biotechnology-based pesticides 'plant-incorporated protectants' (PIPs) and describes procedures specific for PIPs in "Procedures and Requirements for Plant-Incorporated Protectants". The EPA grants experimental use permits for field testing and registrations that permit the sale and use of pesticides in commerce under the Federal Insecticide, Fungicide, and Rodenticide Act (FIFRA). The EPA also issues tolerances or tolerance exemptions that permit pesticide residues in food and/or feed under the Federal Food, Drug, and Cosmetic Act (FFDCA).

3. Environmental Policy in Relation to GMOs

The Cartagena Protocol on Biosafety specifically requires regulation in relation to the trans-boundary transfer of living modified organisms[6] that may have adverse effects on the conservation and sustainable use of biological diversity, taking also into account risks to human health. In addition, the Convention on Biological Diversity requires, through Article 8(g), that Members institute national frameworks in order to "[E]stablish or maintain means to regulate, manage or control the risks associated with the use and release of living modified organisms resulting from biotechnology which are likely to have adverse environmental impacts that could affect the conservation and sustainable use of biological diversity, taking also into account the risks to human health;".

The provisions of the Cartagena Protocol extend only to those organisms resulting from modern biotechnology that might cause potential adverse effects to the conservation and sustainable use of biodiversity. Human health has 'then' to be taken into account. However, when designing a regulatory system for biosafety, it is legitimate to assure safety of the environment and human health in general, with the needs for the Protocol forming a sub-set within the regulatory system. It seems likely that any attempt to link the protection of human health to legislation that primarily addresses biodiversity would not be acceptable to most legislatures. Countries have understood that biosafety in this instance means primarily protection of the environment, and that the release of living modified organisms needs be regulated in order to protect the environment. Safety concerns are not, however, limited to the impact on the environment. Regulatory systems that attempt to assure human and animal health are often different from that set in place for environmental protection.

4. Precaution

Scientific data is able to be collected at many sites around the world, which can provide an insight into the manner in which the product may interact with its environment when released into a particular environment. Where data is not available or where a country believes that its environment is different from that in which the organism was tested it may require field testing before an organisms is released or placed on the market. Where data is 'knowable' further experimentation will provide data that may address concerns and provide answers as to the likely behaviour of the organism in a particular environment.

Because of the inherent variability of biological systems information as to the behaviour may fall into the 'not-knowable' category where no amount of information collected can provide more than increased precision in determining the variability. Further experimentation will not provide any assurance that the organism will (or will not) impact on the environment in an unacceptable manner. The *precautionary approach* is invoked in order to address the absence of data. It is usually taken to refer to Principle 15 of Agenda 21 agreed in Rio de Janeiro in 1992.

The Precautionary Principle (or approach) has attracted many and various interpretations; for many it means that if the science is unknown and there is a risk of environmental damage, do not proceed. Caution dictates that it implies that where there is doubt over the safety of an action, that action should not be taken until evidence is available that the steps to be taken will not have disastrous consequences for the environment. The concern in relation to transgenic organisms is due to the possibility that once an organism is in the environment it is virtually impossible to recall and because of its property of replication, it does not decay over time; indeed numbers may increase disastrously.

Others interpret this as an injunction to proceed with caution, considering each release into the environment on a case-by-case basis and probably also proceeding step-by-step, where small field trials precede larger ones and the results are analysed before proceeding to commercial unfettered release (if ever). Recourse to the approach "presupposes that potentially dangerous effects deriving from a phenomenon, product or process have been identified, and that scientific evaluation does not allow the risk to be determined with sufficient certainty."[6] "Decision making about risks in the context of a precautionary approach is further complicated by the inherent dynamics of science.

Even though scientific information may be inconclusive, decisions will still have to be made to meet society's expectations that risks be addressed and living standards maintained." Scientists may be concerned that the 'principle' is used to stifle research, innovation and competition. "Where action is deemed necessary, measures based on the precautionary principle should be, *inter alia* [6]:

(1) *proportional* to the chosen level of protection,
(2) *non-discriminatory* in their application,
(3) *consistent* with similar measures already taken,
(4) *based on an examination of the potential benefits and costs* of action or lack of action (including, where appropriate and feasible, an economic cost/benefit analysis),
(5) *subject to review,* in the light of new scientific data, and
(6) *capable of assigning responsibility for producing the scientific evidence* necessary for a more comprehensive risk assessment."

The arguments that have ranged around precaution are serious, for they have directly impacted on the policy decisions of many countries. In Europe precaution in relation to transgenic organisms is taken to require case-by-case and step-by step approaches to risk. This way of interpreting precaution is built into the Protocol, which also requires a case-by-case process in assessing risk.

5. Science-based Decisions

There have been many arguments that decisions on the use of these living modified organisms must be based on science; Policy may be defined when designing the system that is applied to individual applications, but that the applications should only be considered in the light of this policy. Decisions are usually made by Government based on advice received from a number of sources. The risk assessment procedure, at the very least, should be science based. This is made very clear in the Cartagena Protocol. Article 15 states that "Risk assessments undertaken pursuant to this Protocol shall be carried out in a scientifically sound manner".

A Royal Society Report in 2002 asserts that "scientific assessments must inform policy decisions but cannot pre-empt them, and that public opinion must be taken into account throughout. We believe that the public debate about GM food must take account of wider issues than the science alone. We also wish to stress the importance of informing debate with sound science" Article 23 of the Cartagena Protocol requires public involvement in the decision making process and article 26 allows for specific socio-economic issues to be taken into account in the process.

Risk assessment of genetically modified organism is largely based on the concept of familiarity, or of 'substantial equivalence'. It is actually difficult to identify other ways of approaching the problem of identifying risk. It assumes that the characteristics of the modified organism except for the specific characteristics introduced are those of the host organism. It is not easy to see any other approach, but the Royal Society of Canada and the Royal Society in the UK have both indicated dissatisfaction with 'substantial equivalence' which is arguably substantially equivalent to familiarity. Can the approach be justified when stress tolerance, modification of metabolism or production of pharmacologically active compounds really begins?

Crop varieties developed through conventional plant breeding techniques not involving modern biotechnologies are not generally tested for their safety. Rather they have to meet plant variety registration requirements that identify whether they are distinct from those currently on the market, uniform and stable and National Seed list requirements of value for cultivation and use. These traditional methods use (primarily) crossing selection and back-crossing processes to select a desired characteristic and remove inadvertently introduced extra characteristics that initially accompany the introduced trait. These mechanisms introduce new and numerous gene combinations.

If there are toxins known to occur in these crops (e.g. glucosinolates in canola, glycol-alkaloid accumulation in potatoes), the new variety is normally tested to ensure that the level of toxin (or allergen) is no greater than the range that is normally observed for that substance. Interactions of traits introduced by traditional methods with other characteristics of the plant are normally ignored until they prove to make the variety unusable. "[T]he implicit assumption behind this methodology is that, even where a breeding-derived novel trait is involved, *new combinations of existing genes operating within highly selected germplasm are not expected to generate harmful outcomes*."[7]

For transgenic crops, the concept of 'substantial equivalence' was introduced. This was first described in a 1993 report of the OECD which stated that "If a new food or food component is found to be substantially equivalent to an existing food or food component, it can be treated in the same manner with respect to safety"[8]. The World Health Organization (1995) published a report in which the "substantial equivalence" as a decision threshold was promoted as the basis for safety assessment decisions concerning GMOs [7].

Substantial equivalence can be considered in three ways [5]:

- the GM foodstuff might be regarded as substantially equivalent to its conventional counterpart both toxicologically and nutritionally. When a product has been shown to be substantially equivalent, no further safety assessment is required.
- it might be substantially equivalent apart from certain defined differences. Sometimes the GM food product includes the components deliberately introduced by genetic modification. In this case the GM food product might be regarded as 'substantially equivalent to its conventional counterpart except for a small number of clearly defined differences'. Assessment is then limited to examining the implications of the difference(s), perhaps by testing the novel components of the GM plant in isolation."
- the GM product might be regarded as not substantially equivalent to its conventional counterpart, or there might not be a suitable reference available for comparison. The product will then need a highly detailed safety assessment, taking all the properties of the modified foodstuff and determining by direct measurement where necessary the impact on human health and the environment.

Many countries are deciding that the use of the term 'substantial equivalence' is misleading. It suggests that if substantial equivalence is demonstrated, no further assessment need be done. There is a 'mistaken perception that the determination of substantial equivalence was the end point of a safety assessment rather than the starting point' [9]. The Royal Society recommended in 2002 [5] "Safety assessments should continue to consider potential effects of the transformation process. The phenotypic characteristics to be compared between foods derived from GM plants and their conventional counterparts should be defined. It may not be necessary or feasible to subject all GM foods to the full range of evaluations but those conditions that have to be satisfied should be defined."

6. Intellectual Property Rights

Over the last few decades the global trading importance of biotechnology has been recognised and as a result there have been concerted and concentrated efforts to protect the results of research and development involving genetic material. The result of this has been the extension of intellectual property protection to most forms of biological material. The trade importance of biological information has been underlined the adoption of the Agreement on the Trade Related Aspects of Intellectual Property Rights (TRIPs) within the World Trade Organisation. This requires states party to the agreement to provide protection for all types of inventions irrespective of the field of technology.

The aim of the Agreement is to ensure that all member states provide effective and appropriate intellectual property protection and protect the intellectual property rights by the appropriate enforcement mechanisms. The Agreement sets down the minimum standards of protection. The TRIPS agreement permits countries to exclude from patentability those inventions whose commercial exploitation may be contrary to *ordre public* or morality. Countries may exclude from patentability "diagnostic, therapeutic and surgical methods for the treatment of humans or animals" and more importantly, Article 27(2) allows Members to exclude from patentability innovations in order to protect animal, plant life or health or to avoid serious damage to the environment. Article 27(3) provides for exclusion from patentability of: "plants and animals other than micro-organisms, and essentially biological processes for the production of plants or animals other than non-biological and microbiological processes.

Members shall provide for the protection of plant varieties either by patents or by an effective *sui generis* system or by any combination thereof..." What constitutes *sui generis* protection for new plant varieties is not defined hence countries are free to adopt a system that ensures intellectual property protection for plants. One option is for countries to implement the International Convention for the Protection of New Varieties of Plants (UPOV). Here again there is a fundamental difference between the US and the EU in the systems used for the protection of new plant varieties.

The US has three main laws that govern the protection of plants, but in reality only one is used. The Plant Patent Act 1930 (PPA) allowed the patenting[7] of new plants which is reproduced asexually which must also be shown to meet to general patent requirements of novelty, inventive step and utility. The Plant Variety Protection Act 1970 (PVPA) allows protection for sexually reproduced plant varieties shown to be novel, distinct, uniform and stable. Since 1980 [10], the Utility Patent Protection has become the norm for protecting new plant varieties in the US. The granting criteria for biological material are the same as the creteria for any other invention: plants or plant material must be novel, have an inventive step and have utility.

The European system for the protection of plants is a little more complicated. All EU Member States (except Belgium and Greece) have ratified either the 1978 or 1991

International Convention for the Protection of New Plant Varieties (UPOV). Membership of UPOV requires countries to implement national legislation based on the UPOV Convention. In the UK the Plant Variety Act 1997, implemented the UPOV 1991 Convention into national legislation. In addition, The Community Regulation on Plant Variety Rights (1994) which closely follows the 1991 UPOV convention governs all EU member states and allows EU wide plant variety protection.

Traditional plant breeders have used either their national system or the Community Plant Variety Right Regulation rather than patents to protection new plant varieties. This, at least in part, has been due to the fact that plant varieties were excluded from patenting under Article 53(b) of the European Patent Convention (EPC), a convention on which all EPC members national patent laws are based and of which all EU member states are signatories. Plant breeders traditionally used plant variety rights not solely because plant varieties were excluded from patenting, but primarily because the right was designed meet the needs of plant breeders.

Plant variety rights have two main derogations from the right, the breeders' exemption and the farmers' privilege. The breeders' exemption is the right to use protected material freely for research purposes provided that the acts are done for the purpose of breeding or discovering and developing varieties. Plant variety rights can be acquired over the resulting variety even if the resulting variety is essentially derived. The second derogation, farmers' privilege permits farmers to use protected material for propagating purposes on their own holding. This right is limited in that the farmer can only retain propagating material for this purpose from a limited number of categories and must pay an equitable remuneration sensibly lower than the amount originally charged.

The situation changed in early 2000, when the Enlarged Board of Appeal (EBA) of the European Patent Office (EPO) overruled the decision of the Technical Board and said that claims which encompass more than one variety are not excluded. The ruling (Decision G01/98) [11] by the EBA of the EPO, a landmark case with regards to the patenting of plants, has raised the greatest concerns amongst European plant breeders. The 'Novartis' case permits the patenting of plants in Europe provided a technical invention can be shown and plant varieties are not claimed specifically. This decision by the EBA has clarified the situation and demonstrated that the European Patent system actually provides protection for plants provided a technical invention can be demonstrated, for example, the plant contains a novel gene. Thus it now appears that GM plants, in which a technical invention can be demonstrated, may be protectable by patents rather than plant variety rights.

It is clear that plant variety rights do not give the equivalent protection as that afforded by a patent. However, many of the so called weaknesses in the plant variety rights system may in fact be the advantage of that system over patents. The derogations in plant variety rights on did so on the basis of public interest. There is no real equivalent derogation to the breeders' exemption and framers' privilege under patent law. Plant variety rights are generally administered by governmental agencies responsible for agricultural matters and not by offices concerned with trade and industry, as is the practice with patents.

In contrast to the paper assessment undertaken to determine whether and invention is patentable, a plant variety is subjected to two years of practical trails before the right is granted. These serve to show whether or not the variety is actually distinct, uniform and stable as opposed to simply relying on a written description provided by the breeder. The trails are undertaken by the granting offices in conjunction with breeding institutes, such as the United Kingdom's National Institute of Agricultural Botany (NIAB).

7. Public Opinion

In the United States, there appeared to be little public rejection of GM crops and food, attributed by the US government to the openness of the American system. Although a recent survey found that the American public's position on the acceptability of genetic modification of food was mixed [12]. 58% of Americans either strongly approve or somewhat approve of creating hybrid plants using genetic modification, while 37% disapprove. The attitude of the European public changed markedly in the late 1990s resulting in a wide spread distrust of modern biotechnology, especially GM crops and food, ultimately leading to the *de facto* moratorium on the commercialisation of plants manufactured using GM techniques.

The Eurobarometer surveys show that there is considerable discrimination amongst the public (at least in Europe) in relation to the various uses that modern biotechnology can be put. Europeans continue to distinguish between different types of applications, particularly medical in contrast to agri-food applications (see Figure 3)[8]. Food, for example, is considered risky, morally unacceptable and not to be encouraged yet crops (much to the surprise of the researchers) are considered useful but risky, morally acceptable and there is a slight majority in favour of their use.

This response in Western Europe to the new technology cannot easily be dismissed through assertions by scientists that there is negligible risk, or that permits to market transgenic foods and crops (in particular) should be based solely on risk assessments that are science based. If all the scientific information had been available and a consensus amongst scientists could be achieved that the impact on the environment is minimal, it would be possible to argue for a totally science based risk assessment process.

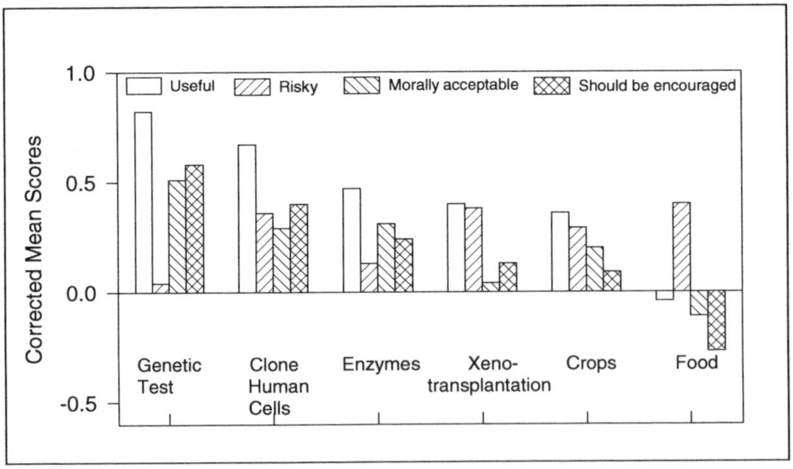

Figure 3. Mean score on a scale from +1.5 to −1.5 for the assessment of use, risk, moral acceptability and should be encouraged for each application.

In an Irish Consultation Paper [13], the concerns about potential environmental and human health effects were thought to arise because of an absence of familiarity with the regulatory systems; the technology was thought to be complex and developing rapidly; "there is little experience on the interaction of GMOs with their surrounding environment"; the information being provided to the public is probably inadequate, particularly in relation to labelling to allow choice; the use of antibiotic resistance marker genes is thought to inimical to their use in human and veterinary medicine and the use of herbicide tolerant crops might increase the use and build-up of herbicides in the environment.

The situation in Europe is significant as it has a direct bearing on that which can be done in developing countries. In the first instance the concerns being expressed by Greenpeace and Friends of the Earth, or even the British Medical Association create a groundswell against the use of this new technology. Can it be right to introduce these 'untested' technologies in developing countries when public 'informed' opinion is so virulently opposed to their use in Europe? When even statutory bodies like the Nature Conservation Organisations in Britain and France reject modern biotechnology because of a predicted negative effect on the environment, are developing countries to embrace them? The rejection of genetically modified foods by many European Supermarkets and food producers has impacted on production and growing of genetically modified crops that have to be exported to one of the largest food markets in the world [14]. To grow rice modified so that it produces vitamin A is a wonderful prospect for nutrition in the many countries that depend on rice as a primary food. If, however, the produce cannot be exported as well, producers will be reluctant to grow it.

8. Conclusions and Outlook

Over the last five to ten years the EU has introduced new Directives and Regulations covering all aspects of GMOs, from their use in the laboratory right through to their commercialisation. GMOs are now possible the most highly regulated sector of any area of biotechnology. Yet the future of GMOs, especially in the area of agriculture, is still unclear due, at least in part to the fact that the Member States of the EU are deeply divided over their commercialisation.

In early December 2003, the EU standing committee for the food chain voted against extending consent for Syngenta's maize transformation event Bt11 to allow cultivation and commercialisation. France, together with Denmark, Greece, Luxembourg, Austria and Portugal voted against approving the maize. Britain, Spain, Ireland, the Netherlands, Sweden and Finland voted to approve the GM corn, while Germany, Italy and Belgium abstained. The final decision on approval for the GM maize will now be made by EU ministers early in the New Year. Syngenta's maize was the first GM to go through the new approval procedure and its approval would have ended the five year *de facto* moratorium on the commercialisation of GM crops. The decision by the EU ministers may well decide the future of GM crops in the EU.

In addition to trying to get their products approved through a system where Member States are split over their policy towards GM crops, agbiotech companies also face the hurdle of changing public opinion. A public consultation exercise carried out in the UK during the summer of 2003 (http://www.gmpublicdebate.org.uk/) concluded that the public might be willing to accept GM crops, but not yet. The public still has grave concerns over such issues as the relationship between GM crops and the environment, food safety and the oligopolistic nature of the industry.

Until the public can be assured that these issues have been fully addressed it will be difficult for any Member State to introduce GM crops. It appears therefore, that the end of the moratorium and the start of commercialisation of GM crops may still be some way off.

Endnotes

[1] Modern biotechnology is defined in the Cartagena Protocol (Article 3).
[2] Variously and at various times called genetic modification, genetic manipulation or genetic engineering.
[3] 'transgenic organism' is used as synonymous with 'living modified organism' or 'genetically modified organism'.
[4] USA, Argentina, Canada, China and Australia, the largest producers of GM crops, have not ratified the Protocol.
[5] Belgium, Finland, Greece, Italy and Portugal have not as yet ratified the Protocol.
[6] Article 1, Objective, of the Cartagena Protocol on Biosafety.
[7] A patent allows the holder to prevent anyone else from using his protected invention (negative right). The right is given in exchange for the applicant demonstrating, in a written document called the patent specification, that his invention is novel (not known in that form anywhere in the world), that it is the result of an inventive, or non obvious, step and that it has utility or is capable of industrial application. The right lasts up to 20 years. There are limited exceptions to the right. Patents are private property rights. This means the cost of acquisition and enforcement is borne by the rights holder.
[8] Europeans and Biotechnology in 2002, Eurobarometer. (www.europa.eu.int/comm./public_opinion/archives/eb/ebs_177_en.pdf).

References

[1] Watson, J. D. and Crick, F. H. C (1953). Molecular structure of nucleic acids. Nature 171, 737-738
[2] United Kingdom (1975). Report of the Working Party on the Experimental Manipulation of the Genetic Composition of Micro-organisms. CMND 5880 (January).
[3] Wright, S. (1994). Molecular Politics: Developing American and British Regulatory Policy for Genetic Engineering 1972-1982 Chicago, USA: University of Chicago Press.
[4] Office of Science and Technology Policy (1986). Coordinated Framework for Biotechnology. Federal Register 51, 23302 – 23350.
[5] The Royal Society (2002). Genetically Modified Plants for Food Use and Human Health – an update, Policy Document 4/02, 20 pp. (www.royalsoc.ac.uk/gmplants/)
[6] Commission of the European Communities (2000). On the precautionary principle. COM(2000) 1 (02.02.2000), Brussels, Belgium: European Commission. 29 pp. (europa.eu.int/comm/dgs/health_consumer/library/pub/pub07_en.pdf)
[7] Royal Society of Canada (2001). Elements of Precaution: Recommendations for the Regulation of Food Biotechnology in Canada. Expert Panel Report on The Future of Food Biotechnology. Ottawa, Ontario: Royal Society of Canada, 265 pp. (www.rsc.ca/foodbiotechnology/indexEN.html)
[8] OECD (1993). Safety Evaluation of Foods Produced by Modern Biotechnology. Concepts and Principles. Paris: OECD, 74 pp. (http://www.oecd.org/dataoecd/57/3/1946129.pdf).
[9] FAO/WHO (2000). Safety Aspects of Genetically Modified Foods of Plant Origin. Report of a Joint Food and Agriculture Organisation/World Health Organisation Consultation. Geneva, Switserland: WHO, 37 pp. (www.fao.org/es/ESN/food/pdf/gmreport.pdf)
[10] Diamond v Chakrabarty (1980). 206 USPTQ 193. (digital-law-online.info/cases/206PQ193.htm)
[11] Novartis/Transgenic Plant (2000). European Patent Office Reports 303.
[12] Hallman, W.K. (2002) Public Perceptions of Genetically Modified Foods: Americans Know Not What They Eat. Food Policy Institute, Rutgers, USA: The State University of New Jersey.
[13] Government of Ireland (1999). Department of the Environment and Local Government Consultation paper on a national policy position on the deliberate release of genetically modified organisms to the environment.(www.environ.ie/DOEI/DOEIPol.nsf/wvNavView/Genetically+Modified+Organisms?OpenDocument&Lang=).
[14] Yates, S.C. (1999). GMO awareness takes flight. Nat. Foods Merchand. July issue. (http://www.newhope.com/nfm-online/nfm_backs/Jul_99/gmos.cfm).

Author Index